Frans Kjellman

The algae of the Arctic Sea

A survey of the species, together with an exposition of the general characters

Frans Kjellman

The algae of the Arctic Sea
A survey of the species, together with an exposition of the general characters

ISBN/EAN: 9783337271329

Printed in Europe, USA, Canada, Australia, Japan

Cover: Foto ©berggeist007 / pixelio.de

More available books at **www.hansebooks.com**

KONGL. SVENSKA VETENSKAPS-AKADEMIENS HANDLINGAR. Bandet 20. N:o 5.

THE ALGÆ OF THE ARCTIC SEA.

A SURVEY OF THE SPECIES, TOGETHER WITH AN EXPOSITION OF THE GENERAL CHARACTERS AND THE DEVELOPMENT OF THE FLORA,

BY

F. R. KJELLMAN.

WITH 31 PLATES.

COMMUNICATED TO THE ROYAL SWEDISH ACADEMY OF SCIENCE 1883, JANUARY 10.

STOCKHOLM, 1883.
KONGL. BOKTRYCKERIET.
P. A. NORSTEDT & SÖNER.

INTRODUCTION.

The description of the Flora of the Arctic Sea here published is chiefly founded on the experience acquired and the collections brought together by myself during voyages in the arctic waters. Since I entered the ranks of Swedish arctic voyagers about a decennium ago as a member of the Spitzbergen expedition of 1872—73, the leader of this as well as all greater Swedish arctic expeditions in later times, A. E. NORDENSKIÖLD, has kindly admitted me as a companion to all his following voyages of exploration in the high North: the expeditions of 1875 and 1876 to Novaya Zemlya and the mouth of the Yenissei and the Vega expedition 1878—80. By these means I have had the advantage of making myself familiar, by studies in the nature, with the marine Flora on the north coast of Norway, where I carried on algological researches during the greater part of the summer 1876, and on the coasts of Spitzbergen, Novaya Zemlya and northern Siberia, accordingly, within a considerable part of the arctic region. Of the vegetation of the rest of the Polar-Sea I have gained knowledge, partly by consulting the collections brought home from there, amongst which there ought to be especially mentioned the collections of algæ from the west coast of Greenland belonging to the Botanical Museum of Copenhagen and placed at my disposal through the kind intercession of Professor J. LANGE and Mr. HJ. KJÆRSKOU, partly by means of the literature written on the subject. The most important works are those of DICKIE and GOBI, and, above all, several treatises by J. G. AGARDH, which are of very high value on account of this eminent algologist having with his usual acumen and accuracy unraveled several of the most complicated and difficult groups of algæ of the arctic Flora. With regard to these and other works which I have made use of, I refer the reader to my list of literature given below.

The definition and division of the Arctic Sea.

In defining the limits of the Arctic Sea there generally prevail two different points of view, the one a purely geographical one, when by the Arctic Sea there is meant the sea north of the north Polar circle, the other a more hydrographical one, when the Arctic Sea denotes the cold glacial sea round the North Pole. By the former view tracts of water are excluded from the Arctic Sea which are perhaps the most rich in ice in all the northern hemisphere and possess, with regard to ice-drift and temperature, a distinctly arctic character, namely the northern Atlantic off south Greenland, while on the other hand, there is included in it the sea on the north coast of Norway, where the temperature of the water, in consequence of warm currents, is far higher than in the other parts of the polar basin and on this account neither in summer nor in winter any greater quantity of ice is formed or sets down from higher latitudes. According to the latter view, on the contrary, the sea off Greenland south of the Polar circle forms part of the Arctic Sea, whereas the sea off the coast of northern Norway is excluded. If thus defined, the Arctic Sea, as will be demonstrated below, can be regarded as a unity with respect to the geographical distribution of plants; which is not the case according to the former definition. It is necessary therefore to establish once for all fixed limits to the Arctic Sea as meaning a distinct region of vegetation, and to make a decided distinction between the Arctic Sea as considered from this purely botanical point of view and from a geographical one, assigning a different name to each of these different regions. I propose that in the geography of plants that part of the northern ocean which stretches along and north of the coasts of the arctic countries, be called the Arctic Sea, and that the name of the North Polar Sea, not uncommon in chartography, be applied to the sea north of the north Polar circle. The following investigation will prove that such a distinction is made necessary by the botanical facts. The present work will treat, accordingly, of the algæ not only in the Polar Sea proper, as here understood, but also in that sea which lies south of the Polar circle off the coast of Greenland.

In dividing the Arctic Sea and denominating its several parts, I have tried to follow, as closely as possible, the maps published in later times. But different geographers, chartographers and arctic voyagers having often applied different names to the same part of the Arctic Sea, or assigned different limits to regions called by the same name, which is especially the case with the Arctic Sea north of the Atlantic, I think I ought to state expressly which names I have decided on employing and which limits I have thought fit to give to the different regions.

The Norwegian Polar Sea. I propose to designate by this name that part of the Polar Sea which extends along the north-west and north coasts of Norway from the

Polar circle in the south to about the 72:nd degree of latitude northwards, and to the longitude of Vardö to the east. Part (The whole?) of this region is called the North Sea — das Nordmeer — on Kiepert's map of the arctic lands [1]), but as this name includes also part of the Atlantic and moreover might possibly be confounded with the North-Sea or German ocean, I have not thought fit to adopt it.

The Greenland Sea is the sea between Greenland and Spitzbergen north of Iceland and of the Norwegian Polar Sea, extending along the east coast of Greenland and the west and north coasts of Spitzbergen. I regard Beeren Eiland as situated in this sea.

The Murman Sea is limited to the north by a line supposed to be drawn from the mouth of Varanger-fjord to Matotshkin Shar at Novaya Zemlya [2]).

The Kara Sea is the sea between Novaya Zemlya and the Taimyr Peninsula, reaching to the longitude of Cape Chelyuskin.

The Spitzbergen Sea is the region north of the Murman and Kara Seas, east of the Greenland Sea.

The Sibirian Sea is the region east of the preceding, to the longitude of Bering Strait.

The American Arctic Sea north of North-America.

I call *Baffin Bay* the region between America and Greenland. I hold it to be bounded to the south by the latitude of Cape Farewell.

---

[1]) KIEPERT, Uebersichts-Karte der Nordpolar-Länder. Neue berichtigte Ausgabe. Berlin 1874.

[2]) Cp. NORDENSKIÖLD, Karta öfver Prövens färd till Jenisej och åter 1875.

[3]) Cp. NORDENSKIÖLD, Vega-Exp. 1. p. 150.

The general character of the vegetation.

Number of individuals. The vegetation richest in individuals is found in the Polar Sea on the coasts of Norway. Here all those parts of at least the litoral and sublitoral zones, which are fit for the growth of algæ, are clothed with dense masses of such plants. It may be stated broadly, that the tracts covered with algæ are comparatively as large in the Norwegian Polar Sea as in the northern Atlantic on the coasts of Norway and Great Britain, and that the density of the vegetation is on the whole the same. If we except the Murman Sea in its most westerly portion and the White Sea, which two regions may most fitly, with regard to vegetation, be regarded as intermediate between the Norwegian Polar Sea and the Arctic Sea as understood here with strict reference to the geography of plants [1]), it is probable that the southern part of Baffin Bay along the west coast of Greenland is that part of the Arctic Sea whose vegetation comes next to that of the Norwegian Polar Sea in number of individuals. I do not know the vegetation here from personal observations, but by the collections I have examined and the informations given by investigators who have visited these regions, I am led to the opinion that on the west coast of Greenland, at least up to Disco Island or about Lat. N. 71°, there is a vegetation of algæ which, though certainly inferior in extent and in number of individuals to that of the Norwegian Polar Sea, comes however next to it, and surpasses by far that of any other larger arctic region. RINK, the foremost knower and most accurate describer of the nature of Greenland, says in his work *Grönland geographisk og statistisk beskrevet.* »The view presented by the sea, where it is clear, close on the coasts of Greenland is no less surprising. The bottom is overgrown with a forest of gigantic algæ with leaves from six to eight ells long by a quarter of an ell broad, which together with the animal world moving between them remind one of the coral reefs of the tropical seas. Besides, the stones on the bottom are covered with corallaceous crusts [2]), and their cavities as well as the clay dredged up teem with animals». The opinion I have, in an earlier work, pronounced on the Flora of the Murman Sea on the west coast of south Novaya Zemlya and Waygats, has not been overthrown by later observations. I said: The vegetation of algæ is here poor in number of individuals, as compared with that on the coasts of Scandinavia. Large tracts of the sea-bottom are completely devoid of algæ, or possess only an extremely poor, thin vegetation, although they are of such a nature that in other seas they would be covered with algæ. The greatest part of those ranges of the bottom which in other seas are

[1]) Cf. GOBI, Algenfl. weiss. Meer. and CIENKOWSKY, Bericht.
[2]) Lithothamnia.
[3]) RINK, Grönland 1. p. 84.

clothed with algæ, are without vegetation, and on those portions of the bottom, where the vegetation is at its densest, it is nevertheless largely inferior in point of individuals to such parts of the Atlantic as are rich in algæ. I think that this statement may be extended even to the Greenland and Spitzbergen Seas, along the east coast of Greenland and the coasts of Spitzbergen, Beeren Eiland and northern Novaya Zemlya. It probably holds good even with respect to the American Arctic Sea, whose algology is as yet only very incompletely known. In the Kara Sea, judging by the few observations hitherto made, the character of the Flora is another at Novaya Zemlya than on the coast of Siberia. If the vegetation along the rest of the east coast of Novaya Zemlya resembles in its general features that of Uddebay, the only point on this coast where it has as yet been made the object of a closer investigation, the Flora of the western part of the Kara Sea is most closely allied, with respect to the number of individuals, to that of the eastern part of the Murman Sea.

Of the whole Arctic Sea, the region along the north coast of Siberia, i. e. the eastern part of the Kara Sea and the Siberian Sea, has the poorest Flora with regard to number of individuals. From the observations hitherto made, it must be considered extremely poor. There is probably no other region of the sea to be found of the same extent with this, whose vegetation presents such a character of poverty and indigence. As is shown by the table and map published by STUXBERG of the dredgings in the Kara Sea and the Siberian Sea made during the Swedish expeditions in 1875, 1876, and 1878, researches have been carried on with the best dredging apparatus of modern time at a considerable number of places along the whole coast of Northern Siberia and generally at such a depth and such a distance from the shore, that in other seas and even in other parts of the Arctic Sea a bottom rich in algæ would quite certainly have been struck very often. However, the data given, which are based on my own notes, show that in the eastern part of the Kara Sea and in the Siberian Sea algæ have been dredged only in 10 places. Only in four of these, viz. at Cape Palander and in Actinia Bay within the Kara Sea, and at Irkaypi and the region about the mouth of Kolyutshin-fjord within the Siberian Sea, algæ were found in any notable numbers. The Flora of the last-mentioned locality may possibly be compared, in point of number of individuals, to that of poorer portions of the Murman Sea and the Greenland Sea; at the other three places the tracts clothed with algæ were of little extent and the number of individuals was small. With regard to the vegetation of the six remaining localities I have noted the following facts, which seem to me worthy of being specially mentioned.

1. Lat. N. 74° 52′ Long. E. 85° 8′. Kjellman's Islands.

Bottom: flat rocks and boulders of granite and gneiss.
Veget.: scanty litoral vegetation of *Urospora penicilliformis.*

2. Lat. N. 76° 8′ Long. E. 90° 25′.

Depth: 16 fathoms. Bottom: stones and shingle.
Veget.: *Lithothamniom fœcundum*, scarce. *Phyllophora interrupta*, extremely small and scarce. *Lithoderma fatiscens*, rather abundant.

3. Lat. N. 76° 18′ Long. E. 92° 20′.

Depth: 40 fathoms. Bottom: stones and clay.

Veget.: *Phyllophora interrupta*, several specimens. *Polysiphonia arctica* one small specimen, attached to the preceding.

4. Lat. N. 77° 36′ Long. E. 103° 25′. Cape Chelyuskin.

Depth: 5—10 fathoms. Bottom: clay with pieces of slate and quarz.

Veget.: almost none; only traces in two places. *Laminaria Agardhii*, one intact specimen and some in a state of dissolution.

Sphacelaria arctica, one specimen.

Pylaiella litoralis, extremely scarce and poor. Not the least trace of a litoral vegetation. Ice-foot remaining almost everywhere.

5. Lat. N. 73° 40′ Long. E. 140° 16′. Blishni Island.

Depth: 4 fathoms. Bottom: hard clay.

Veget.: Some specimens of *Phyllophora interrupta* were found imbedded in the clay which was brought up by the dredges. Their basal parts were torn off, but in general they had a fresh appearance. They had probably been lying loose on the bottom, but not drifted far.

6. Lat. N. 69° 27′ Long. E. 177° 14′.

Depth: 4—5 fathoms. Bottom: sand and pebbles.

Veget.: *Delesseria sinuosa*, one specimen attached to a species of Hydromedusa.

My total judgment on the Flora of the Arctic Sea with regard to its number of individuals may, in accordance with the facts exhibited above, be stated in the following manner.

In about one third of this sea, namely the greater part of the Kara Sea and the Siberian Sea, the vegetation is very poor in individuals; in the Norwegian Polar Sea it is comparable in richness to that of the North Atlantic; in the rest of the Arctic Sea it is considerably more poor, a comparatively lesser surface of the bottom being furnished with algæ and the vegetation even on these portions being less dense than in the Atlantic. The vegetation on the west coast of Greenland (in the western part of the Murman Sea and in the White Sea) approaches most nearly to that of the Norwegian Polar Sea in number of individuals.

The distribution of the vegetation on the different bottom-zones, the litoral, the sublitoral, and the elitoral. I think the limits of these zones may be drawn in the Arctic Sea in the same manner that I have done in my account of the Flora of the Murman Sea. Thus the litoral zone would comprise the bottom-range between tide-marks. The sublitoral zone extends from the lower boundary of the litoral to the depth of 20 fathoms. Still deeper parts of the bottom covered with algæ form the elitoral zone. Its lower limit certainly varies in different parts of the Arctic Sea. In the Greenland Sea on the coast of Spitzbergen algæ are found growing even at the depth of 150 fathoms.

With regard also to the distribution of the vegetation on the several bottom-zones different parts of the Arctic Sea differ essentially. In the Norwegian Polar Sea the litoral zone is overgrown with a rich, luxuriant vegetation of varying composition. More than half of the species known from this sea occur on this region of the bottom, either exclusively or at least sometimes. Some of them, for instance *Rhodomela lycopodioides*, *Gigartina mamillosa*, *Chondrus crispus*, do not certainly attain here the same luxuriant development as farther southwards on the coast of Norway, but the great majority are as flourishing here and form as dense masses as on the coasts of the North Atlantic. On the west coast of southern Greenland there is also to be found a litoral vegetation very rich in individuals, although it is more monotonous. I have been informed by Professor TH. FRIES that the litoral zone is clothed with Fucaceæ, and that these constitute the peculiar character of the litoral vegetation. That however several other species occur, beside these, is shown by the collections made here, containing several purely litoral algæ. With regard to several of these, J. VAHL has, besides, expressly noted on the labels, that they were found growing within the litoral zone. Of such species I mention here: *Rhodomela lycopodioides*, *Polysiphonia urceolata*, *Halosaccion ramentaceum*, *Ilea fascia*, *Pylaiella litoralis*, *Enteromorpha intestinalis*, *E. compressa* and *E. micrococca*, *Monostroma Blyttii*, *Diplonema percursum*, *Spongomorpha arcta*, *Cladophora rupestris*, *Rhizoclonium rigidum* and *Rh. riparium*, *Chætomorpha Wormskioldii*, *Urospora penicilliformis*, *Rivularia atra*. In the by far greater part of the Arctic Sea the litoral zone possesses no vegetation at all or an exceedingly meagre one. On the coasts of Spitzbergen I have found the following species in the litoral zone: *Rhodochorton Rothii* and *Rh. intermedium*, *Fucus evanescens f. bursigera* and *nana*, *Chætopteris plumosa*, *Pylaiella litoralis*, *Chætophora maritima*, *Enteromorpha compressa*, *Spongomorpha arcta*, *Rhizoclonium riparium*, *Ulothrix discifera*, *Urospora penicilliformis*, *Codiolum Nordenskiöldianum*, *Calothrix scopulorum*. This number of species is certainly pretty considerable, but it should be observed, that the majority are either generally very rare or at least very rare in the litoral zone. *Fucus evanescens*, commonly diffused in its larger forms on the coasts of Spitzbergen, seldom ascends into the litoral zone. In the form *bursigera* it has been found here in two places, in the form *nana* in one, always in plots of small extent. *Chætopteris plumosa*, one of the commonest algæ of Spitzbergen, I have found as litoral only once and then only a few specimens. *Pylaiella litoralis* and *Spongomorpha arcta*, both of them rather common in the sublitoral zone, are very rare in the litoral. The two *Rhodochorta*, *Ulothrix discifera* and *Codiolum Nordenskiöldianum* have been met with, each of them, only in one place within the litoral zone of Spitzbergen. Only *Enteromorpha compressa*, *Rhizoclonium riparium*, *Urospora penicilliformis* and *Calothrix scopulorum* occur more commonly or in somewhat greater numbers on the uppermost part of the bottom at Spitzbergen. In the by far greatest part of this zone there is no vegetation at all. The same poverty is exhibited by the litoral zone in the eastern part of the Murman Sea; in the western part of it as well as in the White Sea it probably stands in the middle between the richness of the Norwegian Polar Sea and the poverty of the Greenland Sea [1]). In my account of the vegetation on the west coast of Novaya

[1]) Cp. CIENKOWSKY, Bericht.

Zemlya and Waygats I have mentioned that the greatest part of the litoral bottom-zone here is destitute of any vegetation and that the litoral vegetation occurring here and there is extremely poor in individuals and consists exclusively of small algæ. Eleven species have been found here as litoral: *Rhodochorton Rothii*, *Fucus evanescens*, *Phlæospora pumila*, *Pylaiella litoralis*, *Chætophora maritima*, *Enteromorpha compressa* and *E. minima f. glacialis*, *Rhizoclonium riparium* and *Rh. pachydermum*, *Urospora penicilliformis*, *Calothrix scopulorum*. The most common of these were *Rhodochorton Rothii*, *Pylaiella litoralis* *Enteromorpha compressa* and *Rhizoclonium riparium*. *Fucus evanescens* is more rare, *Phlæospora pumila*, *Urospora penicilliformis*, *Chætophora maritima* and *Calothrix scopulorum* are found only in two places, *Enteromorpha minima f. glacialis* and *Rhizoclonium pachydermum* each in one place. Here, as on the coasts of Spitsbergen, the litoral algæ are small. The particular specimens of *Chætophora maritima* can hardly be distinguished with the naked eye. *Calothrix scopulorum* and *Urospora penicilliformis* cover the stones between tide-marks with a thin coating. *Rhodochorton Rothii* and *Rhizoclonium riparium* are matted together to the thickness of some mm., even *Rhodochorton intermedium*, *Phlæospora pumila*, *Enteromorpha compressa* and the litoral form of *Pylaiella litoralis* are small in size, being only some mm. high, and the forms of *Fucus evanescens*, as generally found within the litoral zone, seldom grow more than about 6 cm. in height [1]). In the Kara Sea traces of a litoral vegetation have been detected only in two places, namely at Kjellman's Islands, where, as mentioned above, there were found small tufts of *Urospora penicilliformis* on the rocks at the shore, and in Actinia Bay, where the litoral bottom-zone was clothed in several places, although sparely, with stunted *Enteromorpha compressa*. No litoral algæ are known from the Siberian and the American Seas.

The main mass of the vegetation in the Arctic Sea may be said to be diffused over the sublitoral zone. But this general statement has a somewhat different signification with regard to different parts of the Arctic Sea. The sublitoral zone certainly possesses everywhere the most vigorous and dense vegetation and that which is most rich in individuals, but with regard to the number of species the sublitoral vegetation in the Norwegian Polar Sea is poorer, in the other parts of the Arctic Sea, on the contrary, richer than that of any other bottom-zone.

With the methods hitherto invented for the exploration of the marine vegetation, insurmountable difficulties are opposed to our gaining any sure knowledge of the nature of the vegetation in the elitoral zone. Those few specimens of algæ which have sometimes been brought up by the dredges from a greater depth than 40 fathoms only suffice to prove, according to my experience, that larger algæ really occur on this part of the bottom. But they afford no information about the number of individuals and the general character of the vegetation. It seems to me to result from the investigations carried on in the Arctic Sea, that by far the greatest part of the elitoral zone is destitute of algæ, and that the vegetation found here and there is poor in species as well as individuals.

I do not know of any species that has been found with certainty in the elitoral bottom-zone of the Norwegian Polar Sea. On the coast of Spitzbergen I have found

[1]) Cp. KJELLMAN, Algenveg. Murm. Meer. pp. 58—59.

two or three, namely *Delesseria sinuosa*, of which some few, but quite fresh specimens were brought up during the Swedish expedition of 1872—1873 from the depth of 85 fathoms north of Spitzbergen, *Ptilota pectinata* found at the depth of 150 fathoms in Smeerenberg Bay on the north-western coast of Spitzbergen, and in 80—100 fathoms north of Spitzbergen. Even *Dichloria viridis* descends here to the same considerable depth. On the west coast of Novaya Zemlya the following species are known from the elitoral zone: *Polysiphonia arctica, Delesseria sinuosa, Euthora cristata*, and *Dichloria viridis* [1]). As I have mentioned before, *Phyllophora interrupta* and *Polysiphonia arctica* were met with during the Vega expedition at one place in the Kara Sea at the depth of 40 fathoms. According to DICKIE, a rather considerable number of species were dredged at considerable depths in Baffin Bay during one of the English Franklin expeditions, viz. *Polysiphonia nigrescens*, 40—50 fathoms; *Ptilota pectinata*, 30—40; *Dictyota fasciola(?)*, 20—50; *Agarum Turneri*, 10—100; *Laminaria saccharina* (probably *L. cuneifolia*), 50—100; *Laminaria (Ilea) fascia*, 40—50; *Chordaria flagelliformis*, 40—100; *Dictyosiphon foeniculaceus*, 50—70; *Desmarestia aculeata*, 80—100; *Desmarestia (Dichloria) viridis*, 50—100; *Chætopteris plumosa*, 25—30; *Ectocarpus (Pylaiella) litoralis*, 50—100; *Ectocarpus Landsburgii (Pylaiella varia?)*, 50—100; *Ectocarpus Durkeei?*, 70—80; *Conferva spec. Youngeana?* i. e. *Urospora penicilliformis*, 25—30 fathoms [2]). The correctness of these statements seems for good reasons to be rather questionable.

Above the upper margin of the litoral zone, there are here and there on the arctic coasts to be found lagoons, connected with the sea and possessing from this cause a salinity that makes it possible for marine algæ to flourish in them. Their vegetation is sometimes very rich in individuals, but always highly monotonous, being composed of one, sometimes two or three, species of Chlorophyllophyceæ.

The outlines of the composition of the Flora. Formations of algæ. Three families may be said to dominate the vegetation of the Arctic Sea: *Laminariaceæ, Fucaceæ* and *Corallinaceæ*. They clothe the largest tracts of the bottom, appear in dense, numerically strong masses, and attain a considerable degree of luxuriancy. In consequence of this the other elements of the vegetation are allowed to assert themselves but little in the total impression produced by the vegetation. The Laminariaceæ make the mightiest effect. In the whole extent of the North Polar Sea, these algæ are the most large-sized and occur in the greatest masses and on the widest tracts. In a strict sense, the Arctic Sea might be called the sea of the Laminariaceæ. The Fucaceæ mark the vegetation of larger regions only in those parts of the Polar Sea which are not arctic or at least not purely arctic: on the western coast of Greenland, in the White Sea, in the most westerly portion of the Murman Sea and, above all, in the Norwegian Polar Sea. In the other parts of the Arctic Sea, where they cannot appropriate the litoral region to themselves, they are either wholly wanting, as is the case for inst. in the greatest part of the Kara and the Siberian Seas, or else occur in so little number and so scattered, that their importance as characteristic of the Flora, is none or next to none. Very extensive parts of the sublitoral zone of the Arctic Sea are occupied

[1]) Cp. KJELLMAN, Algenv. Murm. Meer. p. 67.
[2]) Cp. DICKIE, Alg. Sutherl. 1. p. 140—143.

by *Corallinaceæ*. According to KLEEN a species of *Lithothamnion* and *Chorda filum* are the commonest species in the inner sounds and in shallow water in the southern part of the Norwegian Polar Sea on the coast of Nordlanden [1]). Farther to the North, in Tromsö amt and at Finmarken, I have found extensive spaces of the lower sublitoral region covered with species of the genus *Lithothamnion*. *Lithothamnion glaciale* is abundant at several places on the coast of Spitzbergen; at the mouth of Musselbay it covered the bottom to the extent of 4—5 Engl. square miles, in the form of balls that had a diameter of 15—20 cm. [2]). Even on the west coast of Novaya Zemlya I have found a vegetation of *Lithothamnion glaciale*, rich in individuals, growing in regions af considerable extent. [3])

In the same manner a Corallinea, *Lithophyllum arcticum*, characterized the vegetation of a large tract at Uddebay in the Kara Sea [4]). That Corallinaceæ occur in large masses even at Baffin Bay on the west coast of Greenland, is proved by RINK'S above-mentioned description of the marine life on this coast. It is comparatively few and generally small-sized species of algæ, mostly Florideæ, that thrive together with the Corallinaceæ; but even if they appear in greater numbers, as is sometimes the case, still it is the Corallinaceæ that stamp the vegetation with its general character. The bush-shaped *Lithothamnia* occur in very great numbers. In such places the dredges are often brought up over-full.

Although the litoral vegetation of algæ in the Norwegian Polar Sea may be characterized, as mentioned before, as a vegetation of *Fucaceæ*, it should be remarked, however, that it is not uniform in its composition along the whole reach of the coast. Partly the elements of which it is composed are not the same everywhere, partly the same species enter into it in a greater number at some places than at others, partly and principally the number of the Fucaceæ in proportion to other species varies essentially in different parts of the region in question. In the southern part of the Norwegian Polar Sea, besides species of the genera *Fucus*, *Pelvetia* and *Ozothallia*, *Himanthalia lorea* also takes part in constituting the formation of Fucaceæ; in the northern part, on the coast of Finmarken, it is wanting; *Fucus edentatus* occurs at Nordlanden in lesser numbers than on the coast of Finmarken, where it is, in wide tracts, the most common of all the Fucaceæ. At certain places of the litoral region the vegetation is distinctly marked by *Fucaceæ*, at others these are so much inferior to other algæ, that it may be questioned, whether such ranges should really be included within the formation of Fucaceæ. Nowhere, at least not over any larger space, the Fucaceæ are found quite unmixed. Several other species grow upon them or on the rocks and stones beside them, some of these even in large numbers, although they are never able to influence essentially the general aspect of the vegetation. Amongst these the following ought to be mentioned specially: *Rhodomela lycopodioides*, *Polysiphonia fastigiata*, *Rhodymenia palmata*, *Delesseria alata*, *Ptilota elegans*, *Elachista fucicola*, *Spongomorpha arcta* and *Cladophora rupestris*. Those portions of the litoral zone, where

[1]) KLEEN, Nordl. Alg. p. 9.
[2]) Cp. KJELLMAN, Spetsb. Thall. 1. p. 4.
[3]) Cp. KJELLMAN, Algenv. Murm. Meer. p. 66.
[4]) Cp. KJELLMAN, Kariska hafv. Algv. p. 10—11.

the Fucaceæ have much receded, are covered with a very motley vegetation. In general, the mixture of species is such that it can hardly be decided whether one or more species are predominant. However, it cannot be denied that there appears here and there a certain differentiation, Florideæ occurring most numerous in some places, green algæ in others, together with the Fucaceæ. At Finmarken, and, judging by the statements of KLEEN, also at Nordlanden at exposed parts of the coast, where the bottom of the litoral zone is formed of gradually sloping rocks, Florideæ are often found in considerable number: *Rhodomela lycopodioides*, *Polysiphonia urceolata*, *Rhodymenia palmata*, *Halosaccion ramentaceum*, *Gigartina mamillosa*, *Cystoclonium purpurascens* and *Porphyra laciniata*. Here however several other algæ, that do not belong to the Florideæ, also grow abundant, as *Chordaria flagelliformis*, *Monostroma arcticum*, *Spongomorpha spinescens* a. o. Such places of the litoral zone as are rich in tide-pools, are richly furnished with green algæ, though in a very motley mixture, principally with Fucoideæ, but also with Florideæ. This holds good both of Finmarken and, according to KLEEN, of Nordlanden. He says [1]): »The very greatest part of the species observed (at Nordlanden) are to be found between tide-marks, partly and principally in tide-pools, partly on rocks above low-water mark». The following species may be mentioned as characteristic of these parts of the litoral zone of Finmarken: *Corallina officinalis*, (to which are attached *Myriotrichia filiformis*, *Chantransia Daviesii* and *Ch. secundata*), *Lithothamnion polymorphum*, *Hildbrandtia rosea*, *Chondrus crispus*, *Ceramium rubrum*, *Punctaria plantaginea*, *Ilea fascia*, *Dictyosiphon foeniculaceus*, *Enteromorpha intestinalis*, *Monostroma Blyttii* (with *Ectocarpus confervoides* and *Myrionema strangulans*), *Spongomorpha arcta* and *Sp. lanosa*, *Cladophora glaucescens* and *Cl. gracilis* (with *Myrionema strangulans*, small species of *Ectocarpus* and *Pylaiella*). Besides these, smaller species of Fucus are sometimes found, as *F. distichus*, *F. linearis*, *F. filiformis*, *F. miclonensis*, these being then often predominant. In other cases *Enteromorphæ*, *Cladophoreæ*, and *Monostroma Blyttii* hold the most prominent place on account of their superiority in numbers.

Though the litoral vegetation of the Polar Sea on the coast af Norway cannot thus be said to be uniform, still it is not so far gone in differentiation but that it can be regarded as belonging to only one more sharply marked formation of algæ — that of the Fucaceæ.

In the other parts of the Arctic Sea, where the litoral Flora is more rich, especially more rich in Fucaceæ, the uniformity is certainly greater and the differentiation still lesser than in the Norwegian Polar Sea. [2])

It has already been intimated above, that species which in the Norwegian Polar Sea are litoral or most nearly related to litoral species, occur commonly, in other parts of the Arctic Sea, within the sublitoral zone. This is the case for instance with *Rhodymenia palmata*, *Rhodomela lycopodioides*, generally in the form *tenuissima*, *Halosaccion ramentaceum*, *Fucus evanescens*, *Monostroma Blyttii*, *Spongomorpha arcta* a. o. These grow often scattered in small numbers, entering as elements in the formation of Laminariaceæ; but it happens sometimes that they form

[1]) KLEEN, Nordl. Alg. p. 7.
[2]) Cp. above p. 10—11 and CIENKOWSKY, Bericht.

denser and more numerous masses spreading over considerable tracts, thus constituting a whole that is well defined from other formations. In my exposition of the marine vegetation of the Murman Sea, I have described two such regions of vegetation, that I have called the formations (regions) of *Rhodymenia* and *Dictyosiphon*. The former is found both at Spitzbergen and at Novaya Zemlya at the depth of about three fathoms and is characterized by luxuriant *Rhodymenia palmata*. Together with it, there occurred rather abundant, *Ceramium rubrum* at Spitzbergen, *Polyides rotundus* and *Sarcophyllis arctica* at Novaya Zemlya.[1]) The formation of *Dictyosiphon*, known from the Murman Sea in Pilzbay on the west coast of Novaya Zemlya, was found within the upper part of the sublitoral zone on a bottom composed of pebbles. The predominant species was a form of *Dictyosiphon* that I referred then hesitatingly to *Dictyosiphon hippuroides*, but have now described below under a special name: *D. corymbosus*. Besides, *Rhodomela lycopodioides*, *Chætopteris plumosa*, *Punctaria plantaginea*, *Monostroma Blyttii*, *Spongomorpha arcta* formed elements of this Flora. At several places on the coast of Spitzbergen as well as at one place in the eastern part of the Siberian Sea, I have found a region of vegetation that may be said to be characterized by *Fucus evanescens* and *Rhodomela lycopodioides f. tenuissima*. At Spitzbergen these were most often accompanied by *Polysiphonia arctica*, *Rhodymenia palmata*, *Holosaccion ramentaceum* (with *Elachista lubrica*), *Chordaria flagelliformis*, *Sphacelaria arctica*, *Pylaiella litoralis*, *Phlæospora tortilis* and a few rather small specimens of *Laminaria Agardhii* and *L. solidungula*. In the neighbourhood of the winter quarter of Vega on the north-eastern coast of Siberia, on gravelly and stony bottom, at the depth of 2—3 fathoms, the vegetation of a large tract was composed of the following species: *Rhodomela lycopodioides, f. tenuissima*, the most common species together with *Fucus evanescens; Ahnfeltia plicata*, rather abundant; *Sarcophyllis arctica*, the most common *Floridea*, next *Rhodomela lycopodioides; Antithamnion boreale, Alaria elliptica, Laminaria cuneifolia, Chordaria flagelliformis, Elachista fucicola, Chætopteris plumosa*, all of them scarce; *Sphacelaria arctica*, rather common; *Pylaiella litoralis*, common. These facts show that the Flora of those regions of vegetation consists, for the most part, of species which are either litoral on the coast of Norway at the present time, or most nearly related to species that are litoral here. Proceeding from the supposition, which seems to me necessary and to which I shall return subsequently, that the aspect and composition of the marine vegetation on the coast of Norway agreed, during the glacial period, with the present one at Spitzbergen, the conclusion clearly is that no distinct formation of *Fucaceæ* occurred even at the former place at that remote time. The physical conditions changing gradually, *Fucaceæ*, *Rhodymenia palmata*, *Rhodomela lycopodioides*, and other modern litoral forms were able to ascend to lesser depths, to increase in number of individuals, and to receive amongst themselves forms of more southern origin, thus constituting denser, well-defined masses, in short, to unite into the present formation of Fucaceæ. Under this supposition, the above-mentioned formations in the purely arctic parts of the Polar Sea, composed of *Fucus*, *Rhodymenia*, *Dictyosiphon*, and *Rhodomela*, together

[1]) Cp. KJELLMAN, Algenv. Murm. Meer, p. 67.

with their companions, might be regarded as a sort of litoral formation in course of development, precurser of the present formation of *Fucaceæ* on the coast of Norway. It might be named a *prælitoral formation*.

The formation of *Laminariaceæ* is, as mentioned before, the best marked and most widely distributed vegetation in the whole Arctic Sea, covering also the largest area. Its composition varies in different parts both with regard to the species predominating and the species forming the under-vegetation in these arctic woods of algæ. But although the Laminariaceæ comprise a considerable number of species, agreater one, indeed, than has hitherto been supposed, they are all referable to one and the same general type, and on that account the formation constituted by them makes everywhere, on the whole, the same impression. As the variations in the composition of the formation of Laminariaceæ afford the best characteristics for distinguishing and defining different narrower regions of the arctic marine Flora, they will be set forth in detail, in the description of the regions alluded to. The formation of Laminariaceæ is most strongly developed on a bottom of solid rock or large stones. In the Norwegian Polar Sea and in Baffin Bay on the west coast of Greenland its upper margin coincides with high-tide mark, and from there it descends in all its richness to the depth of 2—5 fathoms. Deeper down it diminishes in number of individuals, but often increases in luxuriancy. Within the rest of the Arctic Sea, it generally keeps to the lower parts of the sublitoral region, in about 3—10 fathoms water. On the coast of Norway, in such places where the bottom is looser, consisting of sand, gravel, and smaller stones, within the upper portion of the sublitoral zone, the formation of Laminariaceæ assumes an aspect very different from the typical one. The common species of *Laminaria* and *Alaria* draw back, diminishing in number and luxuriancy, and instead of these another *Laminariea*, *Phyllaria dermatodea*, increases so as to form the main mass of the vegetation, in conjunction with *Chorda filum*, *Chordaria flagelliformis*, *Dictyosiphon hippuroides*, *Monostroma fuscum*, *Spongomorpha arcta*, *Diploderma amplissimum* a. o.

The formation of *Corallinaceæ* is poor in species and differently composed in different parts of the Arctic Sea. On the coast of Norway *Lithothamnion soriferum* predominates. Fastened to this, there are found in small numbers *Chantransia efflorescens*, *Delesseria sinuosa*, *Rhodophyllis dichotoma*, *Kallymenia septentrionalis*, *Antithamnion Pylaisæi* and *boreale*, *Derbesia marina*, and some few others. On the coasts of Spitzbergen and the west coast of Novaya Zemlya *Lithothamnion glaciale* is the principal species. Together with this, there are found *Ptilota pectinata* in great numbers and *Delesseria sinuosa*, *Rhodophyllis dichotoma*, *Euthora cristata* and *Antithamnion boreale* less abundant. [1]) The formation of *Corallinaceæ* on the west coast of Greenland in Baffin Bay is probably constituted in the same way. In the western part of the Kara Sea, where it is charactèrized by *Lithophyllum arcticum*, it was also found to be uncommonly rich in *Euthora cristata*. One or two *Laminariæ* and a small number of *Antithamnion boreale* were met with besides. [2])

---

[1]) Cp. KJELLMAN, Algenv. Murm. Meer., p. 66.
[2]) Cp. KJELLMAN, Kariska hafvets Algv. pp. 10—11.

In the deepest part of the sublitoral zone, near its lowest margin, I have found, on the arctic coast of Norway, a vegetation which I think may be regarded as a remnant from that period when the sea was filled with ice even on the coast of Norway. It is composed almost exclusively of species that are widely distributed within the present Arctic Sea proper, having probably their centre of development in the high North. I shall speak more fully of this subject below. I have observed this vegetation in several places at Gjesvær and Maasö in Finmarken, at the depth of 10—20 fathoms on gravelly and stony bottom. At Gjesvær it was composed of the following species: *Odonthalia dentata*, *Polysiphonia arctica*, *Delesseria sinuosa*, *Rhodophyllis dichotoma*, *Euthora cristata*, *Ptilota pectinata* and *Pt. plumosa*, *Porphyra abyssicola*. Most of these were common even at Maasö, but they were joined here by a very large-sized, broad-leaved *Laminaria*, strongly resembling *L. Agardhii* common in the Greenland Sea and the Murman Sea. The same plant has been found at Nordlanden by KLEEN, who identifies it with *L. Agardhii* and says that it grows in very deep water. This circumstance, combined with the description made by this algologist of the deep-water vegetation in the southern part of the Norwegian Polar Sea, leads me to suppose that the above-mentioned kind of vegetation, which I should prefer to name the *arctic formation of algæ*, occurs even here, though somewhat altered in composition [1]).

Another kind of vegetation, that appears to stand rather independent and to form a well defined whole, is that formation, found at several places in the Arctic Sea, which I have called the formation of *Lithoderma* [2]) after its preponderant species. It grows on gravelly and stony bottom in 5—15 fathoms water. *Lithoderma fatiscens* clothes every stone with a thin crust. Other characteristic species are *Phyllophora interrupta*, *Rhodochorton Rothii*, *Laminaria solidungula*, *Spongomorpha arcta*, and *Chætomorpha melagonium*. It has been found most richly developed on the north and north-west coasts of Spitzbergen at Smeerenbay, at Fairhaven, at Treurenberg Bay, and on the west coast of Novaya Zemlya in the west mouth of Matotshkin Shar. At all these places the arctic *Laminaria solidungula* formed its chief ornament. Traces of the same formation were also observed, during the voyage of the Vega, in the eastern part of the Kara Sea, Lat. N. 76° 8′ Long. E. 90° 25′. The depth here was 15 fathoms. The bottom consisted of larger and smaller stones, covered with *Lithoderma* and some few crusts of *Lithothamnion foecundum*. *Phyllophora interrupta* occurred poor and scarce.

Possibly there is, besides, to be found in the Arctic Sea some or other particular kind of vegetation that would deserve to be mentioned. This seems to be indicated by the large masses of algæ belonging to one or some few species, that have been found to cover considerable reaches of the sublitoral zone. These have assuredly not grown originally in the places where they were discovered, but have been brought there from other localities. It is possible that they occur attached in great quantities somewhere or other; but nothing is known for certain on this point. They have always been found hitherto growing scattered and in little number in the same neighbourhood. Special attention ought to be called to *Phyllophora interrupta*, that is commonly

[1]) Cp. KLEEN, Nordl. Alg. p. 9.
[2]) Cp. KJELLMAN, Algenv. Murm. Meer. p. 66.

found within the formation of *Laminariaceæ*, and in somewhat greater number within the formation of *Lithoderma*, though it is even here rather scarce. It has been discovered in very large masses, lying loose on the bottom, at several places on the coast of Spitzbergen, and at one place in the eastern part of the Kara Sea, in Actinia Bay. In Spetsb. Thall. 1. p. 22 I have mentioned this fact as follows: »In general, it occurs rare along the whole western and northern coast of Spitzbergen, fastened to shells and smaller stones, at the depth of 5—15 fathoms. But in certain places it is found in large masses lying loose on the bottom. At the southern side of Fairhaven a small glacier discharges itself, before which the bottom consists of clay. The depth was here 12—15 fathoms. Here this species occurred in such a quantity that in a little while several tuns of it were brought up with the dredges. Hundreds of the specimens taken here were examined; but in no one any holdfast could be observed. All showed the lower part of the frond in a state of dissolution. The upper parts, on the contrary, were completely fresh, and some individuals were even furnished with nemathecia. However, these specimens differed in colour and consistency from those which were found attached both at Fairhaven and elsewhere. The frond was thinner, faded, and had a strong greenish tint.» J. G. Agardh says about the same plant: »This species, of which only a few specimens had been found during the preceding expeditions, has been discovered during the last both in Green Harbour and in Liefdebay, occurring in the latter place in great quantity at 5—10 fathoms depth, together with *Delesseria sinuosa* and *Halymenia rosacea*. Many of the specimens brought home seemed to be in a state of decomposition with regard to their lower part, while the upper part was perfectly sound. Only some few were found with root, fastened to smaller stones. These circumstances probably indicate that the specimens had been lying loose, having been brought gradually to the locality where they were found gathered in so great a number, and that they continue to live here, their lower part rotting off, the upper part increasing continually by meens of new prolifications. Even other algæ (*Sargassum bacciferum*) are known to possess the faculty of developing in an analogous manner for a long time». [1]) The *Halymenia (Kallymenia) rosacea* seems also to be a species of this kind. For my own part, I have seen it in the Arctic Sea only in very little number and only a couple of times, and I have found only unattached specimens. During the expedition of 1868, judging from the great number of specimens brought home and the statements of J. G. Agardh, it was found at Spitzbergen in most considerable quantities. J. G. Agardh says: »It was brought up in large quantity from the depth of 5—10 fathoms, together with other algæ (*Delesseria sinuosa*, *Conferva melagonium* and *Phyllophora interrupta*). I cannot state positively that any single specimen of the many brought home was fastened to any other alga; they seemed to have lain loose on the bottom in the shape of somewhat flattened balls, or perhaps rather to have adhered originally to fragments of an older frond, perhaps of the last year, from the margin of which they had prolificated». [2]) The condition of another species, *Desmarestia aculeata*, seems to be

[1]) J. G. Agardh, Spetsb. Alg. Till. p. 47.
[2]) J. G. Agardh, l. cit. p. 46.

almost the same. Though certainly common at several places in the Arctic Sea, it is not known to grow anywhere in any considerable number. But it was found lying loose on the bottom in large masses richly overgrown with the otherwise scarce *Antithamnion boreale*, at Musselbay on the north coast of Spitzbergen.

Monotony of the vegetation of the Arctic Sea. It will be clear already from the preceding pages, that the vegetation of at least the greater part of the Arctic Sea has a very monotonous character. Its main mass is distributed over the sublitoral zone; in the other zones it has almost completely vanished, or at least, on account of its poorness, is thrown almost altogether into the shade by that of the sublitoral zone. This sublitoral vegetation certainly varies in composition and aspect on different parts of the bottom, but partly there exist only few formations thus differentiated, partly one of these, the formation of the Laminariaceæ, occupies the largest space, and on this account as well as by its richness and luxuriancy eclipses all the others. Within that formation it is the large-sized Laminariaceæ that produce the general effect. By far the majority of the other elements are comparatively too small and insignificant and too few in number to come out more strongly. The Laminariaceæ belong certainly to pretty many species, but their types are so few and so little diversified as architectonical elements, that the character of this formation cannot exhibit any richer and more marked variety. The vegetation of the Arctic Sea lacks variety not only in form, but also in colour. The general tone is gloomy, the dark-brown colour of the Laminariaceæ is the prevailing one. The lighter brown shades are almost completely wanting. The red algæ are only little apparent, and their red colour is most often of the darker and graver shades. The Chlorophyllophyceæ are almost altogether suppressed. Those numerous varieties of green, from the most vivid grass-green to the lightest whitish- or yellowish-green, which give such vividness and richness of colour to considerable portions of the vegetation of the Atlantic, are wanting in the Arctic Sea.

This picture applies to the greatest part of the Arctic Sea. In the Norwegian Polar Sea the physiognomy of the vegetation is more diversified in form and colour, chiefly on account of its powerfully developed litoral division, composed of Fucaceæ, more prominent Florideæ, and green algæ. This is also the case, though in a less degree, with the vegetation on the west coast of Greenland, in the White Sea and in the most westerly part of the Murman Sea.

The luxuriancy of the vegetation. No inconsiderable number of the algæ of the Arctic Sea are developed to a very high degree of luxuriancy. Referring the reader for particulars to the special part of my work, I only enumerate here the following species: *Lithothamnion soriferum*, *L. glaciale* and *L. polymorphum*, *Odonthalia dentata*, *Polysiphonia arctica*, *Delesseria sinuosa*, *Rhodymenia palmata*, *Hydrolapathum sanguineum*, *Sarcophyllis arctica*, *Halosaccion ramentaceum*, *Phyllophora interrupta*, *Kallymenia rosacea*, *Ptilota plumosa* and *Pt. pectinata*, *Rhodochorton Rothii*, *Porphyra laciniata*, *Diploderma amplissimum*, several species of *Fucus*, *Ilea fascia*, *Scytosiphon lomentarius*, *Desmarestia aculeata*, *Dichloria viridis*, *Phloeospora tortilis*, *Dictyosiphon corymbosus*, *Chætopteris plumosa*, *Sphacelaria arctica*, *Enteromorpha intestinalis*, *Diplonema percursum*, *Monostroma angicava*, *M. cylindraceum*, *M. fuscum* and *M. Blyttii*, *Spongomorpha spinescens* and *S.*

arcta, Cladophora gracilis, Rhizoclonium rigidum, Chætomorpha melagonium and *Ch. Wormskioldii.* I think I may say that all these are quite as luxuriant in the Arctic Sea as when at their best in the Atlantic, or as their nearest relatives there. *Lithothamnion glaciale* for instance forms, as said above, spherical masses of 15—20 cm. in diameter on the coasts of Spitzbergen; *Odonthalia* is larger in the Murman Sea than on the coast of Bohuslän, *Delesseria sinuosa* in the Greenland Sea reaches not seldom a length of 30 cm. and a breadth of 7 cm.; *Sarcophyllis arctica* in the Murman Sea has a length of above the third part of a meter and a breadth of 20—25 cm. On the north coast of Spitzbergen I have seen specimens of *Halosaccion ramentaceum* more than a quarter of a meter long. *Diploderma amplissimum* on the north coast of Norway is sometimes nearly one meter long by a considerable breadth; bushy specimens of *Dichloria viridis,* half a meter long, are not rare on the coast af Spitzbergen. *Monostroma fuscum* on the north coast of Norway and *Monostroma Blyttii* on the west coast of Greenland have not seldom a surface of about half a square meter. *Chætomorpha melagonium* has sometimes in the American Polar Sea a length of 5 feet a. s. o. As, moreover, those species which chiefly decide the aspect of the vegetation, i. e. the Laminariaceæ, attain in the Arctic Sea such a size and development, that they may be reckoned amongst the largest and most luxuriant algæ of the Ocean, the vegetation of the Arctic Sea must evidently bear a character of uncommon greatness, luxuriancy, and vigour.

Thus the most prominent features in the general aspect of the arctic marine Flora are scarcity of individuals, monotony, and luxuriancy.

Probable causes of the peculiarities in the general character of the arctic Flora.

It has been shown, in the preceding exposition, that the general character of the marine Flora varies in different parts of the Arctic Sea, and that this Flora, considered as a whole, presents several remarkable peculiarities of physiognomy, as compared with that of other parts of the ocean. That these peculiarities are essentially, though not exclusively, caused by certain cooperating physical circumstances peculiar to the Arctic Sea, I regard as a settled truth, although it is impossible at present, on account of our imperfect knowledge of the biology of the marine algæ, to state decidedly which these circumstances are, and in what direction and with what power they act. I think, however, that the chief causes are the state of the ice, the configuration of the coast, the tide, the nature of the bottom, the salinity of the water, the temperature of the sea, the temperature of the air, and the want of light.

The state of the ice. The influence of the ice on the vegetation of the Arctic Sea is decidedly unfavourable, the ice either 1:o) making the growth of algæ impossible, or 2:o) making the period of vegetation too short for the algæ to reach their full development, or 3:o) tearing off algæ in development, or 4:o) making the bottom unfit for the prosperous growth of algæ. The two first-mentioned effects are caused by the fixed unbroken land-ice, the two latter by broken-up ice-masses being carried along the shores by waves and currents. In the greatest part of the Arctic Sea there is formed during the winter a girdle of thick, coarse ice, which nearest the shores is pressed close to the bottom. At certain places this land-ice remains throughout the year, at others it is indeed destroyed, but usually only late in the year. I have already mentioned that the ice-foot was found remaining everywhere along the shore at Cape Chelyuskin during our stay there at the end of August, and I may add here, that this ice was so strong and thick, that the shore could not probably get free from ice in the course of that summer. In 1875 the interior of Karmakul Bay in south Novaya Zemlya was still covered with unbroken masses of ice in the last days of June, and even in the middle of July this was in great part the case with the sound between north and south Novaya Zemlya [1]). It is clear that no algæ can develop as long as this fixed land-ice remains, and it seems to me highly probable that when this ice, as is often the case, is not dissolved or destroyed ere late in summer, — consequently only a short time before new ice is formed again — the time left a marine vegetation to spring up in such regions is insufficient for some algæ to attain their full

[1]) Cp. NORDENSKIÖLD, Pröven p. 14 and 22, and KJELLMAN, Algenv. Murm. Meer p. 59.

development and for others to develop so well as they would do under other circumstances. Besides, on account of this long interruption in the period of vegetation, it becomes impossible for all other algæ to grow on this part of the bottom than those which possess such organs of reproduction as enable them to rest, without injury, during a long part of the year, exposed to a low temperature.

The same region of the bottom, but also its deeper parts, above all the upper part of the sublitoral zone, is subjected to the action of the drift-ice, when it lies still along the coasts, or floats quietly along them, or is rolled on with tremendous violence by the furious waves, mighty masses of ice being thrown or screwed high up on the shores, leaving the bottom, where they have passed, bare and desolate. The drift-ice always exercises a friction now feebler now stronger on the bottom nearest the shores, by which friction the marine vegetation is decimated, masses of mud and small shingle are formed, and rocks and stones are smoothed and, as it were, polished. I believe that the scarcity of algæ within the litoral zone and the upper part of the sublitoral in the greater part of the Arctic Sea depends on this pernicious influence of the ice. The known algologist DICKIE has a priori arrived at the same conclusion. In his description of a collection of algæ brought together in the American Arctic Sea during one of the English arctic expeditions, he says: »The number of litoral species in such regions must be few or in many places altogether absent; the continual abrading influence of bergs and pack-ice would effectually prevent their growth.»[1]) That difference in the distribution of the vegetation among the several bottom-zones, which appears in different parts of the Arctic Sea, seems thus to be essentially connected with a difference in the character of the ice, in such a manner that if the other circumstances are the same, a more equal distribution of the vegetation on the sublitoral and the litoral zones takes place, if the ice is more favourable. With regard to the formation and drifting of ice, the Norwegian Polar Sea is most favourably situated of all parts of the Arctic Sea. Ice is never formed here in greater quantities, nor does the polar ice set down here. In the White Sea ice is formed during the winter, but in the summer the water is free from ice. The state of the ice within the rest of the Arctic Sea may be regarded as being, on the whole, pretty much the same. In the eastern part of the Greenland Sea along the west coast of Spitzbergen, in the eastern part of the Murman Sea, and in the eastern part of Baffin Bay, the state of the ice is relatively favourable in summer, in the two first-mentioned regions in consequence of the Gulf-stream, in the so-called North-Greenland on account of the westerly direction of the ice-current, after it has turned Cape Farewell. However, even here the sea is probably no year free from ice even in summer, though it is not so compact nor packed on to the coasts in so large quantities as to make them inaccessible during any year. The arctic expeditions of the later years have shown that the sea off the east coast of Novaya Zemlya and the north coast of Siberia is rich in ice, though near the coasts it is less thick and more divided. Such is especially the case before the mouths of the great Siberian rivers, where during the summer-months the main mass of the polar ice is kept from the land by currents flowing in an easterly direction, and the coast ice

[1]) DICKIE, Alg. Sutherl. 2, p. 200.

is broken up or melted. [1]) North and east of Spitzbergen the state of the ice is more difficult, as is also the case in the American Arctic Sea, where the English arctic voyagers have been able to perform their grand work of discovery only by small steps and by means of hard, continual struggles against the ice. The part of the Polar region which is the most inaccessible of all in consequence of the ice, is the east and south coast of Greenland, against which the mighty Polar current, coming from the north-east and the east, presses its huge masses of ice. In accordance and, I believe, causal connexion with these conditions, we find the vegetation on the upper part of the bottom in the Greenland Sea, the eastern Murman Sea, the Kara and the Siberian Seas, and the American Arctic Sea extremely poor, in the south-western part of Baffin Bay and in the White Sea richer and more luxuriant, though monotonous, in the Norwegian Polar Sea luxuriant and rich both in individuals and species.

The configuration of the coast. It is a well-known fact that certain algæ choose exclusively, or at least prefer, such parts of the coast as are exposed to the open sea, and that others, on the contrary, attain their most vigorous development and grow in the greatest abundancy in sheltered localities. This certainly applies chiefly to litoral algæ, but even among the sublitoral ones there are to be found forms that are pelagic and such as are not. Thus the composition and general character of the marine vegetation may depend, in a certain degree, on the configuration of the coast. If the other circumstances are alike, a coast ought to be more favourable for the growth of algæ, the more extensive and rich its border of outlying islands (»skärgård») is, and the more the coast is intersected by numerous and deep bays. In this respect, however, the configuration of the coast is no doubt of very little importance for the vegetation of the Arctic Sea, but it is important with regard to its being more or less fit to protect the vegetation against the destructive agency of the drift-ice. A rich »skärgård» forms a fence against the drift-ice, within which algæ may spring up and complete their development in peace, and the ice can hardly penetrate in any great quantity into deep, narrow bays. It seems probable that the richness and luxuriancy of growth which mark the vegetation among the isles off the north-western coast of Spitzbergen are in no slight degree owing to the shelter afforded it by the isles against large, deep-going floes or blocks of ice floating about here. I will also remark that I have never, within the confines of the Arctic Sea, found any noteworthy litoral vegetation on an exposed coast, but only in sheltered places on the inside of the isles, for instance at Fairhaven at Spitzbergen, or in deeper bays, for instance in the interior of Ice-fjord at Goose-islands at Spitzbergen, in the deep Besimannaya Bay, in Karmakul Bay, which is shallow but filled with a number of rocks and islets, on the west coast of Novaya Zemlya, in Actinia Bay, which is sheltered in all directions and outside which is situated the largest »skärgård» in the whole Siberian Sea, except the group of isles at the mouth of the Yenissei. I have mentioned before that Actinia Bay was one of the places richest in algæ that I found during the voyage of the Vega north of Asia. With regard to favourable configuration, the coasts of Norway and Greenland are much better off

[1]) Cp. NORDENSKIÖLD, Vega-exp. 1. p. 23, 154, 155.

than the other coasts of the Arctic Sea. Here there is found a rich »skärgård» and the coast is intersected by numerous larger and smaller bays entering into the land in different directions. At Spitzbergen the »skärgård» is small, and the bays comparatively few, being moreover too wide to afford any greater protection against drift-ice. The coasts of southern Novaya Zemlya and of Waygats are still worse off in this respect, and on all the long Siberian coast, judging from the experience gained during the Vega expedition, there is scarcely to be found any place where a larger ship can lie safe from the waves and the drift-ice during a sea-wind, except Dickson Harbour and Actinia Bay. About half of the American coast on the Arctic Sea appears to be very open. The existence of a richer and more vigorous marine vegetation on the west coast of Greenland in the litoral zone and the upper part of the sublitoral zone may be, and in my opinion ought to be, explained by this coast possessing a large number of isles and bays and affording thereby the necessary protection to the algæ against the drift-ice.

The tides. The tides may be considered to contribute indirectly towards impoverishing or annihilating the vegetation on the upper part of the bottom in the greater part of the Arctic Sea, because by keeping the ice in continual motion they make it perform incessantly its destructive work, and because larger parts of the bottom can be reached and abraded, at certain times, even by more shallow-going ice. Moreover, the litoral zone being laid bare at low-water, the vegetation which may possibly be found in that region, becomes exposed, at least during certain parts of the year, to conditions that must be regarded as unfavourable. This point will be considered below. Even in winter the ice which lies along the coasts is not in repose, whatever may be the size of the pieces. During the winter stay of the Swedish expedition on the north coast of Spitzbergen, the sea off Musselbay was covered by masses of ice, several miles broad and apparently hard frozen together. From these a grating sound was heard incessantly, caused by the friction exercised by the ice-blocks and ice-floes on one another during their ceaseless rising and sinking and their slight progressive and regressive movements. But in these movements, however insignificant, caused by the tidal currents, the ice-masses exercise a continual friction even on the bottom. In summer the motion of the ice, produced by the tides, is sometimes very violent, especially in narrow sounds and bays. Out of a number of instances, I choose one from Novaya Zemlya. In the western part of the strait, Matotshkin Shar, which divides its two main islands, there goes, during the ebb, a very violent current from east to west. During the stay of the Swedish expedition of 1875, the ice in the interior of the strait was breaking up, and the drift-ice, thus formed, rushed westwards during the ebb with such a violence, that the little vessel of the expedition was in great danger and had to be removed incessantly from one anchorage to another in order to be sheltered from the ice. »Once the icemasses, floating vehemently forwards in the strait,» says NORDENSKIÖLD in his relation of this incident, »were on the very point of tearing away our little smack from an anchorage somewhat imprudently chosen and of either pressing it up on the dry land or carrying it out into the open sea» [1]). The rush of

[1]) NORDENSKIÖLD, Pröven, p. 22.

such a mass of ice cannot certainly be withstood by algæ. And it ought to be remarked that the eastern shores of that inconsiderable group of isles which is situated in the western mouth of the strait, were altogether without litoral algæ, while such plants were found to grow, although in little number, on the western shores, which were evidently less exposed to drift-ice.

The height of the tide is comparatively small in the Arctic Sea, and it is, moreover, so equal in general in different parts of it, that even if the influence of the tidal currents were another and a more powerful one than I have tried to show above, the existing difference in the height of the tide cannot be considered as having contributed essentially to the existing difference in the physiognomy of the vegetation within larger tracts of the Arctic Sea proper, that is to say, that sea which abounds in drift-ice during summer. In the eastern part of the Siberian Sea and in the western part of the American Arctic Sea the tides are scarcely perceptible. According to researches made during the voyage of the Vega, the flood-tide at the wintering place Pitlekay [1]) is only 18 cm. At Point Barrow, according to Markham [2]), it is only 7 inches. In the eastern part of the American Arctic Sea it is somewhat higher: on the south coast of Melville Island in Winter Harbour it averages 2 f. 6½ i. in May, 2 f. 7 i. in June, and 2 f. 8½ i. in July [3]). On the east coast of Greenland, at Sabine Island, according to the observations of the second German Polar expedition, the height of the spring-tide is on an average 4,21 Engl. feet, that of the neap-tide 1,86 [4]). At Spitzbergen, according to DUNÉR and NORDENSKIÖLD, the spring-tide is 5—6 f., the neap-tide 3 f. [5]). SPÖRER states the flood-tide to rise 2—3 f. on the west coast of Novaya Zemlya, about 1 f. 4 i. on the east coast [6]). In the south-western part of Baffin Bay the height of the tide is very considerable, about 30 feet [7]), if the statements made to DICKIE are indeed reliable. If such is the case, this tide ought to influence the general character of the vegetation in a remarkable degree.

The nature of the bottom. Whether the chemical composition of the bottom exercises any influence on the marine vegetation or not, is a question that still remains unanswered, and, in my opinion, it cannot be answered by means of the material which we have at our disposition at present. But it is certain and undeniable that the growth of marine algæ, their distribution, richness, variety, and luxuriancy, are essentially connected with and dependent on the physical nature of the bottom. There are tracts of the bottom whose structure is such that algæ do not grow and cannot grow there, however favourable the other conditions may be, while, on the other hand, others are clothed with a rich and luxuriant vegetation, although the physical circumstances are in other respects as unfavourable as possible for the development of a richer Flora. Wherever the bottom is very loose, i. e. formed of mud, sand, and clay, algæ

[1]) See NORDENSKIÖLD, Vega-exp. 1. p. 76.
[2]) MARKHAM, Threshold, p. 221—222.
[3]) See PARRY, Zweite Reise, p. 375.
[4]) See KOLDEWEY, Zweite deutsche Polarf. 2. p. 658.
[5]) DUNÉR, NORDENSKIÖLD, Spetsb. geogr. p. 11.
[6]) SPÖRER, Nov. Semlä, p. 57—58.
[7]) DICKIE, Alg. Cumberl. p. 236.

are wanting, because there are here no larger solid objects to afford the algæ that foot-hold which they need, at least during some part of their existence, in order to attain their full and normal development. On the contrary, no bottom consisting of gravel shells, larger and smaller stones, and hard rocks, especially if furnished with cavities, a. s. o., wants algæ, if the other circumstances are favourable. The rest of the conditions being equal, the marine vegetation is more extensive in proportion as the muddy, sandy, or clayey tracts of the bottom are smaller, and it is richer in individuals and more luxuriant when the bottom is coarser and more solid; it is possible, however, that it may become more diversified, the more the composition of the firmer bottom varies. For it seems, at least in Kattegat, as if certain algæ should stick exclusively, or chiefly, to a certain sort of bottom. So called shelly bottom is especially remarkable for its richness in peculiar species of algæ. I must leave it undecided whether the great scarcity, in the Arctic Sea, of several species which are found most often and in the greatest number in shelly localities on the coast of Bohuslän, is occasioned by the absence of such bottom or by other causes. But this is by no means impossible or improbable. The algæ of the Arctic Sea make larger claims than others on the firmness of the bottom. They need a surer foot-hold in order to be able, on the generally rather exposed coasts, to withstand the drift-ice together with the waves and the violent currents, without being prematurely torn off and destroyed. Very considerable stretches of the bottom of the Arctic Sea are however of an unfavourable structure. Only on the north coast of Scandinavia and the west coast of Greenland [1]), where the ground consists of hard azoic rocks, it can be said to be mainly good. Such rocks predominate, indeed, at comparatively large reaches of the north-west and north coast of Spitzbergen, for instance in the group of isles about Fairhaven, and the bottom, from that cause, is favourable, but along very great stretches of the coast of Spitzbergen the rocks are schistous and of looser consistency. Loose slates and sandstones going down to the sea, the largest space of the bottom is formed there of clay and sand. This is the case also with those parts of the west coast of Novaya Zemlya and Waygats which have hitherto been subjected to algological investigations, and probably also on the east coast. A small part of the east coast of Novaya Zemlya has been examined, at Uddebay. A great part of the east coast northward of this point is occupied by glaciers [2]), and the experience from other arctic countries has shown that outside and near glaciers the bottom of the sea is of loose consistency. The southern and south-eastern part of the Kara Sea along the peninsula of Yalmal has surely a most unfavourable bottom. NORDENSKIÖLD, who landed at one place on the west coast of Yalmal, Lat. N. 72° 18', says: »No solid rock was to be found here. The ground everywhere consisted of sand and sandy clay, in which I was not able to find a single stone of the size of a gun-ball or even of a pea, though I sought for it for several kilometers along the coast-bank. Nor did the dredge ever bring up any bits of rock from the bottom of the sea off the coast» .... [3]) It proceeds from the journal of dredgings, kept during

[1]) Cp. KORNERUP, Grönl. Medd. 1. p. 226.
[2]) See KJELLMAN, Pröven p. 49.
[3]) NORDENSKIÖLD, Pröven p. 40.

the Vega voyage and published by STUXBERG, that in the eastern part of the Kara Sea and in the Siberian Sea the very greatest part of the bottom is formed of sand and clay [1]). Only at some few places of the region west of Taimyr Island, at Irkaypi and at the mouth of Kolyutshin Bay, the bottom was favourable, at another, viz. Actinia Bay, it was found to be tolerably good. In the American Arctic Sea, at least in that part of it where the Archipelago is situated, the bottom probably resembles most nearly that on the coasts of Spitzbergen. Primitive rock is to be found there, but of comparatively little extent. The ground consists chiefly of schistous rocks, lime- and sandstone, belonging to the silurian and coal formations [2]).

I hold the disadvantageous structure of the bottom to be an essential cause why so large stretches of the Arctic Sea possess no algæ, why the eastern Kara Sea and the Siberian sea are furnished with such plants only within so inconsiderable areas, and why the Flora of the greater part of the Arctic Sea is in general so limited in number of individuals.

Salinity of the sea. Another circumstance that contributes certainly to the extraordinary scarcity of algæ in the eastern part of the Kara Sea and the greater part of the Siberian Sea, is the slight salinity of the water, caused by the great quantity of fresh water brought down by the large Siberian rivers and carried along the coast in an easterly direction. With regard to the hydrographical conditions of these seas, I take the liberty to quote the statements of NORDENSKIÖLD, who has particularly studied this subject. About the sea between the mouth of the Yenissei and the New Siberian Islands, he says: »If the depth reaches at least 30 metres, the temperature at the bottom varies between — 1° and 1°,4 C. The specific gravity of the water amounts to from 1,026 to 1,027, corresponding to a salinity little less than that of the Atlantic. At the surface the temperature was exceedingly variable. Thus for instance it was + 10° at Dickson Harbour, + 5°,4 a little south of Taimyr Straits, + 0°,8 amongst the drift-ice immediately off this strait, + 3° off Taimyr Bay, + 0°,1 at Cape Chelyuskin, + 4° off Chatanga Bay, + 1°,2 to 5°,8 between the Chatanga and the Lena. The specific gravity of the surface water in a broad channel along this part of the coast never exceeded 1,023, most often it only amounted to 1,01 or less. The latter figure corresponds to a mixture of about one part of sea water with two parts of river water. These figures show incontestably that a warm and only slightly salt surface current runs from the mouths of the Obi and the Yenissei along the coast in a north-easterly direction, and afterwards, under the influence of the earth's rotation, in a more easterly course. Other similar currents proceed from the Chatanga, Anabar, Olenek, Lena, Jana, Indigirka and Kolyma.» [3]) About the sea east of the New-Siberian Isles he says: »Eastward from this point the sea was ice-free nearest the coast. The water was slightly salt and its temperature rose to + 4° C. . . . . In one respect, there exists a great difference in the nature of the sea off the Siberian coast west and east of Cape Baranow. On the western side a number of large rivers, the Ob, Yenissei, Piasina, Taimyr, Chatanga,

---

[1]) STUXBERG, Vega-Exp. I. p. 684—687. Cp. p. 690.
[2]) HAUGHTON, Fox-Exp. App. 4.
[3]) NORDENSKIÖLD, Vega-Exp. 1. p. 23.

Anabar, Olenek, Lena, Jana, Indigirka, Alasei, and Kolyma are discharged into the Arctic Sea and during the summer produce comparatively warm currents of water along the coast; on the eastern side, on the contrary, no large river issues into the sea. Accordingly, no coast currents conducive to the forming of an ice-free sea occur here, as is the case along the whole coast from the White Sea to the Kolyma.» [1]) Detailed statements on the saltness of the water in the Siberian Arctic Sea will be had, when the hydrographical observations regularly made during the Vega expedition shall have been worked out. I have only to add here, that according to these observations the specific gravity of the sea-water at the surface at the dredging-places between Cape Chelyuskin and Cape Baranow never exceeded 1,023, and generally kept at about 1,01, i. e. did not even reach the gravity corresponding to a mixture of one part of sea-water with two parts of river-water. Off the Lena for instance, the specific gravity amounted only to from 1,0040 to 1,0046, i. e. to about as much as that of the water in the southern part of the Botnic Bay [2]).

It is certainly quite true that in the eastern part of the Kara Sea and in the western part of the Siberian Sea the salinity of the water increases with the depth, as has been stated both by NORDENSKIÖLD and STUXBERG in their above-mentioned works. Observations made during the Vega expedition are quite conclusive on this point; some of them have been set forth by STUXBERG. But on examining the figures contained in the dredging journal of the Vega, it will be clearly seen that at many places from the surface down to that depth where algæ grow in the greatest number in the Arctic Sea proper, the salinity of the water is comparatively slight, essentially less than in many other seas and in the greater part of the Arctic Sea itself. The following figures out of the journal just mentioned may serve to prove this fact [3]).

Lat.	Long.	Depth in fathoms.	Spec. gravity of the water at the bottom.
74° 52' N.	85° 8' E.	6	1,0133.
73° 41' »	114° 58' »	6	1,0151.
74° 9' »	130° 20' »	15	1,0050.
74° 4' »	135° 38' »	16	1,0128.
73° 53' »	138° 0' »	12	1,0165.
73° 40' »	140° 16' »	4	1,0120.
73° 2' »	142° 36' »	9	1,0145.
73° 5' »	144° 20' »	8	1,0144.
72° 20' »	153° 30' »	10	1,0202.
71° 39' »	157° 15' »	10	1,0198.

These figures prove that in the eastern part of the Kara Sea and still more in the western part of the Siberian Sea, namely the extensive stretch between the mouths of the Chatanga and the Kolyma, the sea-water at the coast at the depth of about

[1]) NORDENSKIÖLD, Vega-exp. 1, p. 29 and 154—155.
[2]) Cp. STUXBERG, Vega-exp. 1, p. 684—687 and 694
[3]) Cp. STUXBERG, Vega-exp. 1, p. 684—687.

5—15 fathoms has a most considerably lesser gravity, i. e. salinity, than sea-water in general. It is not to be supposed, indeed, that this salinity is too slight to allow marine algæ in general to grow, but is certainly too little to offer tolerable or suitable conditions for the development of purely pelagic forms, such as the majority of the arctic algæ. It is remarkable that the only localities richer in algæ that have been met with in these seas, are situated in those parts where the influence of the large Siberian rivers is least felt, viz. in the most easterly parts of the Kara Sea about Taimyr Island and in the Siberian Sea east of Cape Baranow.

In the other parts of the Arctic Sea the salinity of the water is nearly equal to that of common sea-water, as is shown by the following table.

Place.	Depth in fathoms.	Salinity of the water.	Source of the statement.
The North Sea (German Ocean)	—	3,26	Cool. Phys. Geogr. p. 269—270.
The Atlantic between Scotland and Newfoundland	—	3,59	» »
North Cape and Spitzbergen	—	3,53	» »
The Norwegian Polar Sea, Fegle Sound	0	3,34	NORDENSKIÖLD, Pröveu. p. 110.
The Morman Sea, Besimannaja Bay	0	3,27	» »
» »	50	3,42	» »
» , western mouth of Matotshkin Shar	0	3,09	» »
» » »	20	3,38	» »
The White Sea	—	3,22	STUXBERG, Vega-exp. p. 694.
The American Arctic Sea, Lancaster Sound	0	3,32	PARRY, Zweite Reise, p. 126.
Baffin Bay, Simiutat on the west coast of Greenland	0	3,41	JENSEN, Gröul. Medd. p. 206—207.
Nagsugtok » »	0	3,30	
» » »	5	3,31	
» » »	10	3,35	
» » »	20	3,36	
The Greenland Sea off the east coast of Greenland	0	3,326	BÖRGEN, Zweite deutsche Polarf. p. 680.

Temperature of the sea. The following statements seem to me to afford sufficient materials for deciding whether the conditions of temperature in the Arctic Sea are to be considered as causing in some degree the peculiarities in the general character of the marine vegetation. In the Norwegian Polar Sea near North Cape, according to MOHN, the average temperature of the sea is

during	December—February	+ 3°,03 C.
»	March—May	+ 3°,0 »
»	June—August	+ 7°,3 »
»	September—November	+ 6°,3 » [1]

Observations made during the Swedish expedition of 1872—1873 north and west of Spitzbergen in the month of July (1—18) gave the following results:

[1]) MOHN, Temp. Verhältn. p. 429.

	Medium.	Maximum.	Minimum.
Temperature of the water at the surface............	+ 3,3	+ 5,2	+ 1,5
» at the depth of 16—280 fathoms	— 0,9	+ 1,9	— 3,2 [1])

According to observations by the same expedition at Musselbay on the north coast of Spitzbergen, the temperature of the sea at the end of September and during the whole of October was about — 1°,0, during November from — 0°,5 to 1°,0, and varied during December to April between — 1°,5 and — 1°,8, rising a very little during the few days when the sea was free of ice. It never rose during the period mentioned to more than — 1°,0 C. [2]).

In the eastern part of the Murman Sea along the west coast of southern Novaya Zemlya and Waygats, the Swedish expedition 1875 found the temperature of the water at the surface of the sea to be:

	Medium.	Maximum.	Minimum.
During June (22—30).....................	+ 1,31	+ 2,4	+ 0,0.
» July.......................................	+ 4,33	+ 8,8	+ 0,6. [3])

With increase of depth the degree of warmth diminishes, as is shown for instance by the following researches off the west coast of Novaya Zemlya Lat. 72° 43' N. Long. 52° 0' E.

Temperature at 0 fathoms.........................	+ 0°,6 C.
» » 10 »	— 1°,4 »
» » 20 »	— 1°,0 »
» » 30 »	— 1°,7 » [4])

The Swedish expedition 1875 found the mean temperature of the water at the surface in the western part of the Kara Sea during 2—3, 24—31 August and 1—2 September to amount to + 3,17 [5]).

That the warmth of the water even here decreases towards the depth, is shown by the following observations from different parts of the sea between Novaya Zemlya and the mouth of the Yenissei.

1. West of Yalmal Lat. 72° 19' N. and Long. 18° 40' E. August 8:th.

Temperature of the water at 0 fathoms....	+ 5°,2 C.
» » » 2 »	+ 3°,0 »
» » » 3 »	+ 2°,4 »
» » » 5 »	+ 1°,5 »
» » » 8 »	+ 0°,0 » (near the bottom).

2. Lat. 73° 30' N., Long. 69° E. August 9:th.

Temperature of the water at 0 fathoms....	+ 7°,8 C.
» » » 2 »	+ 7°,0 »
» » » 4 »	+ 0°,6 »
» » » 8 »	— 1°,0 » (near the bottom).

[1]) Nordenskiöld, Pröven, p. 109.
[2]) See Kjellman, Vinteralgveg. p. 62—63.
[3]) Cp. Nordenskiöld, Pröven, p. 92—98.
[4]) Nordenskiöld, Pröven, p. 106.
[5]) Cp. Nordenskiöld, Pröven, p. 99—103.

3. Off the mouth of the Yenissei Lat. 73° 55′ N. Long. 60° 40′ E. August 15:th.
Temperature of the water at 0 fathoms.... + 7°,4 C.
» » » 15 » .... — 1°,4 » (near the bottom).

4. Off the east coast of northern Novaya Zemlya Lat. 75° 40′ N. Long. 65° E. August 25:th.
Temperature of the water at 0 fathoms.... + 1°,4 C.
» » » 60 » .... — 1°,8 » (near the bottom).

5. Off the east mouth of Matotshkin Shar Lat. 73° 34′ N. Long. 58° E. August 31:st.
Temperature of the water at 0 fathoms.... + 3°,9 C.
» » » 25 » .... + 1°,4 »
» » » 55 » .... — 1°,7 » (near the bottom) [1]).

In the eastern part of the Kara Sea and in the western part of the Siberian Sea the temperature of the water, as I have already intimated, is dependent on the neighbourhood of the mouths of the large rivers. At the depth of more than 5 fathoms the temperature is generally below 0°, except in the proximity of the rivers. At Dickson Harbour the water even at the depth of 5 fathoms has a temperature of + 9°,0 during the first part of August, and off the mouth of the Lena somewhat later in the year its temperature amounted to + 3°,8 at 15 fathoms. The influence of the latter river reaches as far as to the New Siberian Islands. West of one of these, Blishni Island, the temperature at the depth of 4 fathoms was the same as at the surface, + 2°,6. At all the places east of the New Siberian Islands where dredging were carried on during the voyage of the Vega, the temperature at the depth of 3—10 fathoms kept at or below 0°. [2])

A good guidance in judging of the conditions of temperature in the northern part of the American Arctic Sea is afforded by the observations made during Belcher's expedition. According to these [3]), the average temperature of the water in Northumberland Sound is:

	at the surface.	at the bottom.
during September	— 1°,17 C.	— 1°,28.
» October	— 1°,56 »	— 1°,50.
» November	— 1°,50 »	— 1°,50.
» December	— 1°,67 »	— 1°,67.
» January	— 1°,62 »	— 1°,67.
» February	— 1°,67 »	
» March	— 1°,56 »	
» April	— 1°,39 »	
» May	— 1°,39 »	
» June	— 0°,22 »	
» July	+ 0°,11 »	
» August	— 0°,83 »	

[1]) See NORDENSKIÖLD, Pröven, p. 107—108.
[2]) See STUXBERG, Vega-exp. 1, p. 686—687.
[3]) Cp. Contrib. arct. Meteor. 2. p. 172.

From »Meddelelser om Grönland» I borrow the following statements, which show that in bays on the west coast of Greenland the temperature of the water at the surface is considerably lower at the mouth than in the interior of the bay, and at the same time that in the interior of deep bays the temperature of the water decreases strongly and rapidly towards the depth.

The temperature of the water near the mouth of Nagsugtok Bay Lat. 67° 32' N. Long. 53° 28' W. July 11:th at 0 fathom was from 2°,0 to 2°,2; in the interior of the same bay Lat. N. 67° 47' Long. 52° 22' W. in July:

at 0 fathoms	+ 8°,8	C.
» 5 »	+ 2°,8	»
» 10 »	+ 1°,1	»
» 20 »	+ 0°,0	»
» 30 »	+ 1°,0	»[1])

According to the observations of the second German expedition, the temperature of the sea at the surface during the month of August on the east coast of Greenland along Shannon, Pendulum, and Sabine Isles (Lat. 74° 30'—75° 30' N.) varies between ÷ 2°,0 and — 1°,6 C. On examining the temperature of the water beneath the winter ice, it was found to be

1869 October 3:d	— 2°,2	C.	
» » 29:th	— 1°,9	»	(at 27 fms)
» November 11:th	— 2°,2	»	
1870 January 20:th	— 2°,1	»	
» February 18:th	— 2°,5	»	
» May 21:st	— 1°,9	»[2])	

All these facts now set forth may be condensed, I think, as follows: in the Arctic Sea proper, i. e. the Greenland Sea, the eastern Murman Sea, the Siberian Sea, the American Arctic Sea, and Baffin Bay, the average temperature of the surface water in the middle of summer is about as high as or lower than in the Norwegian Polar Sea during winter (Dec.—Febr.), and at that depth where the richest marine vegetation is to be found, it does not rise in general above 0° C. at any time of the year. This difference of temperature between the Norwegian Polar Sea and the other above-mentioned parts of the Arctic Sea, is assuredly the most important cause, though it may not be the only one, of the essential difference shown by the Flora of the Norwegian Polar Sea as to its general character, in comparison with that of the rest of the Arctic Sea.

The temperature of the air is probably also an element that ought to be noticed in accounting for the peculiarities of the arctic marine vegetation.

Of course the temperature of the air cannot exercise any influence but on those portions of the vegetation which may come in contact with the air, that is to say, the vegetation of the litoral zone. Possibly the great poverty and scantiness of this vegetation depend in some part on too cold airs sweeping over the exposed litoral zone at certain times,

[1]) Cp. JENSEN, Grönl. Medd. 2. p. 207.
[2]) Cp. KOLDEWEY, Zweite deutsche Polarf. 2, p. 618—620.

destroying the vegetation that has begun to spring up. This may happen in spring at those parts of the coasts of the Arctic Sea, for instance the west coast of Spitzbergen, where the winter ice breaks up early, and again in autumn, before new ice has been formed along the shore, and even in the middle of winter, if the masses of ice draw off suddenly from the shores, as happens probably not unfrequently from time to time. The account of the Swedish wintering expedition at Spitzbergen 1872—73 shows the sea at the north coast of Spitzbergen to have been open even to the shore several times in the course of the winter [1]). During our wintering in the Vega, there were formed once or twice wide openings in the sea, which may possibly have reached to the shore in the neighbourhood of our wintering station. It appeared from the statements of the natives living there, that the sea opens now and then in winter, though it freezes soon again. It may be objected that, the sea being open at the coast, the temperature of the air cannot be so low as to be injurious to algæ. This may be true, indeed, in general, but it is to be remarked that there occur nowhere else so sudden and strong variations of temperature as in the arctic regions. Amongst many instances of this I select one from the voyage of the Vega. In February 1879 the temperature of the air at noon on the 6:th was — 40°,4 C., at the same hour two days later + 0°,1, on the 12:th — 2°,0, but on the 13:th — 24°,9 and on the 15:th — 29°,0 C. At Musselbay the sea once in the winter froze at a temperature of the air of — 27°,6 [2]), to which low degree it had descended in the course of a few hours. It need not be supposed that a low temperature must continue long, in order to be hurtful. Just as one night of sharp frost suffices to damage the land-vegetation, the extreme degrees of temperature in the arctic regions may act destructively, if their action lasts during one or two tidal periods.

The temperature of the air at different parts of the coasts of the Arctic Sea is set forth in the following table. From this, several conclusions may be drawn with regard to the biological conditions of the algæ of the Arctic Sea, which I shall bring forward below.

Table [3]) showing the average temperature of the air in different parts of the Polar region.

	Tromsö, Norway.	Vardö, Norway.	Musselbay.	Novaya Zemlya.	Pitlekay.	Point Barrow.	Northumberland.	Jakobshavn, West-Greenland.	Sabine, Island, East-Greenland.
January	— 4,20	— 6,00	— 9,89	— 13,72	— 25,06	— 28,20	— 39,22	— 17,40	— 24,15
February	— 4,0	— 6,40	— 22,69	— 18,49	— 25,09	— 30,42	— 33,44	— 17,30	— 23,81
March	— 3,80	— 5,10	— 17,68	— 15,43	— 21,65	— 26,02	— 27,50	— 16,70	— 23,32
April	— 0,10	— 1,70	— 18,12	— 13,94	— 18,93	— 15,72	— 22,89	— 10,40	— 16,51
May	+ 3,20	— 1,80	— 8,26	— 3,79	— 6,79	— 6,61	— 9,44	— 0,10	— 5,42
June	+ 8,70	+ 5,90	+ 1,11	+ 2,41	— 0,60	+ 0,18	— 0,06	+ 4,40	+ 2,26
July	+ 11,50	+ 8,80	+ 4,55	+ 4,80	+ 2,68	+ 2,67	+ 2,61	+ 7,70	+ 3,80
August	+ 10,40	+ 9,80	+ 2,87	+ 4,66	——	+ 7,30	+ 1,22	+ 6,20	+ 0,67
September	+ 7,00	+ 6,40	— 3,86	— 0,28	——	— 3,22	— 7,50	+ 1,10	— 4,32
Oktober	+ 2,00	+ 1,30	— 12,69	— 1,88	— 5,20	— 16,89	— 18,50	— 4,80	— 13,82
November	— 1,70	— 2,10	— 8,13	— 15,67	— 16,58	— 22,47	— 20,33	— 7,50	— 18,32
December	— 3,20	— 4,00	— 14,44	— 26,61	— 22,80	— 25,16	— 34,50	— 11,80	— 17,14

[1]) See NORDENSKIÖLD, Spetsb.-Exp. p. 55—58.
[2]) NORDENSKIÖLD, l. c. p. 58 and WIJKANDER, Obs. météor. p. 20—21.
[3]) Cp. HILDEBRANDSSON, Obs. Météor. p. 578—579 and KOLDEWEY, Zweite deutsche Polarf. p. 536.

Want of light. I have already mentioned the scantiness of green algæ as a characteristic feature of the vegetation of the Arctic Sea proper. The species of this kind existing there occur in very little number, and are generally very, not to say extremely, poorly developed, sometimes stunted to such a degree that they can hardly be recognized. The supposition lies near at hand, that one of the causes of this state of things is the want of light, the majority of the green algæ being known to love light and to prefer in general such localities where they can enjoy it in the greatest quantity. In consequence of several concurrent causes, many of them cannot, on the coast of the icy Polar Sea, spread themselves over the litoral zone, but are obliged to keep within the sublitoral one. The quantity of light here afforded them is certainly very slight in the more northerly parts of the Arctic Sea, as compared with that which they receive for instance on the coast of Scandinavia within the litoral zone. At the north coast of Spitzbergen the sun is below the horizon for many months, and consequently the darkness even above the surface of the sea is during a long time so deep that a man cannot find his way, even at noon, without artificial light. It must be still darker, of course, at the bottom of the sea, in order to reach which the scanty light would have to penetrate masses of ice several feet thick, covered with a fathom of snow, and besides the layer of water above the bottom.

These masses of ice and snow are highly impervious to light, and as long as they remain, only very little light gets to the bottom of the sea, even when the sun is above the horizon during a longer or shorter part of the day. But the numerous Polar expeditions undertaken in this century bear witness that the time is short, indeed, within considerable portions of the Arctic Sea, when the sea is not more or less covered with ice. On the north coast of Norway *Spongomorpha arcta* is met with very abundant and very luxuriant, on the coast of Spitzbergen as well as in the Murman Sea and the Siberian Sea it is scanty and very poor. This is the case also with other green algæ, for instance *Spongomorpha lanosa* and *Monostroma Blyttii.* Other causes, such as the lower temperature of the water, the want of suitable localities a. s. o. may have contributed to produce this difference, and have certainly done so, but in all probability the insufficient supply of light has cooperated and continues to cooperate to the same effect. The quantity of light afforded, for instance on the coast of Spitzbergen, may be sufficient for these algæ to live on, though it is not so large that they can attain any luxuriancy of growth and form so many reproductive organs as to multiply in any considerable numbers.

Survey of the composition of the Flora of the Arctic Sea.

In order to bring clearly into view the composition of the Flora of the Arctic Sea both as a whole and in its differents parts, I have composed the following tables, founded on the statements set forth in detail in the special part of the present work. In the first of these tables, which gives a list of the hitherto known algæ of the Polar Sea and exhibits the outlines of their distribution within the Polar Sea, as understood in the wide sense mentioned before and without any reference to the geography of plants, I have at the same time indicated if a species is known or not from the northern part of the Atlantic and of the Pacific. I intend to make use of the materials thus brought together, in my researches on the origin of the Arctic Flora and the history of its development. An examination of these tables will show that the composition of the vegetation in different parts of the sea-region in question is too disparate to allow this region to be considered as a unity with regard to its Flora. On this account, it seems to me unsuitable and purposeless to attempt drawing up any comparison between this Flora, taken as a whole, and the Flora of other seas. Such a comparison will be instituted below with regard to the Arctic Flora, considered as a whole from the point of view of the geography of plants. Concerning the arrangement of the tables, it may be remarked that I have put together the western part of the Murman Sea and the White Sea, because the vegetations of these two regions present a very great resemblance, as is shown by the description of GOBI [1]).

TABLE 1. **List [2]) of the algæ of the Arctic Sea and the outlines of their distribution in different parts of the Arctic Sea, in the northern Atlantic and in the northern Pacifac.**

	The Norwegian Polar Sea.	The Greenland Sea.	The White Sea and the western Murman Sea.	The eastern Murman Sea.	The Kara Sea.	The Siberian Sea.	The American Arctic Sea.	Baffin Bay.	The northern Atlantic.	The northern Pacific.
Corallinaceæ.										
Corallina officinalis	+	......	+	......	......	......	+	+	+	? [3])
Lithothamnion soriferum	+	......	......	......	......	......	......	......	......	......
" Ungeri	+	......	......	......	......	......	......	......	+	......
" alcicorne	+	......	......	......	......	......	......	......	......	......
" norvegicum	+	......	......	......	......	......	......	......	+	......
" glaciale	+	+	+	+	......	......	+	+	......	......

[1]) Algenfl. Weiss. Meer.

[2]) I have not entered into this list some species that have been reported from the Arctic Sea, but seem to be imperfectly known or most probably incorrectly determined. It will be stated further on which those species are.

[3]) The mark ? indicates that the occurrence of the plant within the region is uncertain.

	The Norwegian Polar Sea.	The Greenland Sea.	The White Sea and the western Murman Sea.	The eastern Murman Sea.	The Kara Sea.	The Siberian Sea.	The American Arctic Sea.	Baffin Bay.	The northern Atlantic.	The northern Pacific.
Lithothamnion intermedium	+	......	......	......	......	......	......	......	......	......
» flavescens	+	......	......	+	......	......	......	......	......	......
» foecundum	......	......	......	......	+	......	......	+	......	......
» compactum	......	......	......	+	......	......	......	......	......	......
» polymorphum	+	......	......	......	......	+	+	+	+	+
Lithophyllum arcticum	......	......	......	......	+	......	......	+(?) [3]	......	......
» Lenormandi	+	......	+	......	......	......	......	......	+	......
Melobesia membranacea	+	......	......	......	......	......	......	......	+	......
» macrocarpa	+	......	......	......	......	......	......	......	+	......
» Lejolisii	......	+(?)	......	......	......	......	......	+	+	......
Rhodomelaceæ.										
Odonthalia dentata	+	+	+	+	+	+	+	+	+	+
Rhodomela lycopodioides	+	+	+	+	+	+	......	+	+	+
» larix	......	......	......	......	......	......	+	......	......	+
Polysiphonia parasitica	+	......	......	......	......	......	......	......	+	?
» urceolata	+	......	+	......	......	......	......	+	+	?
» Brodiæi	+	......	......	......	......	......	......	......	+	......
» fibrillosa	+	......	......	......	......	......	......	......	+	......
» Schübelerii	+	......	......	......	......	......	......	......	......	......
» elongata	+	?	......	......	......	......	......	......	+	......
» fastigiata	+	+	......	......	......	......	......	+	+	?
» arctica	+	+	+	+	+	+	......	+	+	......
» atrorubescens	+	?	......	......	......	......	......	......	+	?
» byssoides	+	......	......	......	......	......	......	......	+	......
» nigrescens	+	......	+	......	......	......	......	+	+	+
Spongiocarpeæ.										
Polyides rotundus	+	......	+	+	......	......	......	......	+	......
Wrangeliaceæ.										
Spermothamnion Turneri	+	......	......	......	......	......	......	......	+	......
Chantransia efflorescens	+	+	+	+	+	......	......	......	+	......
» Daviesii	+	......	......	......	......	......	......	......	+	......
» virgatula	+	......	......	......	......	......	......	......	+	......
» secundata	+	......	......	+	......	......	......	......	+	......
Delesseriaceæ.										
Delesseria rostrata	......	......	......	......	......	......	......	+	......	......
» Bærii	......	+	+	+	......	......	......	......	......	+
» corymbosa	......	......	......	......	......	......	......	+	......	......
» angustissima	+	......	......	......	......	......	......	......	+	......

[3]) The mark +(?) signifies that there is found within the region an alga, probably identical with the species thus marked, and assuredly not identical with any other species reported from the same division of the Arctic Sea.

	The Norwegian Polar Sea.	The Greenland Sea.	The White Sea and the western Murman Sea.	The eastern Murman Sea.	The Kara Sea.	The Siberian Sea.	The American Arctic Sea.	Baffin Bay.	The northern Atlantic.	The northern Pacific.
Delesseria alata	+	......	......	......	......	......	......	......	+	?
» Montagnei	......	......	......	......	......	......	......	+	+	......
» sinuosa	+	+	+	+	+	+	+	+	+	+
Nitophyllum punctatum	......	......	......	......	......	......	......	+	+	......
Hildbrantiaceæ.										
Hildbrandtia rosea	+	+	+	+	......	......		......	+	+
Squamarieæ.										
Peyssonnelia Dubyi	+	......	......	......	......		......	+	+	+
Petrocelis cruenta	+	......	......	......	......	......	......	......	+	......
» Middendorffi	+	......	......	......	......	......	......	......	......	+
Cruoria pellita	+	......	......	......	......	......	......	+	+	......
Hæmescharia polygyna	......	......	......	......	......	+	......	......	......	......
Rhodymeniaceæ.										
Hydrolapathum sanguineum	+	......	......	......	......	......	......	+	+	......
Rhodophyllis dichotoma	+	+	+	+	......	......	......	+	+	+
Euthora cristata	+	+	+	+	+	......	+	+	+	+
Plocamium coccineum	+	......	......	......	......	......	......	......	+	+
Rhodymenia palmata	+	+	+	+	......	......	......	+	+	+
» pertusa	......	+	......	......	......	......	......	+	......	+
Champiaceæ.										
Chylocladia clavellosa	+		......	......	......	......	......	......	+	......
» articulata	+	......	......	......	......	......	......	......	+	......
Dumontiaceæ.										
Sarcophyllis edulis	+	+	+	......	......	......	......	......	+	......
» arctica	......	+	+	+	+	+	......	+	......	+
Halosaccion ramentaceum	+	+	+	+	+	+	+	+	+	+
» saccatum	......	......	+	......	......	......	......	......	......	......
Dumontia filiformis	+	......	+	......	......	......	......	+	+	+
Furcellariaceæ.										
Furcellaria fastigiata	+	+	+	+	......	......	......	+	+	......
Gigartinaceæ.										
Cystoclonium purpurascens	+	......	+	......	......	......	......	+	+	......
Callophyllis laciniata	......	......	......	......	......	......	......	+	+	+
Kallymenia rosacea	......	+	......	......	......	......	......	......	......	......
» septemtrionalis	+	+	......	+	......	......	......	+ ?	......	......
» Pennyi	......	......	......	......	......	......	+	+	......	......
Phyllophora Brodiæi	+	+	+	+	+	......	......	......	+	......
» interrupta	......	+	+	+	+	+	+	+	......	......
» membranifolia	+	......	......	......	......	......	......	......	+	......

	The Norwegian Polar Sea	The Greenland Sea.	The White Sea and the western Murman Sea.	The eastern Murman Sea.	The Kara Sea.	The Siberian Sea.	The American Arctic Sea.	Baffin Bay.	The northern Atlantic.	The northern Pacific.
Ahnfeltia plicata	+	+	+	......	......	+	+	+	+	+
Gigartina mamillosa	+	......	......	......	......	......	......	+	+	?
Chondrus crispus	+	......	......	......	......	......	......	......	+	?
Ceramiaceæ.										
Microcladia glandulosa	+	......	......	......	......	......	......	......	+	......
Ceramium Deslongchampii	+	......	......	......	......	......	......	......	+	......
» circinatum	+	......	......	......	......	......	......	......	+	......
» rubrum	+	+	+	+	......	......	......	+	+	+
» acanthonotum	+	......	......	......	......	......	......	......	+	......
Ptilota elegans	+	......	......	......	......	......	......	......	+	......
» plumosa	+	+	+	+	......	......	+	+	+	+
» pectinata	+	+	+	+	+	......	+	+	+	+
Callithamnion polyspermum	+	......	......	......	......	......	......	......	+	+
» Hookeri	+	......	......	......	......	......	......	......	+	......
» arbuscula	+	......	......	......	......	......	......	......	+	+
» roseum	+	......	......	......	......	......	......	......	+	......
» corymbosum	+	......	......	......	......	......	......	......	+	......
Antithamnion floccosum	+	......	......	......	......	......	......	+	+	?
» Pylaisæi	+	......	......	......	......	......	......	+	+	......
» boreale	+	+	+	+	+	+	......	......	......	+
» americanum	......	......	......	......	......	......	+	+	+	+
Rhodochorton intermedium	......	+	......	......	......	......	......	......	......	......
» spinulosum	......	......	......	......	......	......	......	+	......	......
» Rothii	+	+	......	+	......	......	......	+	+	+
» sparsum	+	......	......	......	......	......	......	+	+	......
» mesocarpum	+	+	......	+	+	......	......	+	+	......
» spetsbergense	......	+	......	......	......	......	......	......	......	......
Porphyraceæ.										
Diploderma amplissimum	+	......	......	......	......	......	......	......	......	......
» miniatum	......	+	......	......	......	......	......	+	......	......
Porphyra laciniata	+	......	+	......	......	......	......	+	+	+
» abyssicola	+	......	+	......	......	......	......	+	......	......
Bangia fuscopurpurea	+	......	......	......	......	......	......	?	+	......
Erythrotrichia ceramicola	+	......	......	......	......	......	......	......	+	......
Fucaceæ.										
Himanthalia lorea	+	......	......	......	......	......	......	......	+	......
Halidrys siliquosa	+	......	......	......	......	......	......	......	+	......
Ozothallia nodosa	+	+	+	......	......	......	......	+	+	......
Fucus serratus	+	+	+	+	......	......	......	+	+	......
» vesiculosus	+	?	+	?	?	......	......	+	+	+
» ceranoides	+	+	......	......	......	......	......	+	+	+
» spiralis	+	......	......	......	......	......	......	......	+	+

	The Norwegian Polar Sea.	The Greenland Sea.	The White Sea and the western Murman Sea.	The eastern Murman Sea.	The Kara Sea.	The Siberian Sea.	The American Arctic Sea.	Baffin Bay.	The northern Atlantic.	The northern Pacific.
Fucus evanescens	......	+	+	+	+	+	+	+	+	+
» edentatus	+	......	......	......	......	......	......	+	+	......
» miclonensis	+	......	......	......	......	......	......	......	+	+
» linearis	+	......	+	......	......	......	......	+	+	......
» filiformis	+	......	+	......	......	......	......	+	+	......
» distichus	+	......	......	......	......	......	......	......	+	......
Pelvetia canaliculata	+	......	+	......	......	......	......	+	+	......
Tilopterideæ.										
Scaphospora arctica	......	......	......	+	......	......	......	......	......	......
Haplospora globosa	......	+	......	+	......	....	......	......	+	......
Laminariaceæ.										
Alaria esculenta	+	+	......	......	......	......	......	......	+	......
» Pylaii	+	......	......	......	......	......	......	+	+	?
» membranacea	+	+	+	+	......	......	......	+	......	......
» grandifolia	?	+	......	+	......	......	......	......	......	......
» dolichorhachis	......	......	......	......	......	+	+ ?	......	......	......
» oblonga	......	......	......	......	......	+	......	......	......	......
» elliptica	......	......	......	......	......	+	......	......	......	......
Agarum Turneri	......	......	......	......	......	......	+	+	+	+
Phyllaria dermatodea	+	+	+	+	......	......	......	+	+	+
» lorea	......	+	......	+	......	......	......	......	+	......
Laminaria solidungula	......	+	......	+	+	+	......	+	......	+
» cuneifolia	......	......	?	......	......	+	?	+	......	+
» saccharina	+	......	+	......	......	......	......	......	+	?
» longicruris	......	......	......	......	......	......	+	+	+	+
» Agardhii	......	+	+	+	+	......	......	......	?	......
» atrofulva	...	......	......	......	......	......	......	+	......	......
» fissilis	......	+	......	+	......	......	......	......	......	+
» nigripes	......	+	......	+	+	+	......	+	......	......
» Clustoni	+	......	......	......	......	......	......	......	+	......
» digitata	+	+	+	+	+	......	......	?	+	?
» stenophylla	+	......	......	......	......	......	......	......	+	+
Chorda filum	+	+	+	+	......	......	+	+	+	+
» tomentosa	+	......	......	......	......	......	......	......	+	......
Encœliaceæ.										
Stilophora Lyngbyei	+	......	......	......	......	......	......	......	+	......
Asperococcus echinatus	+	......	......	......	......	......	......	......	+	......
» bullosus	......	......	......	......	......	......	......	+	+	......
Ralfsia deusta	+	......	+	+	......	......	......	+	+	+
» verrucosa	+	......	......	......	......	......	......	......	+	......

	The Norwegian Polar Sea.	The Greenland Sea.	The White Sea and the western Murman Sea.	The eastern Murman Sea.	The Kara Sea.	The Siberian Sea.	The American Arctic Sea.	Baffin Bay.	The northern Atlantic.	The northern Pacific.
Chordariaceæ.										
Chordaria flagelliformis	+	+	+	+	......	+	......	+	+	+
Castagnea divaricata	+	......	......	......	......	......	......	......	+	......
Eudesme virescens	+	......	+	......	......	......	......	......	+	......
Mesogloia vermicularis	+	......	+	......	......	......	......	......	+	......
Myrionemateæ.										
Leathesia difformis	+	......	......	......	......	......	......	......	+	......
Elachista fucicola	+	+	+	+	+	+	......	+	+	+
» lubrica	+	+	+	+	......	......	......	+	......	+
Myrionema strangulans	+	......	......	......	......	......	......	+	+	......
Lithodermateæ.										
Lithoderma fatiscens	+	+	+	+	+	+	......	+	+	+
» lignicola	+	......	......	......	......	......	......	......	......	......
Scytosiphoneæ.										
Ilea fascia	+	......	+	......	......	......	+	+	+	+
Scytosiphon lomentarius	+	......	+	......	......	......	......	+	+	+
» attenuatus	......	+	......	......	......	......	......	......	......	......
Punctariaceæ.										
Punctaria plantaginea	+	+	......	+	......	......	......	+	+	......
Desmarestiaceæ.										
Desmarestia aculeata	+	+	+	+	+	......	......	+	+	+
Dichloria viridis	+	+	+	+	......	......	......	+	+	+
Phlœospora subarticulata	+	+	+	+	......	......	......	......	+	......
» tortilis	+	+	+	+	+	......	......	+	+	+
» pumila	......	......	......	+	......	......	......	......	......	......
Coilonema Ekmani	+	......	......		......	......	......	......	+	......
» Chordaria	+	......	......	......	......	......	......	......	+	......
Dictyosiphon corymbosus	......	......	......	+	......	......	......	......	......	......
» hippuroides	+	+	+	+	......	......	......	+	+	......
» foeniculaceus	+	+	+	+	......	......	+	+	+	+
» hispidus	......	+	......	+	......	......	......	......	......	......
Lithosiphon Laminariæ	+	......	......	?	......	......	......	......	+	......
Aglaozoniaceæ.										
Aglaozonia parvula	+	......	......	......	......	......	......	......	+	......
Sphacelariaceæ.										
Cladostephus spongiosus	+	......	......	......	......	......	......	+	+	......
Stupocaulon scoparium	......	+	......	......	......	......	......	......	+	......
Chætopteris plumosa	+	+	+	+	+	+	+	+	+	+
Sphacelaria cirrhosa	+	......	......	......	......	......	......	+	+	......
» arctica	+	+	+	+	+	+	......	+	......	......
» olivacea	+	......	......	......	......	......	......	+	+	......

	The Norwegian Polar Sea.	The Greenland Sea.	The White Sea and the western Murman Sea.	The eastern Murman Sea.	The Kara Sea.	The Siberian Sea.	The American Arctic Sea.	Baffin Bay.	The northern Atlantic.	The northern Pacific.
Ectocarpaceæ.										
Isthmoplea sphærophora	+	......	+	......	......	......	......	......	+	......
Ectocarpus confervoides	+	+	+	+	......	......	......	+	+	+
» pygmæus	+	......	......	......	......	......	......	......	+	......
» draparnaldioides	+	......	......	......	......	......	......	......	+	......
» fasciculatus	+	......	......	......	......	......	......	......	+	......
» tomentosus	+	......	......	......	......	......	......	......	+	......
» ovatus	+	+	......	......	......	......	......	......	+	......
» Lebelii	+	......	......	......	......	......	......	......	......	......
» terminalis	+	......	......	......	......	......	......	......	+	......
» reptans	+	......	......	......	......	......	......	......	+	......
Pylaiella litoralis	+	+	+	+	+	+	......	+	+	+
» nana	+	......	......	......	......	......	......	......	......	......
» varia	+	+	......	......	+	......	......	+	......	......
Myriotrichia filiformis	+	......	......	......	......	......	......	......	+	......
Gleothamnion palmelloides	......	......	+	......	......	......	......	......	......	......
Chætophoraceæ.										
Chætophora maritima	......	+	......	+	......	......	......	......	......	......
» pellicula	+	......	......	......	......	......	......	......	......	......
Ulvaceæ.										
Enteromorpha clathrata	+	......	......	......	......	......	......	+	+	+
» intestinalis	+	......	......	......	......	......	......	+	+	+
» compressa	+	+	+	+	+	......	+	+	+	+
» complanata	+	......	......	......	......	......	......	......	+	......
» minima	......	......	......	+	......	......	......	......	+	......
» tubulosa	......	......	......	......	......	......	......	+	+	......
» micrococca	+	+	+	+	......	+	......	+	+	......
Ulva crassa	+	+	......	+	......	......	......	......	......	......
» lactuca	+	......	......	?	......	......	......	+	+	+
Monostroma latissimum	+	......	......	......	......	......	......	......	+	......
» undulatum	+	......	......	......	......	......	......	......	......	......
» lubricum	......	+	+	......	......	......	......	+	......	......
» cylindraceum	+	......	......	......	......	......	......	......	......	......
» saccodeum	+	......	......	......	......	......	......	......	......	......
» angicava	+	......	......	......	......	......	......	......	......	......
» Grevillei	+	......	+	......	......	......	......	+	+	......
» arcticum	+	......	......	......	......	......	......	......	......	......
» leptodermum	......	......	......	+	......	......	......	......	......	......
» fuscum	+	+	+	+	......	......	......	+	+	+
» crispatum	+	......	......	......	......	......	......	......	......	......
» Blyttii	+	+	......	+	......	......	......	+	+	......
Diplonema percursum	+	+	......	+	......	......	......	+	+	......
Prasiola stipitata	+	......	......	......	......	......	......	......	+	......

	The Norwegian Polar Sea.	The Greenland Sea.	The White Sea and the western Murman Sea.	The eastern Murman Sea.	The Kara Sea.	The Siberian Sea.	The American Arctic Sea.	Baffin Bay.	The northern Atlantic.	The northern Pacific.
Confervaceæ.										
Spongomorpha spinescens	+	......	......	......	......	......	......	......	+	......
» arcta	+	+	+	+	+	......	......	+	+	+
» lanosa	+	......	......	+	......	......	......	+	+	......
Cladophora rupestris	+	......	+	+	......	......	......	+	+	......
» diffusa	+	+	......	+	......	......	......	......	+	......
» glaucescens	+	......	......	......	......	......	......	......	+	+
» gracilis	+	......	......	......	......	......	......	+	+	......
» crispata	......	......	+	......	......	......	......	......	+	......
Rhizoclonium rigidum	+	+	+	......	......	......	......	......	+	......
» pachydermum	......	......	......	+	......	+	......	+	......	......
» riparium	+	+	......	+	......	......	......	+	+	+
Chætomorpha melagonium	+	+	+	+	+	......	+	+	+	+
» Wormskioldii	......	......	......	......	......	......	......	+	......	......
» linum	......	......	+	......	......	......	......	......	+	......
» tortuosa	+	......	......	......	......	......	......	+	+	+
» septemtrionalis	+	......	......	......	......	......	......	......	......	......
Ulothrix Sphacelariæ	+	......	......	......	......	......	......	......	......	......
» submarina	......	......	+	......	......	......	......	......	+	......
» discifera	......	+	......	......	......	......	......	......	......	......
Urospora penicilliformis	+	+	......	+	+	......	......	+	+	+
Bulbocoleon piliferum	+	......	+	......	......	......	......	......	+	......
Derbesiaceæ.										
Derbesia marina	+	......	......	......	......	......	......	......	+	......
Bryopsideæ.										
Bryopsis plumosa	+	......	......	......	......	......	......	+	+	+
Characiaceæ.										
Characium marinum	......	+	......	......	......	......	......	......	......	......
Codiolum longipes	+	......	......	......	......	......	......	......	+	......
» pusillum	+	......	......	......	......	......	......	......	+	......
» Nordenskiöldianum	+	+	......	......	......	......	......	......	......	......
Chlorochytrium inclusum	......	+	......	+	+	+	......	......	......	......
Palmellaceæ.										
Chlorangium marinum	......	......	+	......	......	......	......	......	......	......
Rivulariaceæ.										
Rivularia hemisphærica	+	......	......	......	......	......	......	+	+	......
» microscopica	......	......	......	......	......	......	+	......	......	......
Calothrix Harveyi	+	......	......	......	......	......	......	......	+	......
» scopulorum	+	+	......	+	......	......	......	......	+	......
» confervicola	......	......	+	......	......	......	......	......	+	......

	The Norwegian Polar Sea.	The Greenland Sea.	The White Sea and the western Murman Sea.	The eastern Murman Sea.	The Kara Sea.	The Siberian Sea.	The American Arctic Sea.	Baffin Bay.	The northern Atlantic.	The northern Pacific.
Oscillariaceæ.										
Lyngbya semiplena	+	......	......	......	......	......	......	......	+	......
Oscillaria subsalsa	......	......	......	......	......	......	......	+	+	......
Spirulina tenuissima	......	......	......	......	......	......	......	+	+	......
Chroococcaceæ.										
Gleocapsa spec.	......	......	+	......	......	......	......	......	......	......

TABLE 2. **Number of species in the series of algæ in the Flora of the Arctic Sea.**

	The whole of the Arctic Sea.	The Norwegian Polar Sea.	The Greenland Sea.	The White Sea and the western Murman Sea.	The eastern Murman Sea.	The Kara Sea.	The Siberian Sea.	The American Arctic Sea.	Baffin Bay.
Florideæ	104	81	32(34)	32	27	15	11	13	47 (48)
Fucoideæ	92	69 (70)	36(37)	34(35)	33(35)	13(14)	13	8 (9)	42 (43)
Chlorophyllophyceæ	54	40	18	14	18(19)	5	3	2	22
Nostochineæ	9	4	1	2	1	....	......	1	3
Total sum of species	259	194(195)	87(90)	82(83)	79(82)	33(34)	27	24(25)	114(116)

TABLE 3. **Number of genera in the families of the Flora of the Arctic Sea.**

	The whole of the Arctic Sea.	The Norwegian Polar Sea.	The Greenland Sea.	The White Sea and the western Murman Sea.	The eastern Murman Sea.	The Kara Sea.	The Siberian Sea.	The American Arctic Sea.	Baffin Bay.
Gigartinaceæ	7	6	3	3	2	1	2	3	6
Confervaceæ	7	7	6	6	5	3	1	1	5
Ceramiaceæ	6	6	4	3	4	3	1	2	4
Desmarestiaceæ	6	6	4	4	4 (5)	2	......	1	4
Rhodymeniaceæ	5	5	3	3	3	1	......	1	4
Fucaceæ	5	5	2	3	1	1	1	1	3
Laminariaceæ	5	4	4	4	4	1	2	4	5
Ectocarpaceæ	5	4	2	2	2	1	1	......	2
Ulvaceæ	5	5	4	2	4	1	1	1	4
Corallinaceæ	4	4	2	3	1	2	1	1	3
Squamariaceæ	4	3	......	......	......	......	1		2
Porphyraceæ	4	4	1	1	......	......	......	......	2 (3)
Chordariaceæ	4	4	1	3	1	......	1	......	1

	The whole of the Arctic Sea.	The Norwegian Polar Sea.	The Greenland Sea.	The White Sea and the western Murman Sea.	The eastern Murman Sea.	The Kara Sea.	The Siberian Sea.	The American Arctic Sea.	Baffin Bay.
Sphacelariaceæ	4	3	3	2	2	2	2	1	3
Rhodomelaceæ	3	3	3	3	3	3	3	2	3
Dumontiaceæ	3	3	2	3	2	2	2	1	3
Encoeliaceæ	3	3	......	1	1	......	......	......	2
Myrionemaleæ	3	3	1	1	1	1	1	......	2
Characiaceæ	3	1	3	......	1	1	1	......	......
Oscillariaceæ	3	1	......	......	......	......	......	......	2
Wrangeliaceæ	2	2	1	1	1	1	......	......	......
Delesseriaceæ	2	1	1	1	1	1	1	1	2
Tilopterideæ	2	......	1	......	2	......	......	......	......
Scytosiphoneæ	2	2	1	2	......	......	......	1	2
Rivulariaceæ	2	2	1	1	1	......	......	1	1
Spongiocarpeæ	1	1	......	1	1	......	......	......	......
Hildbrandtiaceæ	1	1	1	1	1	......	......	......	......
Champiaceæ	1	1	......	......	......	......	......	......	......
Furcellariaceæ	1	1	1	1	1	......	......	......	1
Lithodermateæ	1	1	1	1	1	1	1	......	1
Punctariaceæ	1	1	1	......	1	......	......	......	1
Aglaozoniaceæ	1	1	......	......	......	......	......	......	......
Chætophoraceæ	1	1	1	......	1	......	......	......	......
Derbesiaceæ	1	1	......	......	......	......	......	......	......
Bryopsideæ	1	1	......	......	......	......	......	......	1
Palmellaceæ	1	......	......	1	......	......	......	......	......
Chroococcaceæ	1	......	......	1	......	......	......	......	...
	111	97	58	60	52(53)	28	23	22	69

TABLE 4. **Number of species in the families of the Flora of the Arctic Sea.**

	The whole of the Arctic Sea.	The Norwegian Polar Sea.	The Greenland Sea.	The White Sea and the western Murman Sea.	The eastern Murman Sea.	The Kara Sea.	The Siberian Sea.	The American Arctic Sea.	Baffin Bay.
Ceramiaceæ	23	19	8	4	6	3	1	8	10
Laminariaceæ	23	10(11)	11	6(7)	10	4	6	4 (5)	10(11)
Ulvaceæ	23	19	7	5	8 (9)	1	1	1	11
Confervaceæ	21	15	7	8	8	3	1	1	10
Corallinaceæ	16	12	2	3	3	2	1	2	5
Ectocarpaceæ	15	14	4	4	2	2	1	......	3
Rhodomelaceæ	14	13	4(6)	5	3	3	3	2	6
Fucaceæ	14	13	4(5)	7	2 (3)	1 (2)	1	1	9
Desmarestiaceæ	12	9	7	6	9(10)	2	......	1	5

	The whole of the Arctic Sea.	The Norwegian Polar Sea.	The Greenland Sea.	The White Sea and the western Murman Sea.	The eastern Murman Sea.	The Kara Sea.	The Siberian Sea.	The American Arctic Sea.	Baffin Bay.
Gigartinaceæ	11	7	5	4	3	2	2	3	7
Delesseriaceæ	8	3	2	2	2	1	1	1	5
Rhodymeniaceæ	6	5	4	3	3	1	......	1	5
Porphyraceæ	6	5	1	2	......	......	......	......	3 (4)
Sphacelariaceæ	6	5	3	2	2	2	2	1	5
Wrangeliaceæ	5	5	1	1	2	1	......	......	......
Squamariaceæ	5	4	......	......	......	......	1	......	2
Dumontiaceæ	5	3	3	5	2	2	2	1	3
Encoeliaceæ	5	4	......	1	1	......	......	......	2
Characiaceæ	5	3	3	......	1	1	1	......	......
Rivulariaceæ	5	3	1	1	1	......	......	1	1
Chordariaceæ	4	4	1	3	1	......	1	......	1
Myrionemateæ	4	4	2	2	2	1	1	......	3
Scytosiphoneæ	3	2	1	2	......	......	......	1	2
Oscillariaceæ	3	1	......	......	......	......	......	......	2
Champiaceæ	2	2	......	......	......	......	......	......	......
Tilopterideæ	2	......	1	......	2	......	......	......	......
Lithodermateæ	2	2	1	1	1	1	1	......	1
Chætophoraceæ	2	1	1	......	1	......	......	......	......
Spongiocarpeæ	1	1	......	1	1	......	......	......	......
Hildbrandtiaceæ	1	1	1	1	1	......	......	......	......
Furcellariaceæ	1	1	1	1	1	......	......	......	1
Punctariaceæ	1	1	1	......	1	......	......	......	1
Aglaozoniaceæ	1	1	......	......	......	......	......	......	......
Derbesiaceæ	1	1	......	......	......	......	......	......	......
Bryopsideæ	1	1	......	......	......	......	......	......	1
Palmellaceæ	1	......	......	1	......	......	......	......	......
Chroococcaceæ	1	......	......	1	......	......	......	......	......

TABLE 5. **Number of species in the genera of the Flora of the Arctic Sea.**

	The whole of the Arctic Sea.	The Norwegian Polar Sea.	The Greenland Sea.	The White Sea and the western Murman Sea.	The eastern Murman Sea.	The Kara Sea.	The Siberian Sea.	The American Arctic Sea.	Baffin Bay.
Monostroma	12	10	3	3	3	......	......	......	4
Polysiphonia	11	11	2 (4)	3	1	1	1	......	4
Laminaria	11	4	5	3 (4)	5	4	3	1 (2)	5 (6)
Lithothamnion	10	8	1	1	3	1	1	2	3
Fucus	10	9	3 (4)	5	2 (3)	1 (2)	1	1	7
Ectocarpus	9	9	2	1	1	......	......	......	1
Delesseria	7	3	2	2	2	1	1	1	4

	The whole of the Arctic Sea.	The Norwegian Polar Sea.	The Greenland Sea.	The White Sea and the western Murman Sea.	The eastern Murman Sea.	The Kara Sea.	The Siberian Sea.	The American Arctic Sea.	Baffin Bay.
Alaria	7	3 (4)	3	1	2	......	3	1	2
Enteromorpha	7	5	2	2	3	1	1	1	5
Rhodochorton	6	3	4	......	2	1	......	......	4
Callithamnion	5	5	......	......	......	......	......	......	......
Cladophora	5	4	1	2	2	......	......	......	2
Chætomorpha	5	3	1	2	1	1	......	1	3
Chantransia	4	4	1	1	2	1	......	......	......
Ceramium	4	4	1	1	1	......	......	......	1
Antithamnion	4	3	1	1	1	1	1	1	3
Dictyosiphon	4	2	3	2	4	......	......	1	2
Melobesia	3	2	1	......	......	......	......	......	1
Kallymenia	3	1	2	......	1	......	......	1	2
Phyllophora	3	2	2	2	2	2	1	1	1
Ptilota	3	3	2	2	2	1		2	2
Phloeospora	3	2	2	2	3	1	......	......	1
Sphacelaria	3	3	1	1	1	1	1	......	3
Pylaiella	3	3	2	1	1	2	1	......	2
Spongomorpha	3	3	1	1	2	1	......	......	2
Rhizoclonium	3	2	2	1	2	......	1	......	2
Ulothrix	3	1	1	1	......	......	......	......	......
Codiolum	3	3	1	......	......	......	......	......	......
Calothrix	3	2	1	1	1	......	......	......	......
Lithophyllum	2	1	......	1	......	1	......	......	1
Rhodomela	2	1	1	1	1	1	1	1	1
Petrocelis	2	2	......	......	......	......	......	......	......
Rhodymenia	2	1	2	1	1	......	......	......	2
Chylocladia	2	2	......	......	......	......	......	......	......
Sarcophyllis	2	1	2	2	1	1	1	......	1
Halosaccion	2	1	1	2	1	1	1	1	1
Diploderma	2	1	1	......	......	......	......	......	1
Porphyra	2	2	......	2	......	......	......	......	2
Phyllaria	2	1	2	1	2	......	......	......	1
Chorda	2	2	1	1	1	......	......	1	1
Asperococcus	2	1	......	......	......	......	......		1
Ralfsia	2	2	......	1	1	......	......	......	1
Elachista	2	2	2	2	2	1	1	......	2
Lithoderma	2	2	1	1	1	1	1	......	1
Scytosiphon	2	1	1	1	......	......	......	......	1
Coilonema	2	2	......	......	......	......	......	......	......
Chætophora	2	1	1	......	1	......	......	......	......
Ulva	2	2	1	......	1 (2)	......	......	......	1
Rivularia	2	1	......	......	......	......	......	1	1
Corallina	1	1	......	1	......	......	......	......	......
Odonthalia	1	1	1	1	1	1	1	1	1

	The whole of the Arctic Sea.	The Norwegian Polar Sea.	The Greenland Sea.	The White Sea and the western Murman Sea.	The eastern Murman Sea.	The Kara Sea.	The Siberian Sea.	The American Arctic Sea.	Baffin Bay.
Polyides	1	1	......	1	1	......	......	......	......
Spermothamnion	1	1	......	......	......	......	......	......	......
Nithophyllum	1	......	......	......	......	......	......	......	1
Hildbrandtia	1	1	1	1	1	......	......	......	......
Peyssounelia	1	1	......	......	......	......	......	......	1
Cruoria	1	1	......	......	......	......	......	......	1
Hæmescharia	1	......	......	......	......	......	1	......	......
Hydrolapathum	1	1	......	......	......	......	......	......	1
Rhodophyllis	1	1	1	1	1	......	......	......	1
Euthora	1	1	1	1	1	1	......	1	1
Plocamium	1	1	......	......	......	......	......	......	......
Dumontia	1	1	......	1	......	......	......	......	1
Furcellaria	1	1	1	1	1	......	......	......	1
Cystoclonium	1	1		1	......	......	......	......	1
Callophyllis	1	......	......	......	......	......	......	......	1
Ahnfeltia	1	1	1	1	......	......	1	1	1
Gigartina	1	1	......	......	......	......	......	......	1
Chondrus	1	1	......	......	......	......	......	......	......
Microcladia	1	1	......	......	......	......	......	......	......
Bangia	1	1	......	......	......	......	......	......	1?
Erythrotrichia	1	1	......	......	......	......	......	......	......
Himanthalia	1	1	......	......	......	......	......	......	......
Halidrys	1	1	......	......	......	......	......	......	......
Ozothallia	1	1	1	1	......	......	......	......	1
Pelvetia	1	1	......	1	......	......	......	......	1
Scaphospora	1	......	......	......	1	......	......	......	......
Haplospora	1	......	1	......	1	......	......	......	......
Agarum	1	......	......	......	......	......	......	1	1
Stilophora	1	1	......	......	......	......	......	......	......
Chordaria	1	1	1	1	1	......	1	......	1
Castagnea	1	1	......	......	......	......	......	......	......
Eudesme	1	1	......	1	......	......	......	......	......
Mesogloia	1	1	......	1	......	......	......	......	......
Leathesia	1	1	......		......	......	......	......	......
Myrionema	1	1	......	......	......	......	......	......	1
Ilea	1	1	......	1	......	......	......	1	1
Punctaria	1	1	1	......	1	......	......	......	1
Desmarestia	1	1	1	1	1	1	......	......	1
Dichloria	1	1	1	1	1		......	......	1
Lithosiphon	1	1	......	......	1?	......	......	......	......
Aglaozonia	1	1	......	......	......	......	......	......	......
Cladostephus	1	1	......	......	......	......	......	......	1
Stupocaulon	1	......	1	......	......	......	......	......	......
Chætopteris	1	1	1	1	1	1	1	1	1

	The whole of the Arctic Sea.	The Norwegian Polar Sea.	The Greenland Sea.	The White Sea and the western Murman Sea.	The eastern Murman Sea.	The Kara Sea.	The Siberian Sea.	The American Arctic Sea.	Baffin Bay.
Isthmoplea	1	1	......	1	......	......	......	......	......
Myriotrichia	1	1	......	......	......	......	......	......	......
Gleothamnion	1	......	......	1	......	......	......	......	......
Diplonema	1	1	1	......	1	......	......	......	1
Prasiola	1	1	......	......	......	......	......	......	......
Urospora	1	1	1	......	1	1	......	......	1
Bulbocoleon	1	1	......	1	......	......	......	......	......
Derbesia	1	1	......	......	......	......	......	......	......
Bryopsis	1	1	......	......	......	......		......	1
Characium	1		1	......	......	......	....	......	......
Chlorochytrium	1	......	1	......	1	1	1	......	......
Chlorangium	1	......	......	1	......	......	......	......	......
Lyngbya	1	1	......	......	......	......	......	......	......
Oscillaria	1	......	......		......	......	......	......	1
Spirulina	1	......	......	......	......	......	......	......	1
Gleocapsa	1	......	......	1	......	......	......	......	......

The history of the Flora of the Arctic Sea.

It is seen from table 1 given above, that of the algæ known from the Arctic Sea the following ones are not to be found south of it, as it has been defined here:

Lithothamnion soriferum,
» alcicorne,
» glaciale
» intermedium,
» flavescens,
» foecundum,
» compactum,
Lithophyllum arcticum,
Polysiphonia Schübelerii,
Delesseria rostrata,
» corymbosa,
Hæmescharia polygyna,
Halosaccion saccatum,
Kallymenia septemtrionalis,
» Pennyi,
» rosacea,
Phyllophora interrupta,
Rhodochorton intermedium,
» spinulosum,
» spetsbergense,
Diploderma amplissimum,
» miniatum,
Porphyra abyssicola,
Scaphospora arctica,
Alaria membranacea,
» grandifolia,
» dolichorhachis,
» elliptica,
» oblonga,
Laminaria atrofulva,
» nigripes,
Lithoderma lignicola,
Scytosiphon attenuatus,
Phloeospora pumila,
Dictyosiphon corymbosus,
» hispidus,
Sphacelaria arctica,
Ectocarpus Lebelii,
Pylaiella nana,
» varia,
Gleothamnion palmelloides,
Chætophora maritima,
» pellicula,
Ulva crassa,
Monostroma undulatum,
» lubricum,
» cylindraceum,
» saccodeum,
» angicava,
» arcticum,
» leptodermum,
» crispatum,
Rhizoclonium pachydermum,
Chætomorpha Wormskioldii,
» septemtrionalis,
Ulothrix Sphacelariæ,
» discifera,
Characium marinum,
Codiolum Nordenskiöldianum,
Chlorochytrium inclusum,
Chlorangium marinum,
Rivularia microscopica,
Gleocapsa spec.

Accordingly no less than 63 species, representing 22 families and 34 genera. A considerable number of these algæ are, indeed, now described for the first time, or have only lately been distinguished as separate species, and it is thus certainly possible that some one or other of them, attention being now drawn to it, may prove to go southward of the limit established for the Arctic See. But on the other hand, the great majority consists of species so well marked by size and specific characters, that it seems rather improbable, if they did occur farther to the south, that they would have escaped observation on the coasts comparatively so accurately and long investigated of the northern Atlantic, where in such a case most of them would be expected to grow. About one third of them belong exclusively to those parts of the Arctic Sea which are not filled with ice: the Norwegian Polar Sea, the western Murman and the White Seas, namely:

Lithothamnion soriferum,
» alcicorne,
» intermedium,
Polysiphonia Schübelerii,
Halosaccion saccatum,
Diploderma amplissimum,
Lithoderma lignicola,
Ectocarpus Lebelii,
Pylaiella nana,
Gleothamnion palmelloides,
Chætophora pellicula,
Monostroma undulatum,
» cylindraceum,
» saccodeum,
» angicava,
» arcticum,
» crispatum,
Ulothrix Sphacelariæ,
Chætomorpha septemtrionalis,
Chlorangium marinum,
Gleocapsa spec.

Accordingly 21 species, representing 11 families and 14 genera. The other 42 species occur all of them in the Arctic Sea proper. Only 11 or 10 of these have been as yet met with also in the Norwegian Polar Sea, in the western Murman Sea, and in the White Sea. These are:

Lithothamnion glaciale,
» flavescens,
Kallymenia septemtrionalis,
Porphyra abyssicola,
Alaria membranacea,
» grandifolia?
Sphacelaria arctica,
Pylaiella varia,
Ulva crassa,
Monostroma lubricum,
Codiolum Nordenskiöldianum.

Thus there remain no less than 31 or 32 species that are endemic in the Arctic Sea proper, namely:

Lithothamnion foecundum,
» compactum,
Lithophyllum arcticum,
Delesseria rostrata,
» corymbosa,
Hæmescharia polygyna,
Kallymenia Pennyi,
» rosacea,
Phyllophora interrupta,
Rhodochorton spinulosum,
» intermedium,
» spetsbergense,
Diploderma miniatum,
Scaphospora arctica,
Alaria grandifolia?
» dolichorhachis,
» elliptica,
» oblonga,

Laminaria atrofulva,
» nigripes,
Scytosiphon attenuatus,
Phloeospora pumila,
Dictyosiphon corymbosus,
» hispidus,
Chætophora maritima,
Monostroma leptodermum,
Rhizoclonium pachydermum,
Ulothrix discifera,
Chætomorpha Wormskioldii,
Characium marinum,
Chlorochytrium inclusum,
Rivularia microscopica.

These species belong to 15 different families and 22 different genera.

This strong endemism points to the purely arctic marine Flora being, contrary to the arctic phanerogamous Flora, no immigrated Flora, but one that possesses its centre of development in the Arctic Sea itself properly so called. Other circumstances lead cogently to the same conclusion, indicating at the same time that the present purely glacial marine Flora must have been formerly more widely spread towards the south than it is now. This results, I think, from a comparison of the Flora of the Arctic Sea with that of the northern Atlantic and the northern Pacific.

The Arctic Sea, taken in a wide sense, possesses, as is shown by table 1, a considerable number of species in common with the Northern Atlantic.

These species amount to 184 (185). By far the greatest part are met with on the Atlantic coast of Europe, only 11 are exclusively American.

Amongst these, the following four are not known south of New-Foundland:

Delesseria Montagnei,
Fucus miclonensis,
Laminaria Agardhii.
Phyllaria lorea.

Besides, I think it doubtful whether that plant from New-Foundland which has been called *Laminaria caperata* is in reality identical with the *Laminaria Agardhii* of the Arctic Sea. If this is not the case, *Laminaria Agardhii* is to be reckoned amongst the endemic species of the Arctic Sea. The other 7 species:

Ptilota pectinata,
Antithamnion Pylaisæi,
Fucus edentatus,
» evanescens.
Laminaria longicruris,
Agarum Turneri,
Phyllaria dermatodea

either are not known south of Cape Cod or at least have their proper area of distribution north of this promontory[1]) and grow there most abundant and most richly developed. It is known already by the researches of Harvey, fully confirmed by later observations, that this promontory on the east coast of America forms the boundary between a more southern Flora and one expressly named arctic by American algologists[2]).

[1]) Cp. Farlow, New Engl. Alg.
[2]) Farlow, l. c. p. 4.

It cannot well be doubted that the real origin of the last-mentioned 7 species is to be placed in the Arctic Sea. Their occurrence on the coast of America south of the Arctic Sea is obviously to be explained by the Labrador current flowing along the coast from the north; it may have carried them down, and it makes moreover the external conditions similar to those under which these algæ live in the Arctic Sea. Some of these species, *Ptilota pectinata, Fucus evanescens, Laminaria longicruris, Agarum Turneri,* and *Phyllaria dermatodea,* belong to those arctic species which are most commonly dispersed and occur in the greatest masses. They are all found in Baffin Bay, even at high latitudes. *Delesseria Montagnei* appears also to be a common species in Baffin Bay. At least it is to be found in quite considerable numbers in the collections from Greenland that I have had the opportunity of examining. *Fucus edentatus* and *Antithamnion Pylaisæi* are also recorded from the same part of the Arctic Sea, but it is still unknown whether they are abundant or not. Thus there remain only two species that have not as yet been found with certainty in the Arctic Sea directly north of their reported American locality: *Fucus miclonensis* and *Phyllaria lorea.* The former species, as I understand it, is no properly arctic alga, the latter is on the contrary known from the most arctic parts of the Polar Sea. It is possible that *Fucus miclonensis,* both that which I have set down under that name from the Norwegian Polar Sea and that which J. G. Agardh reports from New-Foundland, Spitzbergen, and Greenland, is nothing else than one of the numerous forms in which *Fucus evanescens* presents itself, or else that *Fucus miclonensis* from New-Foundland is, in accordance with the opinion of J. G. Agardh, identical with that form of Fucus, brought home from Greenland and Spitzbergen, which I think ought to be regarded as a variety of the *Fucus evanescens* that is commonly dispersed in the Arctic Sea. With regard to *Phylloria lorea,* I can see no probable reason why it should be supposed to have had its centre of development at New-Foundland or to have come there from the south. On the contrary it may be assumed with very great certainty to have had its origin and centre of development in the Arctic Sea, just as the other species in question.

With the exception of these species, the others possessed by the Polar Sea in common with the northern Atlantic are known either exclusively from the European coast of the Atlantic or both from here and from the north-eastern coast of America. Amongst these not a few are found in the purely arctic parts of the Polar Sea, reaching here very high latitudes, being widely distributed, occurring often in large numbers, in short, numbering among the most characteric algæ of the Arctic Sea. On following their distribution southwards, some of them are found to disappear immediately south of the limit of the Polar Sea, others go only a few degrees of latitude south of the Polar Circle, others again, though common and luxuriant in the arctic waters, become southwards more and more rare, small and poor. These circumstances apparently indicate that those species have had their origin in the Arctic Sea and spread from here to the more northern parts of the Atlantic. The following algæ may be quoted as good instances of such species:

Halosaccion ramentaceum, known from all parts of the Arctic Sea, very common in certain regions, for instance in the eastern part of the Greenland Sea. It goes northwards at least to the 80:th degree, and between the 79:th and 80:th degrees is still highly luxuriant. This is the case also with *Polysiphonia arctica*.

In the European part of the Atlantic these two algæ are not known south of Iceland; on the coast of America the former is one of the more prominent elements of the so called arctic Flora, occurring in large numbers at Eastport, and being only occasionally met with so far to the south as the coast of Massachusett [1]).

Ralfsia deusta, though less common in the Arctic Sea, is yet found rather generally in the most northerly part of the Murman Sea. In the Atlantic it has about the same southern limit as the preceding, being known from Iceland, but not south of Eastport (Maine) on the coast of America.

Monostroma Blyttii in the Arctic Sea at Spitzbergen ascends to the 79:th degree of latitude; however, its maxima of frequency are on the north coast of Norway and on the south-western coast of Greenland. It is not known south of Iceland. On the north-western coast of America, it is found most abundant and luxuriant about Eastport, descending southwards to about Boston [2]).

Rhodophyllis dichotoma, another characterical arctic alga, found on the coast of Spitzbergen about 79° N. Lat., is rather common in the north-eastern part of the Murman Sea, and common on the north coast of Norway. It is known from Færoe Isles and the neighbourhood of Bergen on the west coast of Norway, but not farther to the south on the European side. On the west coast of America its southern boundary is at Cape Ann.

The following species:

Odonthalia dentata.
Rhodomela lycopodioides,
Euthora cristata,
Ptilota plumosa,
» pectinata,
Phloeospora tortilis,
Chætopteris plumosa,

may undoubtedly be reckoned among the most characteristic species of the arctic Flora. In the European part of the Atlantic no one of them goes to the south of England, most of them are limited to the west coast of Scandinavia and the most northerly parts of Great Britain. Those of them which are met with on the east coast of America have here a northerly distribution, belonging exclusively or chiefly to the arctic part of that region.

Delesseria sinuosa, *Dichloria viridis*, *Desmarestia aculeata*, *Chætomorpha melagonium*, all of them common species even in the most northerly parts of the Arctic Sea, are indeed to be found even south of England, but here they appear to be rare or even very rare. At least they are so at Cherbourg, according to the statements of LE JOLIS [3]). This is perhaps the case also with several other species, but it impossible at present

[1]) Cp. J. G. AGARDH, Spec. Alg. II. p. 359 and FARLOW, New Engl. Alg. p. 5 and 143.
[2]) Cp. KJELLMAN, Isl. Alg. p. 79 and FARLOW, l. c. p. 5 and 42.
[3]) Cp. Le JOLIS, Liste Alg. Cherb.

to arrive at any higher degree of certainty on this point, because the existing statements on the distribution of algæ south of the Channel are scanty and fluctuating.

Another reason which seems to speak for the Arctic Sea being an independent centre of development, is the fact that in the Arctic Sea and even in the parts of it situated to the north of the Atlantic there are found some species which are wanting in the Atlantic, but occur in the northern part of the Pacific. Such are:

Delesseria Bærii,
Petrocelis Middendorffi,
Rhodymenia pertusa,
Sarcophyllis arctica,
Antithamnion boreale,
Laminaria cuneifolia?
» fissilis,
» solidungula,
Elachista lubrica.

The first of these algæ, as is shown by the table of the distribution of the species, is known from the Greenland Sea, and from the White and the Murman Seas. In the Greenland Sea it ascends to the north coast of Spitzbergen [1]), though it seems to be rare here; in the eastern Murman Sea it is more common and developed luxuriantly [2]), in the White Sea and the neighbouring part of the western Murman Sea it is one of the commonest algæ [3]). Without the limits of the Arctic Sea it is known with certainty only from the Ochotsh Sea. Another locality reported is Kamtshatka [4]). If we should assume that the centre of development of this species were the Pacific, it would become very difficult, not to say impossible, to explain its occurrence in the Arctic Sea north of the Atlantic. It cannot well be supposed to have migrated along the shores, nor to have been carried to its present place by currents. For the current goes from the west towards the east along the coast of Siberia, as has been proved by the researches of the Vega expedition. The alga could possibly have been brought to the American Arctic Sea by the Kurosivo-current, of which at least a feeble branch flows in that direction, but there is no current leading over to the Spitzbergen Sea from the American Sea. If on the contrary we suppose that it has originated in the Arctic Sea, there offers a probable and rather easy explanation, which I shall set forth below, to account for its occurrence in the Ochotsh Sea. It may certainly be objected that the species is known neither from the Siberian nor from the American Arctic Sea; but on the other hand it may be rejoined, that the American Arctic Sea and even the Siberian Sea are little known with regard to algology, and that therefore the species may very well exist in these seas. This is especially probable with regard to the American Arctic Sea, because there is found in Baffin Bay a species very nearly related to and very slightly differentiated from *D. Bærii*, namely *D. corymbosa*, which may pos-

[1]) KJELLMAN, Spetsb. Thall. 1, p. 12.
[2]) Cp. KJELLMAN, Algenv. Murm. Meer. p. 18.
[3]) GOBI, Algenfl. Weiss. Meer. p. 11.
[4]) G. RUPRECHT, Alg. Ochot. p. 239 follow.

sibly have issued from *D. Bærii* or from some parent form common to both. Besides, there can be adduced no reason why the arctic algæ might not be supposed to have been more widely distributed formerly in the arctic regions than they are now, which is assumed on very good reasons to be the case with the arctic phanerogams.

What has been said now about *Delesseria Bærii*, may be applied, though with one or two modifications, even to the other above-mentioned species, that are met with in the Arctic Sea and south of it only in the northern part of the Pacific. By means of the table exhibiting the geographical distribution of the arctic algæ and the more detailed data set forth in the special part of this work, every one may easily convince himself of this fact. It is neither necessary nor suitable to enter here into details with regard to each species. But I will, however, call attention to one of them, *Laminaria solidungula*, one of the most peculiar and characteristic algæ of the Arctic Sea. It is found in the Arctic Sea almost circumpolar: in the Greenland Sea, where it is commonly distributed along the whole west and north coasts of Spitzbergen, and not seldom attains such a considerable size as to become one of the most magnificent algæ of the ocean, in the Murman Sea, the Kara Sea, the Siberian Sea not far from the mouth of Behring Strait, and in Baffin Bay on the west coast of Greenland. It probably grows also in the American Arctic Sea. South of the Arctic Sea it is known only from one place, namely the Ochotsh Sea, from where RUPRECHT reports a young »abnormal specimen of *Laminaria saccharina* with undivided scutiform root», which is in all probability a young *Laminaria solidungula* [1]).

The present Flora in the northern part of the Pacific differs so essentially in composition from that of the Northern Atlantic, that is to say, it contains many species that are so sharply distinguished from those of the Atlantic, even belonging to quite different types, that in order to account in any way for this fact, one is necessarily obliged to assume that these two divisions of the ocean appertain to different areas of development, within which different forms have continued to be evolved during a very long time. However, on the other hand, it is a well-known fact that the northern Atlantic has no inconsiderable number of species in common with the northern Pacific. Though it is highly probable, as J. G. AGARDH has rightly remarked, that »a great part of the statements about algæ occurring in widely distant seas is attributable to imperfect knowledge and wrong determinations caused thereby, and that the number of such species as are supposed to grow in widely distant seas will be diminished in proportion as the accuracy of the scientific determinations increases», and though this general judgment may be true even with regard to the reported number of species reputed common to the northern Atlantic and the northern Pacific, still there are undoubtedly to be found in these widely separated sea-regions, as the same algologist expressly points out, several forms that can be proved to be identical [2]). The present hydrographical conditions being so essentially unlike, it can hardly be assumed that these species should have been developed both in the northern Atlantic and in the

[1]) Cp. RUPRECHT, Alg. Ochot., p. 351 and J. G. AGARDH, Lamin. p. 8 and Grönl. Lamin. och Fuc. p. 11.
[2]) Cp. J. G. AGARDH, Spetsb. Alg. Progr. p. 1. Spetsb. Alg. Bidr., p. 10. Grönl. Lamin. och Fuc. p. 8—9, 11, etc.

northern Pacific, after these seas had received their present limits and physical nature [1]). It seems as if J. G. AGARDH inclined to the opinion that these species common to both the regions in question have had their origin in either of them and have been brought from this region into the other by a current continued through all the Arctic Seas. AGARDH says: »Though it appears to be true in general with regard to the algæ, that the distribution of a species is confined to the limits of the same sea-current, still the fact of many larger algæ occurring both in the Atlantic and the Pacific may possibly indicate the existence of a current continued through the Arctic Seas, which might carry algæ from New-Foundland and Spitzbergen to Kamtshatka and the most northern isles of Western America» [2]). The hydrographical researches, carried on in the Arctic Seas by the Polar expeditions of later times, have not, however, demonstrated the existence of such a continued current, having rather proved, on the contrary, that there is in the Arctic Sea a whole net-work of currents. Only by supposing a species to have been removed from one current into another, its occurrence both in the Pacific and the Atlantic could be explained from the influence of sea-currents, and such a combination would be so complicated with regard to several species as to be hardly admissible. It might be supposed, indeed, that such a current existed during previous periods; but I do not know af any reasons for such a supposition. Without entering into this question in general, I will only remark, that in my opinion the occurrence of several algæ in the northern Atlantic at the same time as in the Pacific may probably be explained by the former distribution of water and land and the different physical conditions of the seas in former times as compared with their present state. The study of the present distribution of the algæ of the Arctic Seas leads to such a conclusion. According to pretty reliable statements, the following arctic algæ are to be found in the northern Atlantic as well as in the northern Pacific.

Corallina officinalis?
*Lithothamnion polymorphum,
*Odonthalia dentata,
*Rhodomela lycopodioides,
Polysiphonia parasitica?
» urceolata?
» fastigiata?
» atrorubescens?
» nigrescens,
Delesseria alata?
* » sinuosa,
*Hildbrandtia rosea,
Peyssonnelia Dubyi,
*Rhodophyllis dichotoma,
*Euthora cristata,
Plocamium coccineum,
*Rhodymenia palmata,
*Halosaccion ramentaceum,
Dumontia filiformis,
Callophyllis laciniata?
*Ahnfeltia plicata,
Gigartina mamillosa?
Chondrus crispus?
*Ceramium rubrum,
*Ptilota plumosa,
* » pectinata,
Callithamnion polyspermum,
» arbuscula,
Antithamnion floccosum?
* » americanum,

[1]) Cp. ENGLER, Pflanzenw. p. IX—X.
[2]) J. G. AGARDH, Spetsb. Alg. Bidr. p. 10.

*Rhodochorton Rothii,
Porphyra laciniata,
*Fucus vesiculosus,
* » ceranoides,
» spiralis?
* » evanescens,
» miclonensis?
*Alaria Pylaii?
*Agarum Turneri,
Phyllaria dermatodea,
Laminaria saccharina?
* » longicruris,
» digitata?
» stenophylla,
*Chorda filum,
*Ralfsia deusta,
*Chordaria flagelliformis,
*Elachista fucicola,
*Lithoderma fatiscens,
*Ilea fascia,
*Scytosiphon lomentarius,
*Desmarestia aculeata,
*Dichloria viridis,
*Phloeospora tortilis,
*Dictyosiphon foeniculaceus,
*Chætopteris plumosa,
*Ectocarpus confervoides,
*Pylaiella litoralis,
Enteromorpha clathrata,
» intestinalis,
* » compressa,
Ulva lactuca,
*Monostroma fuscum.
*Spongomorpha arcta,
Cladophora glaucescens?
*Rhizoclonium riparium,
*Chætomorpha melagonium,
» tortuosa,
*Urospora penicilliformis,
Bryopsis plumosa.

Of these 70 species, no less than 41, viz. those marked with an asterisk, accordingly 58½ per cent, are at present known with certainty from the arctic parts of the Polar Sea, and amongst these 41 species there are several of the most commonly distributed and most characteristic forms of that region of the Polar Sea which is rich in ice. As moreover many of them, as has been shown above, are chiefly distributed northwards at least in the Atlantic — there are no detailed statements to be had for the Pacific — we are justified in placing their origin in a glacial sea and in assuming that they have passed from there to the northern Atlantic and the northern part of the Pacific. The percentage of arctic forms amongst those species which are reported common to the Arctic Sea and the northern part of the Atlantic and of the Pacific, is in all probability larger than what is indicated by the figures mentioned. For there is some reason to suppose that a rather large number of species from the northern part of the Pacific, which have been considered identical with forms from the Atlantic, will prove on closer examination to be specifically distinct or wrongly determined. This is probably the case with those marked with a sign of interrogation in the above list [1]). Their number amounts to 16. If these are deducted, the arctic forms would constitute about 75 % of the whole number of species that the Arctic Sea has in common with the northern part of the Atlantic and the northern Pacific. Of the 13 species then remaining, *Porphyra laciniata, Enteromorpha intestinalis auct., E. clathrata auct.* and *Ulva lactuca*, which are however met

[1]) In support of this supposition, I refer to the notices on these species which are given by J. G. AGARDH in Spec. Alg., Epicr. and his treatises on the arctic marine Flora, by RUPRECHT in Alg. Ochot., by FARLOW in New Engl. Alg., by HARVEY in Ner. Am. and Alg. Vanc., by POSTELS and RUPRECHT in Ill. Alg.

with on the west coast of Greenland, are cosmopolitan, if they are understood in that extensive sense still usual with algologists, and on this account afford us no clue for deciding the question of their origin. One species, *Antithamnion floccosum*, whose form in the Atlantic differs from the Pacific form, might possibly be regarded as having issued from *A. boreale* [1]), commonly dispersed in the Arctic Sea. Some species, *Polysiphonia parasitica, P. nigrescens* and *Plocamium coccineum*, that are also stated to occur in the southern hemisphere, have possibly passed from the one sea into the other south of America. The present distribution of *Callithamnion arbuscula* is so limited that no conclusion can be drawn from it as to the original native country of the alga; moreover it appears in the Pacific in another form than in the Atlantic. Concerning the few remaining species, I believe I must refrain at present from uttering any suppositions.

Thus it has been shown that the arctic marine Flora is rich in endemic species; that several species which go far northwards in the Arctic Sea and are widely distributed there, are only slightly spread southwards in the Atlantic; that there are a number of species in the northern part of the Pacific which occur also in the Arctic Sea, even in those parts of it which lie north of the Atlantic, but are wanting in the Atlantic itself; and that of the species comparatively very numerous, which the arctic Flora possesses in common with the northern Atlantic as well as with the northern Pacific, a very large proportion consists of such species as are met with in the Arctic Sea at high latitudes, amongst them several of the most characteristic forms of the Arctic Sea. I think I am justified in drawing from these facts the conclusion that the Flora of the arctic part of the Polar Sea is an old Flora and that it has developed within the Arctic Sea. But this being so, it ought to be explained why several arctic forms are met with at present south of the limits of the Arctic Sea, both in the Atlantic and in the northern part of the Pacific. The occurrence of several arctic algæ on the north-eastern coast of America is easily explained by their having been carried there by the cold Labrador current, which moreover makes the external conditions on the last-mentioned coast very similar to those in which the algæ live in the Arctic Sea. But along the European coast of the Atlantic no current runs down from the Arctic Sea; on the contrary, a current flows upwards into the Arctic Sea from the Atlantic. Even if the occurrence of several arctic algæ on this coast might be explained by their having wandered southwards along the coast, from the Spitzbergen Sea and the Murman Sea along the land of the Cisuralian Samoyedes a. s. o., this explanation cannot be applied to the species met with at Iceland, and on the coasts of Great Britain and France. The currents between the northern Pacific and the Arctic Sea are favourable for carrying algæ into the Arctic Sea, but not for transporting them from the Arctic Sea into the Pacific. However, the causes of the present distribution of the algæ in the great divisions of the Ocean need not nor ought to be sought for in the conditions now existing on the earth, any more than the causes of the distribution of the land-plants. Just as the phanerogamous Flora of Scandinavia contains several elements which are remnants from that period when the glacial formation extended farther southwards than in our days, those arctic algæ occurring in the

[1]) Compare this species in the special part.

northern Atlantic and the northern Pacific may have stayed there from those times when northern Europe was surrounded with a sea filled with ice stretching down to the present northern coast of France. It is not at all improbable that the Flora of this sea may have been similar to that of the present Arctic Sea. When the glacial formation diminished in the south, southern plants immigrated in the sea as well as on the land, dislodging the main mass of the glacial ones. However, some of these, being able to hold out the struggle against the new-comers, maintained themselves in their original home and have done so ever since. Even though, as I have tried to show above, the present poverty in algæ in the eastern part of the Kara Sea and in the Siberian Sea may depend essentially on the unfavourable nature of the bottom and the slight salinity of the water, still it is probable that during the oldest glacial period, when these seas extended farther southwards than now, the configuration of the coast and the condition of the water were more advantageous for algæ and that the marine vegetation of these parts of the Arctic Sea was not then so poor [1]). The occurrence of several species, as *Delesseria Bærii, Rhodophyllis dichotoma, Petrocelis Middendorffi, Ptilota pectinata* a. o., in the Ochotsh Sea, the Murman Sea and the Spitzbergen Sea, and their absence in the intermediate region of the arctic waters, may be explained by the hypothesis that these species formerly grew also in the last-mentioned sea-region, but succumbed afterwards, because the shore was moved northwards, partly through the formation of deltas, partly through a general rising of the land elevating the sea-bottom, which was formed of sand-banks, and because large rivers began to pour ever increasing masses of fresh water into the sea. If *Halosaccion saccatum*, reported from the White Sea, is really to be found there, it should probably be numbered among those species which have formerly been more widely spread along the coast of Siberia; for, even though it be not identical with any of the species of the Pacific, *H. fucicola, H. hydrophora*, or *H. firmum*, these are nevertheless its nearest relations. The change suffered by the Flora in the northern Atlantic and the northern part of the Pacific, in proportion as the glacial formation retreated, took place even in the part of the Arctic Sea surrounding the coast of Norway. By the immigration of southern forms, glacial species were superseded or deprived of their predominating influence. The elements of the Flora were considerably increased, and its character was no longer marked by glacial forms, but by such as belong to the Atlantic. However it is not only into this part of the Polar Sea, most closely allied hydrographically to the present northern Atlantic, that southern species have probably immigrated in later times, or are now in course of immigration; but this is the case also with the Arctic Sea proper. Such a transplantation of algæ into certain parts of the Arctic Sea is very much favoured by currents flowing in that direction. Thus, for instance, algæ may be easily transported into the eastern part of the Greenland Sea to the coast of Spitzbergen by the Gulf Stream, and into the eastern part of the Murman Sea to the west coast of Novaya Zemlya partly by means of this stream, partly along the almost continuous coast from the Norwegian Polar Sea and the western Murman Sea. Between Spitzbergen and Norway at different latitudes I

[1]) NORDENSKIÖLD, Pröven, p. 70—71.

have seen *Ozothallia nodosa* and *Fucus vesiculosus* drifting, and on the south coast of Spitzbergen I have found *Ozothallia nodosa* washed ashore, overgrown with *Polysiphonia fastigiata*.

I have not myself found any of these algæ growing in any of the numerous places on the coast of Spitzbergen, that I have had the opportunity of examining. They are however stated by others to occur there. If this be really true, they may be assumed to have been transferred there in later times by the Gulf Stream. Whether any species have made use of the convenient route along the coast to Waygats and Novaya Zemlya cannot indeed be decided with certainty. It is possible, however, that *Cladophora rupestris*, found at the south-western extremity of Waygats, and *Spongomorpha lanosa*, met with at southern Novaya Zemlya, may have done so. It is possible also that a transfer of algæ has taken place formerly and is perhaps going on even at the present time by means of those vessels which depart yearly in great numbers from northern Norway for Spitzbergen and Novaya Zémlya and stay during the summer in the waters surrounding these islands, as well as by those which start regularly since a long time for Baffin Bay from more southerly regions. I have evidence of a transportation of this kind taking really place, but I do not know whether the algæ thus brought over into the Arctic Sea have been able or are able to maintain themselves there. One of the vessels employed in bringing over the Swedish expedition of 1872 to Spitzbergen, some time after the arrival of the expedition, was found to be richly clothed with small-sized *Enteromorpha compressa*, at and below the water-line. This plant had probably been engendered by spores, which had attached themselves to the vessel in more southern regions on the coast of Sweden or Norway and developed afterwards during the course of the voyage. It is possible also that water-fowls, especially such as inhabit lagoons on the coast, may carry with them some alga or other from the south. Perhaps *Rhizoclonium rigidum*, which grows abundantly in the lagoons at Advent Bay in Spitzbergen, has come to the high North in this manner.

The marine Flora on the coast of Greenland includes a pretty considerable number of species whose origin lies in a southerly direction, in the Atlantic. The number is so considerable and the habitat of the algæ is such that they cannot be supposed to have been transported into those regions by the agency of either men or animals. Nor can these species have come into the arctic waters about Greenland by means of seacurrents directly from the south, i. e. from the east coast of America, as the current here goes from the north southwards, and, moreover, several species reported from Greenland, are wanting on the American coast. This is the case with *Hydrolapathum sanguineum*, *Pelvetia canaliculata*, *Nitophyllum punctatum*, *Furcellaria fastigiata*, *Callophyllis laciniata*, *Asperococcus bullosus*, *Stupocaulon scoparium*, *Enteromorpha tubulosa*. I do not certainly feel quite sure that all these species occur really on the coast of Greenland. But as I have myself seen specimens of some species, stated to have been collected at Greenland, and as experienced algologists allege having seen specimens of the other species from the same regions, I could not but quote them for the present among the Greenland algæ. I cannot explain their occurrence at Greenland otherwise than by the hypothesis that they have arrived there from the east by Iceland, where at least some of them have been hitherto met with. To be sure, there is no current leading directly from Iceland to Greenland, but by that current which passes along the coast of Iceland in a north-westerly direction, they may possibly have been carried into the region of the great Polar current or Greenland current and then brought on by means of that to the coasts of Greenland. But if this can have happened with these species, it can, of course, have happened also with other southern forms occurring on the coast of Greenland.

The immigration to Greenland would thus have taken place not from the south, but from the east, from the European part of the Atlantic by way of Iceland. A closer investigation of the marine Flora of Iceland would surely give several important results on this point.

Even from the northern part of the Pacific algæ have surely immigrated at later times. However, I can hardly cite more than one reliable instance of it. No species is known with certainty from the Siberian Sea, that must be supposed to have immigrated there from the Pacific, if we do not assume, as certain or probable, that those algæ, *Ptilota asplenoides* and a species of *Laminaria* nearly related to *L. longipes*, which are said to be found in the Siberian Sea at the mouth of the Lena, do really grow in the place from where they have been reported [1]). If they belong to the Siberian Sea, which I think rather doubtful, we cannot but suppose that they have come there from the Pacific, where at least *Ptilota asplenoides* is widely distributed and grows in great abundancy at several places. The direction of the currents between these two seas does not favour the importation of algæ into the Siberian Sea; it is more favourable for an immigration into the western part of the American Arctic Sea. However, this sea is as yet very little known with regard to its algology. Nevertheless, among the species reported from there, there is one that may be assumed, on good grounds, to have immigrated from the south, namely *Rhodomela larix*, which is extensively spread in the Pacific and common even in the northern part of the Behring Sea.

Whoever visits the coasts of Spitzbergen, cannot but notice the objects of different kinds washed ashore, a great part of which have apparently been carried there from the south, from the coast of Norway. At some parts of the coast they are more rare, at others they are to be found in very large masses. NORDENSKIÖLD relates, for instance, that at one place on the coast of Spitzbergen pieces of pumice were found in so great quantities that a small sack could be filled with them. Many of these objects, as floats of tree, glass, and cork a. o., are of such a nature that algæ, at least smaller ones, can germinate on them, and in the earlies stages of their development be brought by means of them from the coasts of Norway to the North [2]). Larger algæ may be transferred floating on the water by the Gulf Stream. Thus it seems as if the marine Flora of Spitzbergen might be expected to comprise a rather considerable number of southern algæ, immigrated from the north-west and north coasts of Norway at later times. This however is not the case. By far most elements of the Spitzbergen Flora have at present such a geographical distribution that they must be considered to have their proper home in the Arctic Sea. I have thought fit to call attention to these facts, because

[1]) Cp. J. G. AGARDH, Grönl. Lamin. och Fuc., p. 7.
[2]) Cp. NORDENSKIÖLD, Spetsb.-exp., p. 39—41.

they show that sea-regions belonging to the area of the same sea-currents do not always possess an essentially similar vegetation, and that one must not accordingly, when attempting to divide the marine Flora into different geographical regions, fix the limits of these in all seas chiefly according to the sea-currents [1]). There may be found external conditions opposing insurmountable obstacles to the leveling tendencies of the currents and keeping regions isolated from each other, whose vegetations sea-currents are ever striving to level and to assimilate. The circumstances which counteract and annihilate the action of the Gulf Stream and which have therewith maintained the marine Flora of Spitzbergen at such a high degree of difference to that of the Norwegian Polar Sea, I consider to be chiefly the unsuitableness of the litoral zone of Spitzbergen for a richer vegetation, the low temperature of the water, and the insufficiency of the light. Most of the algæ on the arctic coast of Norway belong to the litoral zone, and apparently it is chiefly algæ of this kind that may be expected to be transported to Spitzbergen by the Gulf Stream and such objects as are carried by it towards the North. When they have arrived at Spitzbergen, they cannot thrive or spread there in their wonted places. The bottom is unfit for them, the ice destroys them, the temperature of the water is too low for them, and the layers of air with which they are brought in contact during ebb-tide, are often so cold as to injure the algæ, stopping their development and diminishing their power of resistance. The temperature of the water in the sublitoral region is at or below 0° during the whole year, accordingly very much lower than that to which on the coast of Norway they are exposed in the coldest season. How could these Norwegian species, which love light and are accustomed to it, be expected to live here in darkness or dimness during a great part of the year, enjoying even during the rest of the year, on account of the different place where they grow, a less quantity of light than in those regions from where they have come? What has been said here of the Flora of Spitzbergen, holds good essentially even of the Flora of the eastern Murman Sea.

Thus the out-lines of the history of the arctic marine Flora, as based on the results of the preceding investigation, may be stated as follows. The Flora has had its centre of development in the Arctic Sea. Its area was more extensive during the glacial period than at present. It has been recruited in later times by more southern species. The immigration has been larger into certain parts of the Arctic Sea than into others; in the Norwegian Polar Sea it has been so large that the Flora has here lost its arctic character; in other parts it has, on the contrary, been much smaller than might be expected on account of the currents prevailing, which are highly favourable for an immigration from more southern seas.

---

[1]) Cp. J. G. AGARDH, Spetsb. Alg. Progr. p. 1.

Provinces of the arctic marine Flora.

I have attempted in the preceding pages to support the opinion that a strong change of the Flora has taken place on the north-west and north coasts of Norway, so that it has assumed here a character different from its previous one and from that of the present Flora in the Arctic Sea properly so called. I am now going to treat this subject more in detail, and to show forth on the ground of the investigations made hitherto, in what degree the Flora on this coast of Norway differs from the Flora of other parts of the Arctic Sea.

The family types are almost altogether the same. Only three families that are represented on the coast of Norway, want representatives in other parts of the Arctic Sea, namely, *Champiaceæ*, *Aglaozoniaceæ*, and *Derbesiaceæ*. The number of genera peculiar to the Norwegian arctic marine Flora is greater, namely, 21(19). These are:

Spermothamnion,
Petrocelis,
Plocamium,
Chylocladia,
Chondrus,
Microcladia,
Callithamnion,
Bangia?
Erythrotrichia,
Himanthalia,
Halidrys,
Stilophora,
Castagnea,
Leathesia,
Coilonema,
Lithosiphon?
Aglaozonia,
Myriotrichia,
Prasiola,
Derbesia,
Lyngbya.

Of the species hitherto known in the Norwegian Polar Sea, the following are not known with certainty from any other part of the Polar Sea:

Lithothamnion soriferum,
» Ungeri,
» alcicorne,
» norvegicum,
» intermedium,
Melobesia membranacea,
» macrocarpa,
Polysiphonia parasitica,
» Brodiæi,
» fibrillosa,
Polysiphonia Schübelerii,
» byssoides,
Spermothamnion Turneri,
Chantransia Daviesii,
» virgatula,
Delesseria angustissima,
» alata,
Petrocelis cruenta,
» Middendorffi,
Plocamium coccineum,

Chylocladia clavellosa,
» articulata,
Phyllophora membranifolia,
Chondrus crispus,
Microcladia glandulosa,
Ceramium Deslongchampii,
» circinatum,
» acanthonotum,
Ptilota elegans,
Callithamnion polyspermum,
» Hookeri,
» arbuscula,
» roseum,
» corymbosum,
Diploderma amplissimum,
Erythrotrichia ceramicola,
Himanthalia lorea,
Halidrys siliquosa,
Fucus spiralis,
» miclonensis,
» distichus,
Laminaria Clustoni,
» stenophylla,
Chorda tomentosa,
Stilophora Lyngbyei,
Asperococcus echinatus,
Ralfsia verrucosa,
Castagnea divaricata,
Leathesia difformis,
Lithoderma lignicola,
Coilonema Ekmani,
Coilonema Chordaria,
Aglaozonia parvula,
Ectocarpus pygmæus,
» draparnaldioides,
» fasciculatus,
» tomentosus,
» Lebelii,
» terminalis,
» reptans,
Pylaiella nana,
Myriotrichia filiformis,
Chætophora pellicula,
Enteromorpha complanata,
Monostroma latissimum,
» undulatum,
» cylindraceum,
» saccodeum,
» angicava,
» arcticum,
» crispatum,
Prasiola stipitata,
Spongomorpha spinescens,
Cladophora glaucescens,
Chætomorpha septemtrionalis,
Ulothrix Sphaceclariæ,
Derbesia marina,
Codiolum longipes,
» pusillum.
Calothrix Harveyi,
Lyngbya semiplena.

Thus:

Florideæ	36	species	representing	10	families,	17	genera,
Fucoideæ	26	»	»	9	»	16	»
Chlorophyllophyceæ	17	»	»	5	»	10	»
Nostochineæ	2	»	»	2	»	2	»
Total Sum	81	species	representing	26	families,	45	genera.

If we leave out the White Sea and the western Murman Sea, the other regions of the Arctic Sea that, on account of their situation, ought to present the greatest resemblance to the Norwegian Polar Sea with regard to the composition of the Flora, should be the eastern Greenland Sea and the eastern Murman Sea. In these sea-regions there are to be found the following species that are not known from the Norwegian Polar Sea.

Lithothamnion compactum,
Delesseria Bærii,
Rhodymenia pertusa,
Sarcophyllis arctica,
Kallymenia rosacea,
Phyllophora interrupta,
Rhodochorton intermedium,
» spetsbergense,
Diploderma miniatum,
Fucus evanescens,
Scaphospora arctica,
Haplospora globosa,
Phyllaria lorea,
Laminaria solidungula,
» Agardhii,
Laminaria fissilis,
» nigripes,
Scytosiphon attenuatus,
Phloeospora pumila,
Dictyosiphon corymbosus,
» hispidus,
Chætophora maritima,
Enteromorpha minima,
Monostroma lubricum,
» leptodermum,
Rhizoclonium pachydermum,
Ulothrix discifera,
Characium marinum,
Chlorochytrium inclusum.

Florideæ	9	species representing	7	families,	8	genera,
Fucoideæ	12	»	»	5	»	8	»
Chlorophyllophyceæ	8	»	»	4	»	7	»
Total Sum	29	species	representing	16	families,	23	genera.

One family, *Tilopterideæ*, four genera, *Scaphospora*, *Haplospora*, *Characium*, and *Chlorochytrium*, have no representatives on the arctic coast of Norway.

If we presuppose that the Flora on the coast of Norway did once possess the same composition as that which exists now in the sea on the coasts of Spitzbergen and Novaya Zemlya, and that it has afterwards, in consequence of altered external conditions, assumed its present character, the change suffered by it, according to the figures set forth above, would consist in its having lost 1 family, 4 genera, 29 species, receiving in compensation as new elements 3 families, from 19 to 21 genera, 81 species. However, it would surely be precipitate to form such a conclusion. In the first place, we have no right to presuppose that the Flora on the coast of Norway should have been so similar in all particulars to that of the Greenland and the Murman Seas, that some species or other could not have been found within the latter regions, though it did not exist within the former. Then it is quite possible that some of those species which belong to the Flora of Spitzbergen and Novaya Zemlya may really occur on the coast of Norway, although they have not as yet been observed. Moreover, it is highly probable, that there are old glacial species even amongst those algæ which are known in the Arctic Sea only from the north and north-west coast of Norway. In all probability, *Petrocelis Middendorffi* is such a species. But, above all, the increase in new species has most certainly been very much greater than that indicated by the figure of 81. For this figure comprehends only those species which are supposed to have immigrated only into the Norwegian Polar Sea, not those which have probably immigrated not only here, but also into the other parts of the Arctic Sea. Amongst these there are surely to be numbered several algæ that are met with in the western part of the Murman Sea and the White Sea, as well as in the Norwegian Polar Sea. The Flora of the two former seas, as has been demonstrated by GOBI'S detailed and accurate description of it, is a mixed Flora, composed of elements from the Atlantic as well as the Arctic Sea. Among these elements the following species are probably to be reckoned:

Corallina officinalis,
Lithophyllum Lenormandi,
Laminaria saccharina,
Eudesme virescens,
Mesogloia vermicularis,
Isthmoplea sphærophora,
Bulbocoleon piliferum.

The same category includes probably several species which are known, besides from the Norwegian Polar Sea, also from Baffin Bay on the west coast of Greenland or from here as well as from the White Sea and the western Murman Sea. Such are:

Polysiphonia urceolata,
» nigrescens,
Peyssonnelia Dubyi,
Cruoria pellita,
Hydrolapathum sanguineum,
Dumontia filiformis?
Cystoclonium purpurascens,
Gigartina mamillosa,
Rhodochorton sparsum,
Porphyra laciniata,
Pelvetia canaliculata,
Myrionema strangulans,
Scytosiphon lomentarius,
Cladostephus spongiosus,
Sphacelaria cirrhosa,
» olivacea,
Enteromorpha clathrata,
» intestinalis,
Monostroma Grevillei,
Cladophora gracilis,
Chætomorpha tortuosa?
Bryopsis plumosa,
Rivularia hemisphærica.

It appears probable also that some species which occur on the coast of Norway, and at the same time are more widely distributed in the Arctic Sea, ought also to be regarded as immigrated from the south, viz.:

Polysiphonia fastigiata,
» atrorubescens,
Sarcophyllis edulis,
Ozothallia nodosa,
Fucus vesiculosus?
» serratus,
Spongomorpha lanosa,
Cladophora rupestris,

Lastly, there are to be mentioned here several species which probably, like some of those peculiar to the Norwegian Polar Sea, neither have immigrated into the Polar Sea, because they do not occur farther to the south or else have a very slight southerly distribution, nor can be regarded as old glacial species, on account of their present distribution within the Arctic Sea. To this class the following species seem to belong:

Phyllophora Brodiæi,
Antithamnion floccosum,
» Pylaisæi,
Fucus edentatus,
» miclonensis,
Fucus linearis,
» filiformis,
» distichus,
Alaria Pylaii.

Thus the whole augmentation gained by the Flora on the arctic coast of Norway during the last part of the glacial period, and after its end, would amount to about 128 species, i. e., about 66 per cent of the total number of its species, as known at present.

The majority of these have surely immigrated from the south. But such a supposition cannot be made with regard to all. The last-mentioned 9 species must be excluded from it as well as the following ones, which, as far is hitherto known, are peculiar to the Norwegian Polar Sea, namely,

Lithothamnion soriferum,	Monostroma undulatum,
» alcicorne,	Monostroma cylindraceum,
» intermedium,	» saccodeum,
Polysiphonia Schübelerii,	» angicava,
Diploderma amplissimum,	» arcticum,
Lithoderma lignicola,	» crispatum,
Pylaiella nana,	Chætomorpha septemtrionalis,
Chætophora pellicula,	Ulothrix Sphacelariæ.

There is at present no other way of explaining the occurrence of these species than to assume that they have developed themselves within the Norwegian Polar Sea. The law for the development of new forms being that one type is originally varied, not that new types are formed, the fact of so many of these species being modifications of two generic types, which are, besides, rich in other forms within the Norwegian Polar Sea, may be said to speak for the hypothesis that many of these species, if not all, have had their centre of development in the region where they grow now. It should be remarked too, that *Polysiphonia Schübelerii* and *Lithoderma lignicola* are nearly related to other species occurring in the Norwegian Polar Sea, the former to *Polysiphonia fibrillosa*, the latter to *Lithoderma fatiscens*. This is the case even with many of the nine first-mentioned species. *Phyllophora Brodiæi* is hardly anything else than a southern form of *Phyllophora interrupta*, *Alaria Pylaii* is very closely allied to *Alaria membranacea*, and *Antithamnion floccosum* and *A. Pylaisæi*, as has already been intimated and will be shown in detail below, may be regarded as species developed out of *Antithamnion boreale*, commonly distributed in the Arctic Sea. The above-mentioned species of Fucus constitute a pretty closely connected series, and there may be shown reasons why this series should possibly be considered as having issued from the arctic *Fucus evanescens*. But the change of the Flora of the Norwegian Polar Sea did not stop at the expulsion of old species, the immigration of new ones, and the development of new forms; even the general character of the vegetation has altered in an essential degree. With regard to the distribution of the species, this Flora differs, as has been shown before, from that of other parts of the Arctic Sea, by the predominating species being different. I have mentioned above, which species are to be regarded as characteristic of the Norwegian Polar Sea, i. e., are most commonly disseminated, occur in the greatest numbers, and contribute most to stamp the vegetation with a peculiar mark [1]). I have only to add here that of the Fucaceæ and Laminariaceæ the following species ought to be regarded as characteristic of this Flora:

Himanthalia lorea [2]),
Halidrys siliquosa [2]),
Ozothallia nodosa,
Fucus serratus,
» vesiculosus,
» spiralis,
» edentatus,
» filiformis,

Pelvetia canaliculata,
Alaria esculenta [2]),
» membranacea,
Phyllaria dermatodea,
Laminaria saccharina,
» Clustoni,
» digitata,
Chorda filum.

The relationship of the present Flora of the Norwegian Polar Sea to that of the other parts of the Arctic Sea, the Northern Atlantic, and the northern Pacific, will be seen from this table:

	Florideæ.	Fucoideæ.	Chlorophyllophyceæ.	Nostochineæ.	Total sum.	Per cent of the whole number of species of the Flora.
The Norwegian Polar Sea has in common with the Greenland Sea......	23(25)	25(27)	13	1	62(66)	32(33) %
» » with the White Sea and the western Murman Sea......................	28	31	9	0	68	35 »
» » with the eastern Murman Sea......	23	21(24)	13(14)	1	58(62)	30(32) »
» » with the Kara Sea......................	11	9(10)	4	0	24(25)	12 »
» » with the Siberian Sea.................	8	6	1	0	15	7,5 »
» » with the American Arctic Sea......	9	4	2	0	15	7,5 »
» » with Baffin Bay..........................	32(33)	34(35)	18	1	85(87)	43,5(44,5) »
» » with the northern Atlantic...........	69	62	29	4	164	84 »
» » with the northern Pacific.............	23(32)	21(24)	12	0	56(68)	29(35) »

These figures clearly show that the Flora of the Norwegian Polar Sea has the greatest affinity to that of the northern Atlantic.

Thus it has been demonstrated by the examination now carried out, that this Flora possesses of old a number of polar or arctic elements, that other elements have been added partly by the forming of new species partly by immigrations from the south, and that these younger elements, being considerably preponderant, determine chiefly the general character of the present Flora. On this account, the Norwegian Polar Sea ought not to be comprehended within the region of the arctic Flora, but, the greater part of the elements of its Flora being natives of the northern Atlantic and having been transplanted from there to the north and north-west coast of Norway, it should be referred to the region of the Atlantic Flora. Though the other parts of the Arctic Sea certainly have no perfectly uniform Flora, their marine vegetation agrees in these respects, viz. that a considerable number of its elements occur everywhere, that

[1]) Cp. p. 14 follow.
[2]) In the southern part of the region.

most species, as far as can be decided, are of arctic origin, or, at least, that the general character of the vegetation is marked by arctic elements. This is true, as has been shown by GOBI [1]), of the Flora of the White Sea and the waters immediately adjacent to it, and still more with respect to the Flora of those regions of the Arctic Sea that are situated farther northwards and eastwards. A more considerable immigration of southern forms has apparently taken place into Baffin Bay on the west coast of Greenland, but even here these forms are inferior to the arctic ones, with regard to the number both of species and of individuals, and with respect to their influence on the general character of the vegetation. On this account, these parts of the Polar Sea and of the northern Atlantic ought to be considered as a whole with regard to the marine vegetation, and may be denominated the region of the arctic marine Flora. This region accordingly comprises those parts of the Ocean which lie along the coasts of the countries occupied by the arctic land-flora, in other words, the whole Polar Sea, except the Norwegian Polar Sea, and besides, the northern Atlantic on the coast of Greenland, and the southern part of Baffin Bay. Possibly there ought to be added to it the sea on the north-eastern coast of America along Labrador, New-Foundland, and the coast southwards to the latitude of Boston. But it is more correct perhaps to consider this part as a transitional region between the arctic Flora and that of the northern Atlantic. It is reserved for future investigations to decide this question. For the present, the latitude of Cape Farewell may be regarded as the southern boundary of the arctic region in the western part of the Atlantic.

The arctic region has accordingly a very large extent, stretching round the Pole and reaching southwards at least down to Lat. N. 60. It might have been expected that within so extensive a region differences in the hydrographical and purely geographical conditions as well as in development, should have produced distinct differences of vegetation, so as to form in the course of time a rather considerable number of narrower regions with dissimilar Floras. There exist indeed such narrower regions distinguishable from one another, but they are fewer than might be supposed a priori. In my description of the vegetation of the Murman Sea, I have shown it to agree so essentially with the Flora of the Greenland Sea on the coasts of Spitzbergen, that these two parts of the Arctic Sea must be regarded as belonging to the same province of the arctic Flora [2]). GOBI has afterwards demonstrated that the Flora of the White Sea and the contiguous parts of the Murman Sea is very closely related to that on the coasts of Spitzbergen and Novaya Zemlya [1]). From the Kara Sea only two species are known that have not been found hitherto in either of the last-mentioned Seas, which it resembles otherwise in the general aspect and characteristic species of the Flora [3]) The scanty notices that we possess on the marine vegetation on the east coast of Greenland, point to its being like that of Spitzbergen. As far as I know, no algæ have been as yet collected in the Spitzbergen Sea, but probably its

[1]) Cp. GOBI, Algenfl. Weiss. Meer, p. 13 and foll.
[2]) Cp. KJELLMAN, Algenv. Murm. Meer. p. 72 and foll.
[3]) Cp. KJELLMAN, Kariska hafvets Algv., p. 9—10.

vegetation agrees with that of the Greenland and the Kara Seas. Thus the whole of that considerable region of the Arctic Sea, which is composed of the Greenland, the Murman, the White, and the Kara Seas, and probably the Spitzbergen Sea, would constitute one division of the arctic Flora, agreeing in the principal characters of its vegetation. Spitzbergen being situated almost at its centre, this province of the arctic Flora may be called the province *of Spitzbergen*, its marine Flora the Spitzbergen marine Flora. The Flora of the Siberian Sea resembles the preceding in many respects. The majority of the elements are the same, but most of the predominating species, i. e., the Laminariaceæ, are specifically different from those of the province of Spitzbergen and of the remaining part of the arctic region. From this reason, the Siberian Sea ought to be regarded as a separate province of the arctic Flora. I propose for it the name of the *Siberian* province. The marine vegetation of Baffin Bay, to which the Flora of the American Arctic Sea seems to approach nearly, is still more isolated. It possesses a considerable number of elements wanting in the rest of the arctic Flora; its characteristic species are partly different, and one of them belongs to a generic type that is not represented at all in the Floras of the other provinces. This province may be properly named the *American*.

The following tables afford a survey of the composition of the marine Flora within the Arctic region regarded as a whole, and within those three provinces:

TABLE 1. **The species of the arctic Flora and the outlines of their distribution.**

	The province of Spitzbergen.	The Siberian province.	The American province.	The northern Atlantic.	The northern Pacific.
Corallinaceæ.					
Corallina officinalis	+	......	......	+	?
Lithothamnion glaciale	+	......	+	......	......
» flavescens	+	......	......	N	......
» foecundum	+	......	+	......	......
» compactum	+	......	......	......	......
» polymorphum	......	+	+	+	+
Lithophyllum arcticum	+	......	+?	......	......
» Lenormandi	+	......	......	+	......
Melobesia Lejolisii	+?	......	+	+	......
Rhodomelaceæ.					
Odonthalia dentata	+	+	+	+	+
Rhodomela lycopodioides	+	+	+	+	+
» larix	......	......	+	......	+
Polysiphonia urceolata	+	......	+	+	?
» elongata	?	......	......	+	......
» fastigiata	+	......	+	+	?
» arctica	+	+	+	+	......
» atrorubescens	?	......	......	+	?
» nigrescens	+	......	+	+	+
Spongiocarpeæ.					
Polyides rotundus	+	......	......	+	......
Wrangeliaceæ.					
Chantransia efflorescens	+	......	......	+	......
» secundata	+	......	......	+	......
Delesseriaceæ.					
Delesseria rostrata	......	......	+	......	......
» Bærii	+	......	......	......	+
» corymbosa	......	......	+	......	......
» Montagnei	......	......	+	+	......
» sinuosa	+	+	+	+	+
Nithophyllum punctatum	......	......	+	+	......
Hildbrandtiaceæ.					
Hildbrandtia rosea	+	......	......	+	+
Squamariaceæ.					
Peyssonnelia Dubyi	......	......	+	+	+
Cruoria pellita	......	......	+	+	......
Hæmescharia polygyna	......	+	......	......	......

	The province of Spitzbergen.	The Siberian province.	The American province.	The northern Atlantic.	The northern Pacific.
Rhodymeniaceæ.					
Hydrolapathum sanguineum	......	......	+	+	......
Rhodophyllis dichotoma	+	......	+	+	+
Euthora cristata	+	......	+	+	+
Rhodymenia palmata	+	......	+	+	+
» pertusa	+	......	+	......	+
Dumontiaceæ.					
Sarcophyllis edulis	+	......	......	+	......
» arctica	+	+	+	......	+
Halosaccion ramentaceum	+	+	+	+	+
» saccatum	+	......	......	......	......
Dumontia filiformis	+	......	+	+	+
Furcellariaceæ.					
Furcellaria fastigiata	+	......	+	+	......
Gigartinaceæ.					
Cystoclonium purpurascens	+	......	+	+	......
Callophyllis laciniata	......	......	+	+	+
Kallymenia rosacea	+	......	......	......	......
» septentrionalis	+	......	+?	N	......
» Pennyi	......	......	+	......	......
Phyllophora Brodiæi	+	......	......	+	......
» interrupta	+	+	+	......	......
Ahnfeltia plicata	+	+	+	+	+
Gigartina mamillosa	......	......	+	+	?
Ceramiaceæ.					
Ceramium rubrum	+	......	+	+	+
Ptilota plumosa	+	......	+	+	+
» pectinata	+	......	+	+	+
Antithamnion floccosum	......	......	+	+	?
» Pylaisæi	......	......	+	+	......
» boreale	+	+	......	N	+
» americanum	......	......	+	+	+
Rhodochorton intermedium	+	......	......	......	......
» spinulosum	......	......	+	......	......
» Rothii	+	......	+	+	+
» sparsum	......	......	+	+	......
» mesocarpum	+	......	+	+	......
» spetsbergense	+	......	......	......	......
Porphyraceæ.					
Diploderma miniatum	+	......	+	......	......
Porphyra laciniata	+	......	+	+	+
» abyssicola	+	......	+	......	......
Bangia fuscopurpurea	......	......	?	+	......

	The province of Spitzbergen.	The Siberian province.	The American province.	The northern Atlantic.	The northern Pacific.
Fucaceæ.					
Ozothallia nodosa	+	......	+	+	......
Fucus serratus	+	......	+	+	......
» vesiculosus	+	......	+	+	+
» ceranoides	+	......	+	+	+
» evanescens	+	+	+	+	+
» edentatus	......	......	+	+	......
» linearis	+	......	+	+	......
» filiformis	+	......	+	+	......
Pelvetia canaliculata	+	......	+	+	......
Tilopterideæ.					
Scaphospora arctica	+	......	......	......	......
Haplospora globosa	+	......	......	+	......
Laminariaceæ.					
Alaria esculenta	+	......	......	+	......
» Pylaii	......	......	+	+	?
» membranacea	+	......	+	......	......
» grandifolia	+	......	......	N?	......
» dolichorhachis	......	+	+?	......	......
» elliptica	......	+	......	......	......
» oblonga	......	+	......	......	......
Agarum Turneri	......	......	+	+	+
Phyllaria dermatodea	+	......	+	+	+
» lorea	+	......	......	+	......
Laminaria solidungula	+	+	+	......	+
» cuneifolia	?	+	+	......	+
» saccharina	+	......	......	+	?
» longicruris	......	......	+	+	+
» Agardhii	+	......	......	N?	......
» atrofulva	......	......	+	......	......
» fissilis	+	......	......	......	+
» nigripes	+	+	+	......	......
» digitata	+	......	?	+	?
Chorda filum	+	......	+	+	+
Encoeliaceæ.					
Asperococcus bullosus	......	......	+	+	......
Ralfsia deusta	+	......	+	+	+
Chordariaceæ.					
Chordaria flagelliformis	+	+	+	+	+
Eudesme virescens	+	......	......	+	......
Mesogloia vermicularis	+	......	......	+	......

	The province of Spitzbergen.	The Siberian province.	The American province.	The northern Atlantic.	The northern Pacific.
Myrionemateæ.					
Elachista fucicola	+	+	+	+	+
» lubrica	+	......	+	......	+
Myrionema strangulans	......	......	+	+	......
Lithodermateæ.					
Lithoderma fatiscens	+	+	+	+	+
Scytosiphoneæ.					
Ilea fascia	+	......	+	+	+
Scytosiphon lomentarius	+	......	+	+	+
» attenuatus	+	......	......	......	......
Punctariaceæ.					
Punctaria plantaginea	+	......	+	+	......
Desmarestiaceæ.					
Desmarestia aculeata	+	......	+	+	+
Dichloria viridis	+	......	+	+	+
Phloeospora subarticulata	+	......	......	+	......
» tortilis	+	...	+	+	+
» pumila	+	......	......	......	......
Dictyosiphon corymbosus	+	......	......	......	......
» hippuroides	+	......	+	+	......
» fœniculaceus	+	......	+	+	+
» hispidus	+	......	......	......	......
Lithosiphon Laminariæ	?	......	......	+	......
Sphacelariaceæ.					
Cladostephus spongiosus	......	......	+	+	......
Stypocaulon scoparium	+	......	......	+	......
Chætopteris plumosa	+	+	+	+	+
Sphacelaria cirrhosa	......	......	+	+	......
» arctica	+	+	+	N	......
» olivacea	......	......	+	+	......
Ectocarpaceæ.					
Isthmoplea sphærophora	+	......	......	+	......
Ectocarpus confervoides	+	......	+	+	+
» ovatus	+	......	......	+	......
Pylaiella litoralis	+	+	+	+	+
» varia	+	......	+	N	......
Gleothamnion palmelloides	+	......	......	......	......
Chætophoraceæ.					
Chætophora maritima	+	......	......	......	......

	The province of Spitzbergen.	The Siberian province.	The American province.	The northern Atlantic.	The northern Pacific.
Ulvaceæ.					
Enteromorpha clathrata	......	......	+	+	+
» intestinalis	......	......	+	+	+
» compressa	+	......	+	+	+
» minima	+	......	......	+	......
» tubulosa	......	......	+	+	......
» micrococca	+	+	+	+	......
Ulva crassa	+	......	......	N	......
» lactuca	?	......	+	+	+
Monostroma lubricum	+	......	+	......	......
» Grevillei	+	......	+	+	......
» leptodermum	+	......	......	......	......
» fuscum	+	......	+	+	+
» Blyttii	+	......	+	+	......
Diplonema percursum	+	......	+	+	......
Confervaceæ.					
Spongomorpha arcta	+	......	+	+	+
» lanosa	+	......	+	+	......
Cladophora rupestris	+	......	+	+	......
» diffusa	+	......	......	+	......
» gracilis	......	......	+	+	......
» crispata	+	......	......	+	......
Rhizoclonium rigidum	+	......	......	+	......
» pachydermum	+	+	+	......	......
» riparium	+	......	+	+	+
Chætomorpha Wormskioldii	......	......	+	......	......
» melagonium	+	......	+	+	+
» linum	+	......	......	+	......
» tortuosa	......	......	+	+	+
Ulothrix submarina	+	......	......	+	......
» discifera	+	......	......	......	......
Urospora penicilliformis	+	......	+	+	+
Bulbocoleon piliferum	+	......	......	+	......
Bryopsideæ.					
Bryopsis plumosa	......	......	+	+	+
Characiaceæ.					
Characium marinum	+	......	......	......	......
Codiolum Nordenskiöldianum	+	......	......	N	......
Chlorochytrium inclusum	+	+	......	......	......
Palmellaceæ.					
Chlorangium marinum	+	......	......	......	......

	The province of Spitzbergen.	The Siberian province.	The American province.	The northern Atlantic.	The northern Pacific.
Rivulariaceæ.					
Rivularia bemisphærica	......	......	+	+	......
» microscopica	......	......	+	......	......
Callothrix scopulorum	+	......	......	+	......
» confervicola	+	......	......	+	......
Oscillariaceæ.					
Oscillaria subsalsa	......	......	+	+	......
Spirulina tenuissima	......	......	+	+	......
Chroococcaceæ.					
Gleocapsa spec.	+	......	......	......	......

TABLE II. **The number of species in the series of algæ.**

	The whole region of the Flora.	The province of Spitz-bergen.	The Siberian province.	The American province.
Florideæ	65 (68)	47 (49)	11	48 (49)
Fucoideæ	65 (66)	51 (53)	13	43 (44)
Chlorophyllophyceæ	37	29 (30)	3	22
Nostochineæ	7	3	—	4
Total number of species	**174(178)**	**130(135)**	**27**	**117(119)**

TABLE III. **The number of species in the families.**

	The whole region of the Flora.	The province of Spitzbergen.	The Siberian province.	The American province.
Laminariaceæ	20	12(13)	6	11(12)
Confervaceæ	17	14	1	10
Ulvaceæ	14	10(11)	1	11
Ceramiaceæ	13	8	1	10
Corallinaceæ	9	8	1	5
Gigartinaceæ	9	6	2	7
Fucaceæ	9	8	1	9
Desmarestiaceæ	9(10)	9(10)	—	5
Rhodomelaceæ	7 (9)	6 (8)	3	7
Delesseriaceæ	6	2	1	5
Sphacelariaceæ	6	3	2	5
Ectocarpaceæ	6	6	1	3
Rhodymeniaceæ	5	4	—	5
Dumontiaceæ	5	5	2	3
Rivulariaceæ	4	2	—	2
Squamariaceæ	3	—	1	2
Porphyraceæ	3 (4)	3	—	3 (4)
Chordariaceæ	3	3	1	1
Myrionemateæ	3	2	1	3
Scytosiphoneæ	3	3	—	2
Characiaceæ	3	3	1	—
Wrangeliaceæ	2	2	—	—
Tilopterideæ	2	2	—	—
Encoeliaceæ	2	1	—	2
Oscillariaceæ	2	—	—	2
Spongiocarpeæ	1	1	—	—
Hildbrandtiaceæ	1	1	—	—
Porcellariaceæ	1	1	—	1
Lithodermateæ	1	1	1	1
Punctariaceæ	1	1	—	1
Chætophoraceæ	1	1	—	—
Bryopsideæ	1	—	—	1
Palmellaceæ	1	1	—	—
Chroococcaceæ	1	1	—	—

TABLE IV. **The number of genera in the families.**

	The whole region of the Flora.	The province of Spitzbergen.	The Siberian province.	The American province.		The whole region of the Flora.	The province of Spitzbergen.	The Siberian province.	The American province.
Confervaceæ	7	7	1	5	Porphyraceæ	2 (3)	2	—	2 (3)
Gigartinaceæ	6	4	2	6	Tilopterideæ	2	2	—	—
Laminariaceæ	5	4	2	5	Encoeliaceæ	2	1	—	2
Corallinaceæ	4	4	1	3	Myrionemateæ	2	1	1	2
Rhodymeniaceæ	4	3	—	4	Scytosiphoneæ	2	2	—	2
Ceramiaceæ	4	4	1	4	Rivulariaceæ	2	1	—	1
Desmarestiaceæ	4 (5)	4 (5)	—	4	Oscillariaceæ	2	—	—	2
Sphacelariaceæ	4	3	2	3	Spongiocarpeæ	1	1	—	—
Ectocarpaceæ	4	4	1	2	Wrangeliaceæ	1	1	—	—
Ulvaceæ	4	4	1	4	Hildbrandtiaceæ	1	1	—	—
Rhodomelaceæ	3	3	3	3	Furcellariaceæ	1	1	—	1
Squamariaceæ	3	—	1	2	Lithodermateæ	1	1	1	1
Dumontiaceæ	3	3	2	3	Punctariaceæ	1	1	—	1
Fucaceæ	3	3	1	3	Chætophoraceæ	1	1	—	—
Chordariaceæ	3	3	1	1	Bryopsideæ	1	—	—	1
Characiaceæ	3	3	1	—	Palmellaceæ	1	1	—	—
Delesseriaceæ	2	1	1	2	Chroococcaceæ	1	1	—	—

TABLE V. **The number of species in the genera.**

	The whole region of the Flora.	The province of Spitzbergen.	The Siberian province.	The American province.		The whole region of the Flora.	The province of Spitzbergen.	The Siberian province.	The American province.
Laminaria	9	6 (7)	3	5 (6)	Lithophyllum	2	2	—	1
Fucus	7	6	1	7	Rhodomela	2	1	1	2
Alaria	7	3	3	3	Chantransia	2	2	—	—
Rhodochorton	6	4	—	4	Rhodymenia	2	2	—	2
Enteromorpha	6	3	1	5	Sarcophyllis	2	2	1	1
Lithothamnion	5	4	1	3	Halosaccion	2	2	1	1
Delesseria	5	2	1	4	Phyllophora	2	2	1	1
Monostroma	5	5	—	4	Ptilota	2	2	—	2
Polysiphonia	4 (6)	4 (6)	1	4	Porphyra	2	2	—	2
Antithamnion	4	1	1	3	Phyllaria	2	2	—	1
Dictyosiphon	4	4	—	2	Elachista	2	2	1	2
Cladophora	4	3	-	2	Scytosiphon	2	2	—	1
Chætomorpha	4	2	—	3	Ectocarpus	2	2	—	1
Kallymenia	3	2	—	2	Pylaiella	2	2	1	2
Phloeospora	3	3	—	1	Ulva	2	1 (2)	—	1
Sphacelaria	3	1	1	3	Spongomorpha	2	2	—	2
Rhizoclonium	3	3	1	2	Ulothrix	2	2	—	—

	The whole region of the Flora.	The province of Spitzbergen.	The Siberian province.	The American province.		The whole region of the Flora.	The province of Spitzbergen.	The Siberian province.	The American province.
Rivularia	2	—	—	2	Asperococcus	1	—	—	1
Calothrix	2	2	—	—	Ralfsia	1	1	—	1
Corallina	1	1	—	—	Chordaria	1	1	1	1
Melobesia	1	1	—	1	Eudesme	1	1	—	—
Odonthalia	1	1	1	1	Mesogloia	1	1	—	—
Polyides	1	1	—	—	Myrionema	1	—	—	1
Nithophyllum	1	—	—	1	Lithoderma	1	1	1	1
Hildbrandtia	1	1	—	—	Ilea	1	1	—	1
Peyssonnelia	1	—	—	1	Punctaria	1	1	—	1
Cruoria	1	—	—	1	Desmarestia	1	1	—	1
Hæmescharia	1	—	1	—	Dichloria	1	1	—	1
Hydrolapathum	1	—	—	1	Lithosiphon	1?	1?	—	—
Rhodophyllis	1	1	—	1	Cladostephus	1	—	—	1
Euthora	1	1	—	1	Stupocaulon	1	1	—	—
Dumontia	1	1	—	1	Chætopteris	1	1	1	1
Furcellaria	1	1	—	1	Isthmoplea	1	1	—	—
Cystoclonium	1	1	—	1	Gleothamnion	1	1	—	—
Callophyllis	1	—	—	1	Chætophora	1	1	—	—
Ahnfeltia	1	1	1	1	Diplonema	1	1	—	1
Gigartina	1	—	—	1	Urospora	1	1	—	1
Ceramium	1	1	—	1	Bulbocoleon	1	1	—	—
Diploderma	1	1	—	1	Bryopsis	1	—	—	1
Bangia	1?	—	—	1?	Characium	1	1	—	—
Ozothallia	1	1	—	1	Codiolum	1	1	—	—
Pelvetia	1	1	—	1	Chlorochytrium	1	1	1	—
Scaphospora	1	1	—	—	Chlorangium	1	1	—	—
Haplospora	1	1	—	—	Oscillaria	1	—	—	1
Agarum	1	—	—	1	Spirulina	1	—	—	1
Chorda	1	1	—	1	Gleocapsa	1	1	—	—

It appears from these lists, that the arctic marine Flora possesses no family that is peculiar to it, but one genus, viz. that of *Hæmescharia*, and a considerable number of species. These are:

Lithothamnion fœcundum,
» compactum,
Lithophyllum arcticum,
Delesseria rostrata,
» corymbosa,
Hæmescharia polygyna,
Halosaccion saccatum,
Kallymenia rosacea,
» Pennyi,
Phyllophora interrupta,
Rhodochorton intermedium,
» spinulosum,
» spetsbergense,
Diploderma miniatum,
Scaphospora arctica,
Alaria grandifolia?
» dolichorhachis,
» elliptica,

Alaria oblonga.
Laminaria Agardhii?
» atrofulva,
» nigripes,
Scytosiphon attenuatus,
Phloeospora pumila,
Dictyosiphon corymbosus,
» hispidus,
Gleothamnion palmelloides,
Chætophora maritima,
Monostroma lubricum,
» leptodermum,
Rhizoclonium pachydermum,
Chætomorpha Wormskioldii,
Ulothrix discifera,
Characium marinum,
Chlorochytrium inclusum,
Chlorangium marinum,
Rivularia microscopica,
Gleocapsa spec.?

Thus:

Florideæ	14	species representing	7	families,	9	genera,
Fucoideæ	13(12)	»	»	5	»	7	»
Chlorophyllophyceæ	9	»	»	5	»	8	»
Nostochineæ	2	»	»	2	»	2	»

Total sum 38(37) species representing 19 families, 26 genera.

The number of endemic species, 38 or perhaps only 37, is about 22 per cent of the total number of species of the Flora. The following table presents the outlines of the distribution of the arctic Flora:

	Species occurring on the north coast of Norway, besides in the region of the arctic Flora.	Species occurring both in the northern part of the Atlantic and in the northern part of the Pacific.	Species known only from the northern Atlantic.	Species known only from the northern Pacific.	Species peculiar to the arctic Flora.
Florideæ	4	20 (26)	17 (18)	4	14
Fucoideæ	3 (4)	20 (23)	22 (24)	3	13 (12)
Chlorophyllophyceæ	2	11	15	—	9
Nostochineæ	—	—	5	—	2
Total sum	10 (11)	51 (00)	00 (02)	7	38 (37)

If we assume, that the Arctic Sea has been the centre of development not only of the endemic species, but also of those species which the arctic Flora has in common with the northern Atlantic and the northern Pacific, and besides of those species which are known from the Polar Sea on the coast of Norway at the same time as from the Arctic region proper, the number of arctic species that may be considered on good grounds to have developed themselves within the glacial sea, would amount to 98 by the lowest estimate. One more species must, however, surely be added to these, viz. *Antithamnion boreale*, known from the north coast of Norway and the northern part of the Pacific. Besides, I believe it can hardly be doubted with regard to those 7 species which the arctic Flora has in common with the northern part of the Pacific, that all except one have their original home in the Arctic Sea. As respects those species which are found in the northern part of the Atlantic, besides in the present glacial region, it is difficult to form any decisive conclusion. It seems probable, however, that a considerable number of these must or may be thought to have developed in the Arctic Sea and to have migrated from there southwards. In any case, the number of those species of the arctic marine Flora whose origin must be placed within the Arctic Sea, cannot be estimated at less than about 100 species, i. e. about 60 per cent of the total number of the species.

In comparison with other northern Floras, the arctic Flora is proportionally poor in Florideæ, rich in Chlorophyllophyceæ and especially in Fucoideæ. This is shown by the following list:

	The arctic Flora.	The Flora of the Norwegian Polar Sea.	The Flora of Scandinavia.[1]	The Flora of New-England.[2]	The Flora of Great Britain.[3]	The Flora at Cherbourg.[4]
Florideæ are, of the total number of the species	37 %	41 %	42 %	43 %	49 %	46 %
Fucoideæ	37 »	35 »	35 »	25 »	25 »	28 »
Chlorophyllophyceæ	21 »	21 »	19 »	18 »	17 »	13 » [5]

Of the Nostochineæ I shall not speak here, their marine species being still too little studied, and the determination of the species being more fluctuating in this group of algæ than in any other. Nearly a third part of the Fucoideæ belongs to the same family, the *Laminariaceæ*, the richest in species of all the families of the arctic Flora, with 20 species, among which one Chorda. In the Scandinavian Flora this family possesses with certainty no more than 10 species, amongst which are three species of Chorda, on the coasts of England only 7, at Cherbourg 5, including at the two last-mentioned places one species of Chorda. On the north-eastern coast of America the Laminariaceæ are more numerous. But it is impossible at present to state with certainty the number of their species there. However, it does not probably amount to more than half of the arctic.

The average number of species in the families is somewhat less in the arctic Flora than in the Scandinavian. In the former there are about 5,1 species in each family, in the latter 6,1. However, the Scandinavian Flora contains comparatively more (38 %) families with only a single species, than the arctic (26 %). Of the 34 families of the arctic Flora 2 are monotypical, of the 47 families of the Scandinavian Flora only 3

[1]) Cp. Enum. Plant. Scand.
[2]) Cp. Farlow, New-Engl. Alg. p. 184.
[3]) Cp. Harvey, Phyc. Brit.
[4]) Cp. Le Jolis, List. Alg. Cherb.
[5]) To these figures, of course, only a relative value can be assigned, as I have used the statements given in the works quoted, without allowing for differences in the determination of species or the alterations suffered by the figures exhibited, in consequence of new species having been added after the publication of the works. But notwithstanding this, the proportion between the series in question is no doubt correctly expressed in the main by the figures given, as a reduction or addition may be supposed to have taken place in an equal degree with regard to all.

are of such a kind. The other families which are represented in the Flora of Scandinavia by only one species, are richer in species farther to the South, and may therefore be supposed to have immigrated into the Scandinavian Flora from that direction.

Table IV shows that the majority (54) of the 90 (92?) genera of the arctic Flora are represented by only one species; the greatest number of species possessed by any genus is 9 (*Laminaria*), and there are only few genera with any considerable number of species. Thus the average number of species in each genus becomes small, namely 1,9. This is less than in the Flora of Scandinavia (2,3), of New-England (2,1), of Great Britain (3,5), of Cherbourg (2,4), and even less than in the Flora of the Norwegian Polar Sea (about 2,0). Among the genera of the arctic Flora the following are monotypical: *Polyides, Hæmescharia, Hydrolapathum, Dumontia, Furcellaria, Haplospora, Dichloria, Isthmoplea, Gleothamnion, Diplonema,* and *Bulbocoleon;* two of these are unknown without the arctic region. These circumstances, viz. the little number of species in the genera and the richness in monotypical genera, tend to prove the high age of the region of the arctic Flora.

The following species of the arctic Flora are common to all the three provinces of the region:

Odonthalia dentata,
Rhodomela lycopodioides,
Polysiphonia arctica,
Delesseria sinuosa,
Sarcophyllis arctica,
Halosaccion ramentaceum,
Phyllophora interrupta,
Ahnfeltia plicata,
Fucus evanescens,
Laminaria solidungula,
Laminaria cuneifolia (?),
» nigripes,
Chordaria flagelliformis,
Elachista fucicola,
Lithoderma fatiscens,
Chætopteris plumosa,
Sphacelaria arctica,
Pylaiella litoralis,
Enteromorpha micrococca,
Rhizoclonium pachydermum.

Accordingly 19 or possibly 20 species, i. e. about 10 % of the total number of species in the whole region, about 15 % of that of the province of Spitzbergen, 70 % of that of the Siberian province, and about 16 % of that of the American province. A survey of the relation of the provinces to each other, with regard to the number of species, is exhibited in the following table.

	The province of Spitzbergen.		The Siberian province.		The American province.	
	Total number.	% of the number of species of the province.	Total number.	% of the number of species of the province.	Total number.	% of the number of species of the province.
Peculiar species	51	39 %	4 (3)	15 (11) %	37 (38)	32 %
Species common with the province of Spitzbergen	—	—	21	78 %	73 (77)	62 (66) %
» » with the Siberian province	21	16 %	—	—	21 (22)	18 %
» » with the American province	73 (78)	56 (60) %	21 (22)	78 %	—	—

Amongst the species common to all the provinces of the arctic Flora, those of which the names are printed in italics in the list are amongst its commonest and most abundant ones. Besides, the following species are both common and more or less abundant in the whole province of Spitzbergen or in the greatest part of it.

Lithothamnion glaciale,
Rhodymenia palmata,
Ptilota pectinata,
» plumosa,
Rhodochorton Rothii,
Alaria grandifolia,
Laminaria Agardhii,
» digitata,
» nigripes,
Desmarestia aculeata,
Dichloria viridis,
Phloeospora tortilis,
Pylaiella litoralis,
Enteromorpha compressa,
Rhizoclonium riparium,
Chætomorpha melagonium,
Urospora penicilliformis.

The following species I believe may be regarded as abundant within more limited areas of the same province:

in the White Sea [1]):
Lithophyllum Lenormandi,
Polysiphonia nigrescens,
Chantransia efflorescens,
Ahnfeltia plicata,
Ozothallia nodosa,
Fucus serratus,
Alaria membranacea;

in the White Sea and the western Murman Sea [1]):
Delesseria Bærii and Rhodophyllis dichotoma;

in the eastern Murman Sea:
Phyllophora Brodiæi and Dictyosiphon corymbosus;

in the Greenland Sea:
Elachista lubrica and Rhizoclonium rigidum.

The following species are especially characteristic of the Siberian province:
Alaria dolichorhachis,
» elliptica,
Alaria ovata,
Laminaria cuneifolia.

Within the American province the following species may be regarded as the most abundant ones:

Lithothamnion glaciale,
» polymorphum,
Polysiphonia urceolata,
Rhodophyllis dichotoma,
Rhodymenia pertusa,
» palmata,
Ptilota pectinata,
» plumosa,
Diploderma miniatum,
Fucus vesiculosus,
Fucus filiformis,
Agarum Turneri,
Laminaria longicruris,
» atrofulva,
» cuneifolia.
Chorda filum,
Desmarestia aculeata,
Dichloria viridis,
Dictyosiphon foeniculaceus,
Enteromorpha compressa,

[1]) See GOBI, Algenfl. weiss. Meer. p. 11.

Monostroma fuscum,	Rhizoclonium pachydermum,
» Blyttii,	Chætomorpha Wormskioldii,
Diplonema percursum,	» melagonium,
Spongomorpha arcta,	Urospora penicilliformis,
Cladophora rupestris,	Rivularia hemisphærica.

The character of the vegetation is chiefly marked in the province of Spitzbergen by:

Alaria grandifolia,	*Laminaria digitata,*
» *membranacea,*	» *nigripes,*
Laminaria Agardhii,	» *solidungula;*

in the Siberian province by:

Alaria dolichorhachis,	*Laminaria solidungula,*
» *elliptica,*	» *cuneifolia;*
» *ovata,*	

in the American province by:

Fucus vesiculosus,	*Laminaria atrofulva,*
Agarum Turneri,	» *cuneifolia,*
Laminaria longicruris,	*Alaria spec.? (membranacea?)*

This difference of the vegetation of Laminariaceæ within the different provinces can hardly be explained otherwise than by the supposition that secondary centres of development have been formed within the great arctic centre. Especially in that part of the Arctic Sea denominated the American province, the development of the Flora seems to have taken a direction independent of the others in certain respects. This is, besides, indicated by several other peculiarities of its vegetation, as the occurrence of such species as *Kallymenia Pennyi*, *Antithamnion americanum*, *Rhodochorton spinulosum*, and *Chætomorpha Wormskioldii*, but, above all, by the species of *Delesseria* of which no less than three belong exclusively to this province. One of these, *D. corymbosa*, is nearly related to *D. Bærii*, pretty common in the province of Spitzbergen, and is perhaps to be considered as having branched off from this; another, *D. Montagnei*, is most closely allied to and only slightly different from *D. alata* of the Atlantic, whose mother form it may possibly be; and the third, *D. rostrata*, resembles in habit most nearly *Delesseria Bærii*, but in anatomical structure agrees more closely with *D. alata* or *D. Montagnei*, from the latter of which it may be thought to have issued.

The general conclusion to which my investigation of the Flora of the Arctic Sea, taken in a wide sense, has led me, may be briefly stated thus: the arctic part of the Ocean comprises two separate regions differing with regard to their history of development, one situate on the north coast of Norway and closely connected with the Atlantic, the other arctic, comprehending the rest of the Arctic Sea; in different parts of the arctic region the development of the Flora has been in a certain degree independent, which is especially indicated by the distribution of the Laminariaceæ; on this account, the arctic region may be devided into three provinces: that of Spitzbergen, the Siberian, and the American.

The conditions of life of the arctic marine algæ.

On the western coast of Sweden the composition of the marine vegetation changes in a remarkable degree at different seasons. Besides a number of species occurring and developing all the year round, in summer as well as in winter, there is to be found a pretty considerable number which are constantly met with during a fixed period, but are wanting during the rest of the year. Again, other species are met with, indeed, during the whole year, but are in course of development only during part of it. Some species, belonging to these two categories, occur or are in course of development during the warmer part of the year, in spring or summer; others during the colder part, in late autumn or winter; some belong to the litoral zone, others on the contrary to the deeper parts of the sea. I hope to return soon to these facts, which I can only allude to here, and to expose them in detail in a separate paper.

The facts mentioned show that there are amongst the Scandinavian species such as need not, under the external conditions prevailing in the sea on the western coast of Sweden, a whole year for the purpose of completing their development from spore to spore, or, if they are perennial, to perform those vital functions whose object is the maintenance of the individual and the species. Thus the external conditions are here such as to make the occurrence of annual species possible. As far as my experience goes, acquired by examining the marine vegetation in different parts of the arctic region and at two different occasions, each time almost throughout a whole year, there are not to be found among the sublitoral and elitoral algæ of the arctic Flora any species whose whole development is limited to less than one year. But in a more southerly part of the region, in the Siberian Sea, near Behring Strait, accordingly near the Polar Circle, I found one species, *Rhodomela lycopodioides*, whose development was interrupted during part of the year, namely, during the winter, in order to be resumed again afterwards, that is to say in other words, a perennial species which did not need the whole year to develop the necessary number of vegetative and reproductive organs. The same species occurs also about thirteen degrees farther northward, on the north coast of Spitzbergen. Here its development is extended to the whole year. It bears a profusion of propagative organs at that season when it is in rest on the north-eastern coast of Siberia. I have not had an opportunity of investigating the litoral algæ during the winter. It is possible, indeed, that some of them, for instance, *Urospora penicilliformis*, *Codiolum Nordenskiöldianum*, *Enteromorpha compressa*, *E. minima* a. o. live only during that part of the summer when the litoral zone is free from land-ice, and that they are accordingly annual, being able even in those regions to complete their development in a short part of the year. But all the species are surely not annual. This seems especially improbable in respect of such algæ as *Rhodochorton Rothii*, *Rhizoclonium riparium*, and *Fucus evanescens*. For on the coast of Novaya Zemlya I have seen all these algæ fully developed early in the year, before the land-ice broke up or just as it was dispersing. Thus I cannot but assume that they had lain frozen and covered with ice during winter, and when delivered at last from their cold cover resumed at once their development which had been interrupted by the ice. I have in the arctic region seen quite evident instances of algæ being frozen and inclosed in ice and continuing their development when delivered. In a lagoon at Pitlekay, which during the winter froze to the bottom, there was found in abundance an *Enteromorpha* which I have below named *E. micrococca f. subsalsa*. When the ice of the lagoon melted at the end of June, this plant remained at the bottom in large seemingly lifeless masses. But in a short time it began to develop vigorously by producing new shoots from the parts which had persisted through the winter. This being so, it is quite possible that all the species which are litoral within the arctic region, persist all the year round and do not complete their development in one year, but in two or three years, with longer or shorter interruptions. Accordingly there is as yet no sure evidence of the existence of annual species among the arctic algæ. I believe, however, that such species are to be found on the west coast of Greenland, and that, besides, the first-mentioned Chlorophyllophyceæ are probably also annual. But in any case, to state it broadly, we may say with regard to the arctic region, that the conditions under which the algæ live there, are such that annual algæ cannot endure or at least cannot occur in any larger number either of species or of individuals, and that the perennial species, in the very most cases, need the whole year to reach the development designed for each period of growth.

But though the development of the arctic algæ is thus extended throughout the whole year, there appears a certain periodicity at least in certain species. For it may be regarded as a pretty general fact that the purely vegetative development is livelier and more energetic during the favourable season, while the development of propagative organs, on the contrary, is stronger and richer in the latter part of the autumn, in winter, and in early spring. But nevertheless I cannot but maintain the statement I have made once before, that a development of vegetative parts takes place on a very large scale on the coast af Spitzbergen during mid-winter, when the sun is at its lowest beneath the horizon and consequently the darkness is intense. Spores germinated and grew into pretty well developed embryonic plants. Of several species, for instance, *Delesseria sinuosa*, young plants were common towards the end of the dark season. It could not well be doubted that these had been developed during the winter-months. Older plants of the same species, and also of *Halosaccion ramentaceum*, *Rhodymenia palmata*, *Phyllophora interrupta*, *Rhodomela lycopodioides*, *Sphacelaria arctica*, *Phlœospora tortilis* put forth new shoots in winter, which were in course of growth and more or less developed at the entrance of the light season. However, it must be admitted that the development of such organs did not set in with greater vigour nor produce any considerable results before the beginning of May.

On the other hand, it is during the winter-months that the reproductive function attains its maximum of energy in the arctic algæ. At least this was the case on the north coast of Spitzbergen, and I suppose that the same state of things prevails also in other parts of the arctic region.

To begin with, it may be remarked that with regard to this branch of vital activity there is a sharp difference between the marine vegetation and the arctic land-flora, especially the cryptogamic. It has been asserted that the phanerogams seldom, nay only quite exceptionally, produce ripe fruit within the arctic regions. This assertion is certainly quite unfounded, but it is true, indeed, that the production of seeds is less rich here than farther southwards, and that the arctic phanerogams are endowed with a peculiar conformation of their own, in order to get time, during the short season, to form reproductive organs. Mosses [1]) and lichens [2]) rarely fructify, but increase in the vegetative way. It must therefore be concluded that the plants of the arctic sea live under more advantageous conditions than the land-plants in the formidable climate of the Polar countries.

On the north coast of Spitzbergen, during the winter 1872—73, I had the opportunity of following 27 species in their development, almost day by day. As I have mentioned in my account of these researches, 22 of these species, belonging to various classes and various families, were furnished with organs of propagation during the whole or some part of the winter. Carpospores, tetraspores, egg-cells, brown and green zoospores were produced and ripened. Some species, as *Rhodomela lycopodioides*, *Laminaria solidungula*, *Elachista lubrica*, and *Chætopteris plumosa* formed reproductive organs in a surprisingly great number, at least as great as the same or nearly related species farther to the south.

What has been said now, must not be thus understood as if the development of reproductive organs were in all the arctic algæ relegated to the winter-months. In this respect great variety prevails. There are to be found species, as *Rhodomela lycopodioides* f. *tenuissima*, *Delesseria sinuosa*, *Rhodymenia palmata*, *Phyllophora interrupta*, *Ptilota pectinata*, *Fucus evanescens*, *Laminaria Agardhii*, *Laminaria nigripes*, *Chordaria flagelliformis*, *Elachista lubrica*, *Pylaiella litoralis* which bear such organs of some kind or other at all times of the year, although, in many of them, this function is most energetic in winter. In other species as *Lithoderma fatiscens*, *Chætopteris plumosa*, *Sphacelaria arctica*, *Laminaria solidungula*, *Alaria grandifolia* a. o., the development of reproductive organs is decidedly limited to or chiefly carried on in late autumn and in winter; again others have been found hitherto with propagative organs only during the summer-months, as *Odonthalia dentata*, *Chantransia efflorescens*, *Ceramium rubrum*, *Antithamnion boreale*, *Dictyosiphon foeniculaceus*, *Ectocarpus confervoides*, *Monostroma fuscum*, *M. Blyttii* etc.

I have shown above, that arctic species occur also in the northern Atlantic. It is a startling fact that most of them, when growing within the arctic region, are found

[1]) Cp. BERGGREN, Musci Spetsb., p. 19.
[2]) Cp. TH. FRIES. Lich. Spetsb., p. 5.
[3]) KJELLMAN, Vinteralgveg.

to be perfectly similar in habit to such as have grown farther southwards, where they have yet been exposed to external conditions essentially different from those of the arctic regions. Specimens of *Rhodymenia palmata* or *Rhodomela lycopodioides* f. *typica* that have lived in deep water on the coasts of Spitzbergen or Novaya Zemlya, are found to agree so completely as to external and internal anatomical characters with specimens of the same species growing within the litoral zone on the west coast of Norway, that even the most sharpened eye cannot detect other than merely individual differences between them. This holds good also of several other species. Hence it follows, that the algæ in general, and particularly the arctic forms, have a great ability to adapt themselves to different external conditions without being influenced by them in any sensible degree. The pressure to which a *Rhodymenia palmata* is exposed in the Greenland Sea, the temperature at which it lives here, and the quantity of light that is afforded to it, are all most essentially different from those on the coast of Norway, without any alteration being discernible in the exterior of the plant. With other species the case is different. *Spongomorpha arcta* and green algæ in general, as well as several others, certainly agree in morphological characters with their southern co-species, but they never attain the same luxuriancy, strength, and richness as farther to the south. Again, other species agree with their co-species in the south as to the form and development of the organs, but differ from them in biology, or the differences of conditions have effected even morphological differences. For instance, *Odonthalia dentata* from the coasts of Spitzbergen resembles the same species from the coast of Bohuslän with regard to all exterior parts, but while developing in the former locality its tetrasporangia at the middle of summer, viz. at the end of July, it is found with such organs at Bohuslän in the winter-months. *Polyides rotundus* offers a pretty similar instance. At Bohuslän its fructification takes assuredly place chiefly in winter, on the coast of Novaya Zemlya, where it occurs in a less luxuriant form, in summer. *Rhodomela lycopodioides* f. *tenuissima* on the north coast of Spitzbergen needs continue its development throughout the whole year, in the Ochotsh Sea as well as on the north-eastern coast of Siberia part of the year suffices for it. After having here at the end of the season thrown off part of the side-organs formed, it rests for some time. Then it begins again to develop new parts from the surviving rests of the stem and the branches. This difference in the mode of life causes such a considerable difference in the external form, that one would not hesitate to regard the Spitzbergen form as specifically different from the Siberian, if the falseness of such a view were not demonstrated by following the plant from latitude to latitude. Such is the case also with *Chætopteris plumosa*, so common in the arctic region. On the coasts of Spitzbergen as well as on the west coast of Sweden, the period when it forms its zoospores (gamets) is in winter. At this time the aspect of the plant in the two localities is very different. At Spitzbergen it has preserved all its assimilating external organs, that is to say, it resembles the summer form at Bohuslän; at the latter place, on the contrary, the formation of those side-parts by which the zoosporangia (gametangia) are supported and particularly developed, is preceded by a far gone decomposition of all the organs — we may call them leaves — developed during the period of vegetation more especially for the purpose of assimilation. These facts can and certainly ought to be explained in the following way. In its original home, the Arctic Sea, this plant has need of the whole year, and, during that time, of all its assimilating organs in order to accomplish its development; whereas on the coast of Bohuslän, where it has come into more favourable conditions, it is able, by carrying on assimilation for only a part of the same time, to form such a quantity of nutrient substances as suffices not only to develop the reproductive organs, but also to supply the assimilating organs, that it has cast off after they have functioned during the necessary time.

With regard to the physiology of nutrition, the arctic algæ are in several respects most instructive. They may during very long periods be inclosed in ice and exposed to high degrees of cold, without being killed or losing their power to resume vigorously their development, when the hindering fetters have been broken. Still more, they prove that plants can germinate at a temperature of from — 1° to — 2° C., and are able, without being checked in their vital functions by the temperature scarcely ever rising to the freezing-point, to develop into magnificent forms producing endless masses of reproductive cells throughout all the year or during the greater part of it. We have thus in these algæ vegetative organisms whose optimum of temperature may be stated to be about or below zero C. Besides, the energy of assimilation requisite for this rich and vigorous development seems to comport very ill with the slight quantity of light afforded to these plants. As far as I can judge, this cannot be explained otherwise than by the assumption that the arctic algæ in general are content with a very inconsiderable measure both of light and of warmth.

With the modern theories on the nature of the process of assimilation, it is certainly difficult to assume that the algæ should continue uninterruptedly their assimilation at the 80:th degree of latitude during the winter when there prevails an almost absolute darkness to the human eye; but such an assumption becomes almost necessary on account of the rich and vigorous development of new parts that was proved to take place during the winter. Otherwise one would be obliged to assume that the considerable quantity of plastic substance used up by the algæ in forming new organs on a large scale during the dark season, are nutriments stored up in reserve during the preceding period of light. I cannot affirm decisively that this was not the case. But on the materials that I have had at my disposition, such an assertion cannot be founded. Certainly, several Florideæ contained a remarkably large quantity of solid substances in their cells during the winter. But neither in the Fucoideæ nor in the *Chlorophyllophyceæ* such stores were to be detected. However, nutrient substances in reserve may have occurred in them in a liquid form. I had no means of investigating this [1]). If the raw materials are assumed to have been gathered during the light season, this implies, on the other hand, that nutritive substances must be prepared then to an extraordinary extent, as not only all the material is to be formed of which vegetative organs are built, but also a sufficient quantity is to be reserved for the developing of

[1]) It ought to be remarked here that the observations to which I refer chiefly, were carried on during an involuntary and unpremeditated wintering on the north coast of Spitzbergen.

new parts during the winter. With regard to this, it must be taken into account, that in consequence of the masses of ice and snow only a slight quantity of light penetrates to the arctic algæ growing at a greater depth, even during a considerable part of the so called light season. Thus I think it may be said that the arctic sea-algæ, as independently assimilating plants, need uncommonly little warmth and light.

Species and forms of the Flora of the Arctic Sea.

Series FLORIDEÆ (LAMOUR.) BERTH. [1])

Bangiaceen. LAMOUR. Essai p. 115; lim. mut.

Fam. CORALLINACEÆ (LAMOUR.) HAUCK.

Meeresalg. p. 19; LAMOUR. Hist. Polyp. p. 224.

Gen. Corallina (TOURN.) LAMOUR.

Hist. Polyp. p. 275; TOURN. Inst. Herb. p. 570; char. mut.

Corallina officinalis L.

Fauna Suec. p. 539.

f. *typica.*

Descr. Corallina officinalis ARESCH. in J. G. AG. Spec. Alg. 2, p. 562.
Fig. » » HARV. Phyc. Brit. t. 222.
Exsicc. » » ARESCH. Alg. Scand. exsicc. N:r 8.

f. *flexilis* nob.

f. dense cæspitosa, 10 cm. alta; fronde quam in forma typica graciliore et flexiliore, ramosissima, ramis ramulisque plus minus fasciculato-congestis, ramulis oppositis, alternis, subsecundis, ultimorum ordinum æqualibus, elongatis, flexilibus, sæpe in callo reniformi desinentibus; articulis, summis exceptis, plus minus compressis, diametro 2:plo—3:plo longioribus; conceptaculis tetrasporangiferis ramulos vulgo brevissimos, articulo singulo compositos, interdum longiores, articulatos terminantibus; tetrasporangiis, divisione peracta, obovato-oblongis vel oblongis, 185—190 μ. longis, 60—65 μ. crassis.

f. *robusta* nob.

f. quam C. officinalis typica major et fere duplo crassior, ponderosa, læte rosea; fronde parce et irregulariter ramosa, ramis ultimi ordinis elongatis, strictis, alternis vel subsecundis, vix oppositis; articulis teretibus, subcylindricis, raro subcompressis, obconicis, diametro sesqui- ad 2:plo longioribus.

Syn. Corallina officinalis GOBI, Algenfl. Weiss. Meer. p. 21.
» » KLEEN, Nordl. Alg. p. 11.
» » Nyl. et Sæl. Herb. Fenn. p. 74.

Remark on the definition of the form. I know perfectly well that in specimens of *C. officinalis* on the west coast of Sweden several of the branches of the last order are in certain cases elongated and assume a cylindrical form. The form that I have characterized above under the name of *flexilis* certainly resembles such specimens. But they

[1]) I quote BERTHOLD as having given the Florideæ the limits here accepted, on the ground of his being the first that has referred decidedly the Porphyraceæ (Bangiaceæ) to this series.

differ, however, from it by being more robust, by the branches being very scanty in general and especially more regularly feathered, and by most of the branchlets presenting the appearance characteristic of the typical *C. officinalis*. From this, f. *flexilis* distinguishes itself by the greater slenderness and flexibility of all its parts, and by its irregular and profuse branching, in consequence of which the upper branches form dense bundles. At least upwards, in their upper portions, the branches are cylindrical, terminating not seldom in a kidney-shaped fastening-disc, that is convex on the upper side, plane on the under-side, and the upper cortical cells of which are isodiametrical, the lower ones parallelopipedical, arranged in rows radiating like a fan. On the coast of Bohuslän I have seen no form identical with this, although the plant is here very variable in appearance. It often grows here in such localities where on the coast of Norway I have found *f. flexilis*. In such a case, it often assumes an aspect differing from the typical one, becoming stunted, and irregularly but at the same time scantily branched. It accords with *C. elongata* ELLIS in point of slenderness, but differs from it by its branching and by the joints being chiefly round. On the other hand, it appears identical with that form of *C. officinalis* which RUPRECHT in Alg. Och. p. 354 reports from the coasts of Russian Lapland and of Cisuralian Samoyede-land, possibly also with the dwarfish form, richly and finely branched, observed by MAGNUS at Glesvær near Bergen (Magnus Nordseef., p. 70).

The form for which I have proposed the name of *robusta*, is in most respects the opposite of the preceding one. It differs from this as well as from the typical form by being larger in size and especially more robust and by having more irregular and scanty branches. The joints are round, cylindrical or slightly tun-shaped. The main axes and the secondary axes of the first order, those with branches as well as those without branches, are thickest at the middle, tapering towards the top as well as towards the base, but more strongly towards the base. The branches of the last order are, on the contrary, of equal thickness, and do not taper strongly towards the base, as in the typical *C. officinalis*. The color is more vividly rosy red than that of the principal form, and in preservation remains longer than in this. I have not found any reproductive organs.

Perhaps this plant is more rightly to be regarded as a northern species of the genus. But as I know it myself but incompletely, and as such great authorities as ARESCHOUG (J. E. Ag. Spec. Alg. II p. 563) and HARVEY (Phyc. Brit. t. 222) state that *C. officinalis* occurs in a number of different forms, it is possible that that one which I have called *f. robusta*, not having found it recognizably described in the literature, is only a form of the common Scandinavian species of *Corallina*. I have not seen it on the coasts of Sweden.

Habitat. The species occurs, fastened to stones, rocks, or, more rarely, to algæ, as species of Laminaria, most commonly in rock-pools in the litoral zone or at low-water mark, sometimes in 1—2 fathoms water, seldom at a greater depth. It grows generally scattered, or in small close groups, and prefers sheltered places. Of the form *gracilis* I have found specimens with tetrasporangia at the end of August.

Geographical distribution. It belongs properly to that region of the Arctic Sea which lies within the northern limits of the Atlantic, but it is known also from the White Sea and the western part of the Murman Sea. Its northernmost locality is Gjesvær near North Cape about Lat. N. 71°. It attains its maximum of number of individuals in the southern part of the Norwegian Polar Sea. In Ill. Alg. p. II. the species is said to have been found at Novaya Zemlya by K. v. BAER. Not having seen it there myself, and GOBI, who has examined the Russian collections of algæ, not reporting it from that locality, I suppose this statement to have arisen from some mistake. ARESCHOUG in J. G. Ag. Spec. Alg. II. p. 785 mentions the species as occurring »ad oras maris glacialis cum lapponicas tum sibiricas». The latter region ought surely to be excluded. I could not detect any trace of it along the coast of Siberia. There are no sure statements about the occurrence of the species in other parts of the Arctic Sea [1]).

Localities: The Norwegian Polar Sea: Nordlanden, *f. typica*, common and plentiful; Finmarken, local and scarce at Öxfjord, (*f. robusta*), Maasö and Gjesvær, (*f. flexilis*), the south coast of Magerö, (*f. typica*).

The White Sea, probably rare and scarce.

The Murman Sea on the coast of Cisuralian Samoyede-land, (*f. flexilis?*).

Gen. **Lithothamnion Phil.**

Wiegm. Arch. I. op. 387.

Lithothamnion soriferum nob.

L. fronde pilam in fundo liberam jacentem, sphæricam vel subsphæricam, diametro usque 8 cm., colore roseo-purpuream formante, decomposito-subdichotome ramosissima; ramis e centro solido, exiguo, undique egredientibus, vel omnino liberis vel in planta adulta inferne plus minus coalitis, teretibus vel subcompressis, lævibus, extremis elongatis, æqualibus vel apicem versus subattenuatis, apicibus rotundatis; conceptaculis sporangiferis superficialibus, numquam innatis, minutis, convexiusculis at parum prominentibus, infra apices ramulorum regiones fere definitas occupantibus, perpaucis vel numerosis; sporangiis quaternas sporas foventibus, 95 μ. longis, 20 μ. crassis. Tab. 1.

Syn. Lithothamnion fasciculatum KLEEN, Nordl. Alg. p. 11.

Description. The frond forms rather regularly spherical masses, that attain even 8 cm. in diameter, with a strong colour, between purplish and rosy. I have examined a great many specimens, but I have not found any one that had developed itself on or around a stone or any other hard object. The solid central mass both in young individuals, pl. I. fig. 1, 2, 3, 5, and in older fully developed ones fig. 4, is insignificant, which shows that ramification sets in at the very earliest stage of growth. The frond is repeatedly subdichotomously branched, with axes of at least three orders. The branch-systems issue in all directions from the centre of the frond, and can be followed in their whole length, although they are more or less anastomosing below. This anastomose has taken place during the growth of the plant. The branch-systems, some diffe-

[1]) For further particulars on the distribution of the species the reader is here as well as below referred to the works of the authors quoted under each species in the list of synonyms.

rent forms of which are represented by fig. 6—10, are sometimes flattened, with the branchlets arranged almost palmately, sometimes obpyramidal, in larger specimens about 2—3 cm. long. In typically developed specimens the branches are erect, fastigiate, and straight, in others they are spreading and more or less curved, those of the last order being 5—10 mm. long, terete, or somewhat compressed, either cylindrical, or tapering, or slightly enlarged towards the tip, with the ends rounded, 1,5—2,5 mm. in diameter.

The structure of the frond. At the broken end of a branch there always appears a more solid, central part of greater or lesser circumference. A pellucid, transverse section, obtained by grinding, shows this central part to be composed of a very dense tissue of angular, iso-diametric cells with very small cell-rooms and very thick walls marked by double contours; fig. 12 and 15. This is surrounded with numerous, pretty regularly concentric layers, distinctly marked against one another and resembling the yearly rings of a dicotyledonic stem, every one of which is found on the transverse section to be formed of comparatively large-roomed cells, which are arranged in pretty regularly concentric and radiating rows. Of these cells the inner ones appear longer, rectangular, the outer ones shorter, almost square. In the layer nearest the central part the cells are less regular and in the outermost layers the difference in length is rather imperceptible. A cut made thin by grinding, parallel to the longitudinal axis of a branch, fig. 13 and 16, shows the branches to consist of superposed, generally distinctly separate, very regular, cup-shaped layers of tissue, whose cells are arranged in rows radiating in the shape of a fan. The lower, inner cells of each of these layers, on an optical longitudinal section, are rectangular, 20 μ. long at the most; the upper ones are square, with somewhat thicker walls than the others, and like these, 5—8 μ. thick.

Organs of propagation. Sporocarpia unknown. The conceptacles of the sporangia occupy a generally sharply defined zone below the tips of the branches, and commonly occur in great numbers, forming what may be called a sorus; fig. 11. Hence the name of the present species. They are always superficial, never growing down into the frond nor becoming overgrown by it; so that traces of old organs of that kind are never to be seen in the interior of the frond (cf. fig. 12—13). This fact, in this species as well as in several others, together with which it appears to form a well marked group, depends apparently on the thickening meristema of the frond lying below the basal surface of the conceptacles. They are circular in circumference, very little prominent, small, scarcely perceptible to the naked eye. The roof is slightly convex, traversed by numerous canals, which are transversely 5—6-angular and filled with a gelatinous substance, and the orificial cells of which are somewhat different in shape from the other cortical cells of the roof; fig. 18. The sporangia are tetrasporic, oblong or club-shaped, somewhat variable in size, but generally, after the formation of spores has begun, about 95 μ. long and 20 μ. thick; fig. 19.

Remark on the relation of this species to other species described. L. soriferum described as a new species, is more nearly related to *L. fasciculatum* LAM. than to any other known species. However, it cannot in my opinion be identified with this species, which in the present acceptation of the name probably comprises several specifically distinct forms. LAMARK's description of *L. fasciculatum* Hist. Anim. 2, p. 203 is very summary. Nevertheless he mentions a character which does not accord with my species, namely, »ramis apice incrassatis, obtusis». The plant, delineated (Phyc. Brit. Plate 74) and described, under the name of *Melobesia fasciculata*, by HARVEY, who quotes as synonymous the *Millepora fasciculata* of LAMARK, is obviously distinct from the one in question, as is easily seen on comparing HARVEY's figures with mine. *Melobesia fasciculata* HARVEY is distinguished from *L. soriferum* mihi by its strongly developed »solid, central stony mass», and by its short branches that are »remarkably truncated at the tips, which are moreover depressed in the centre». *L. fasciculatum* ARESCHOUG (in J. G. AG. Spec. Alg. 2, p. 522), with which HARVEY's *M. fasciculata* is cited as synonymous, certainly coincides in several particulars with my species, but it differs from it by the frond being »circa lapillum plerumque undique effusa» and by the branches being sometimes simple, sometimes compound, thickened upwards, with truncate tips. Even in very young specimens of *L. soriferum* the branches issuing from the centre are decompound. *L. fasciculatum* ARESCHOUG (Obs. Phyc. III, p. 5) differs most essentially in development and ramification from *L. soriferum*, and, as far as I can see, it comprises both *M. fasciculata* HARVEY and that species which I describe below under the name of *L. glaciale*. *L. fasciculatum* SOLMS-LAUBACH (Corall. Monogr.), with which name the author very hesitatingly designs a species occurring in the Mediterranean at Naples, can hardly be identical with HARVEY's *M. fasciculata*. In ramification and in the shape and arrangement of the conceptacles of the sporangia it agrees, in its most developed form, with *L. soriferum*, but it differs from this by the lower part of its frond spreading over stones in the form of a crust. But, on the other hand, I think it is possible or probable that the fragment from Iceland mentioned by the author belongs to *L. soriferum*.

The species occurring in Bohuslän, called *L. fasciculatum*, agrees with *M. fasciculata* HARVEY and *L. fasciculatum* ARESCHOUG (Spec. Alg.) and through the form of the branches is distinctly known from the plant here described. Professor J. E. ARESCHOUG, the well-known monographer of the Corallineæ, has kindly allowed me to look over his collections of such plants. I have not found in these any species, to which *L. soriferum* could be considered to belong. Thus no other course was left me than to describe under a specific name this species, which is very abundant in the southern part of the Polar Sea.

Hab. According to my own experience, the present species grows on sandy and shingly bottom in quiet bays or on protected coasts, in 10—15 fathoms water. Dead it is found at greater depths, and it is probably such specimens that KLEEN has brought up from the elitóral region (cf. KLEEN, Nordl. Alg. p. 11). It is gregarious, covering large spaces of the bottom in great masses, and serves as substratum for various smaller algæ, such as *Antithamnia*, *Rhodophyllis dichotoma*, *Derbesia marina* a. o. I have found specimens with ripe sporangia in July and at the end of September.

Geogr. Distr. As far as is known hitherto, it belongs only to the Atlantic region of the Polar Sea. The most northerly place where it has been found, is Maasö, about Lat. N. 71°.

Localities. The Norwegian Polar Sea: Nordlanden, common and abundant; Tromsö amt for inst. at Tromsö and Carlsö, at the latter place plentiful; Finmarken, abundant in several places, as Maasö and the southern coast of Magerö. FOSLIE has communicated to me specimens collected at Honningsvaag and Lebesby.

Lithothamnion Ungeri, novum nomen.

Descr. et Fig. Lithothamnion byssoides Unger Leithakalk p. 19—20, tab. 5, fig. 1—8.

Remark on the determination of the species. From specimens collected on the coast of Norway and preserved in the Museum of Bergen, UNGER has, under the name of *Lithothamnion byssoides* (LAM.) Phil., described and figured a *Lithothamnion*, also found in the Arctic Sea by Mr M. FOSLIE and kindly sent to me. The coincidence of the specimens I have at my disposal, with the description and figures of UNGER is palpable. But on the other hand, it is evident that this plant cannot by any means be identical with that which goes now commonly under the name of *Lithothamnion byssoides* (LAM.) Phil. Therefore I propose that this Norwegian species take the name of *L. Ungeri* after him who first described and figured it in a recognizable manner. To the description of UNGER I have nothing essential to add. The specimens I have had an opportunity of examining, are sterile.

The relation of the present species to other species. With reference to its structure, *L. Ungeri* approaches nearly to the preceding species. Like this, it never possesses occluded conceptacles of sporangia, and, besides, the arrangement of the cells is the same in both. It differs distinctly by its strongly developed, crustaceous hypothallus, by its denser ramification, with shorter, much finer, and less compound branches. It can scarcely be confounded with any known species of the genus.

Hab. Unknown to me. The specimens, taken in September, are sterile.

Geogr. Distr. The Atlantic region of the Polar Sea.

Locality. The Norwegian Polar Sea: Tromsö amt near the town of Tromsö (FOSLIE)

Lithothamnion Alcicorne nob.

L. fronde initio affixa, demum libera in fundo jacente, flavescente, apicibus pulchre roseis, 4—5 cm. alta, decomposito-subpalmatim-ramosa; ramis ex axi primario brevissimo flabellatim egredientibus, vel omnino liberis vel in planta provectiore ætate plus minus coalitis, subcompressis, lævibus, extremis brevioribus, cylindricis vel compressis, apicibus rotundatis; conceptaculis sporangiferis superficialibus, numquam innatis, planato-hemisphæricis, sat magnis, infra apices ramulorum sparsis, conceptaculis sporocarpiferis, elevatis, conicis, acutis, apice perforatis intermixtis. Sporangiis quaternas sporas foventibus, 250 μ. longis, 100 μ. crassis. Tab. 5, fig. 1—8.

Description. External shape of the plant. I have not seen any attached specimens of this species, but those I have at my disposal through the kindness of Mr M. FOSLIE, show plainly that the plant is attached at first to some hard object, although afterwards, at least in certain cases, it detaches itself and lies free on the bottom. The frond has a short, either flattened and upwards broadening, or extremely short and almost terete, mainstem, from which there issue repeatedly palmately branching branch-systems, spreading like a fan almost in one plane. These are also situated in the same general plane, in consequence whereof the plant gets the appearance of a low, flat bush. The branch-systems contain axes of at least four degrees, all of which, except the ultimate ones, are downwards almost terete, but upwards at the branching-point strongly triangularly expanded; fig. 1. The branch-systems are sometimes almost free, as is shown by the above-quoted figure, sometimes, in older specimens, more or less confluent with one another. It even happens, that branch-systems belonging to different individuals growing close to one another, become confluent. The tips of the branches, except when they are on the point of dividing, are scarcely enlarged, but rounded, or almost truncate.

The structure of the frond. In this respect the plant coincides so closely with *L. soriferum*, that I need not enter into any detailed description. The figures 3, 4, 5, 6, 7 exhibit the structure better than words could do, and on comparing them with the corresponding figures of *L. soriferum*, the similarity of the two species in this respect is easily perceived. They agree even with regard to the size of the cells. In the cup-shaped layers of tissue on a thin median section the cells are 6—10 μ. thick and 20 μ. long at the most.

Organs of propagation. Conceptacles of sporocarpia and sporangia are to be found on the same individual and on the same branch. They are thinly scattered on and below the tips of the branches (fig. 2) and differ very much in shape from each other. The conceptacles of the sporocarpia present the form of acute cones, upwards traversed by a canal. The conceptacles of the sporangia are flatly hemispherical, about 0,5 μ. in diameter at the base, larger and more elevated than in *L. soriferum*. They are never found to have grown down into older portions of the frond, which proves the growth of the frond to take place in the same manner as in the preceding species. The roof of the conceptacles of sporangia is intersected by numerous canals, generally 6-sided in transverse section, which are filled with a gelatinous substance and whose orificial cells differ somewhat in shape from the other surface-cells of the roof. The sporangia, fig. 8, are tetrasporic, considerably larger than in *L. soriferum*, about 250 μ. long and 100 μ. thick, cylindrical, cylindrically spindle-shaped or cylindrically claviform.

The relation of the present species to others. Amongst the Lithothamnia that I know, this species stands nearest to *L. soriferum*, from which it seems, however, to be clearly distinguished by its peculiar ramification, its few scattered conceptacles of tetrasporangia and its large sporangia.

Hab. According to the kind communication of Mr M. Foslie, the finder of the species, it grows in 20 fathoms water, and, as it seems, in sheltered places of the coast. Specimens, collected at the beginning of August, have carpospores in development and ripe tetraspores.

Geogr. Distr. The Atlantic region of the Polar sea. Within the arctic region it has not as yet been found.

Locality. The Norwegian Polar Sea: Tromsö amt near the town of Tromsö, the only place where it is known to occur. Here it has been found by M. FOSLIE.

Lithothamnion Norvegicum ARESCH. (nob).

Descr. Lithothamnion calcareum var. norvegicum ARESCH., Obs. Phyc. 3, p. 4—5.
Fig. Lithothamnion norvegicum tab. nostra 5, fig. 9—10.

Remark on the determination of the species. J. E. ARESCHOUG l. c. has described, under the preceding combination of names, a plant from the southwest coast of Norway, that has also been found in the Polar Sea. On pl. 5, I have given figures of one specimen from the south of Norway (fig. 9) and of another from the Polar Sea (fig. 10). Judging from the fact that no overgrown sporangia are to be found on sections of older frondal portions, the plant with regard to the inner structure belongs to that group of Lithothamnia which possesses external conceptacles of sporangia. ARESCHOUG considers this plant to be a local form of *Melobesia (Nullipora) calcarea* ELL. et SOL., occurring on the coasts of England and figured by HARVEY (Phyc. Brit.) and JOHNSTON (Brit. Spong. Lith.) It is certainly possible that this may indeed be the case, but the statement of ARESCHOUG »forma M. calcarea quam depinxit et descripsit HARVEY (est forsan maxima et magnopere evoluta) a nostra valde abhorrens» being clearly quite correct, and the Norwegian specimens being unlike those figures of *Nullipora calcarea*, much differing from one another and possibly designating different species, which JOHNSTON gives, I consider myself entitled to regard the above-mentioned plant, found on the coast of Norway, as a distinct species separated from *Melobesia calcarea* HARVEY and *Nullipora calcarea* JOHNSTON.

Hab. It grows, spreading stratiformly, in 10—15 fathoms water. Cf. ARESCHOUG l. c. Only sterile specimens are known.

Geogr. Distr. This species belongs only to the Atlantic region of the Polar Sea, and seemingly only to its most southerly part. It has not been found to the north of Nordlanden.

Locality. The Norwegian Polar Sea: Nordlanden at Lödingen (FOSLIE).

Lithothamnion glaciale nob.

L. fronde demum crustam formante validam circum lapides vel conchas effusam e roseo flavescentem, plus minus lobatam, ramos simplices, conicos, obtusos vel subcylindricos, usque 7—8 mm. altos, inferne diametro usque 5 mm., scabriusculos undique emittentem; conceptaculis sporangiferis demum innatis, minutis, convexiusculis at parum prominentibus, creberrimis, nullo ordine in crusta ramisque dispositis; sporangiis binas sporas foventibus, 80—135 μ. longis, 40—60 μ. crassis; Tab. 2 et 3.

Syn. Lithothamnion calcareum KJELLM., Vinteralg. p. 64.
» » KLEEN, Nordl. Alg. p. 11.
» fasciculatum ARESCH. Obs. Phyc. 3, p. 5 (saltem ex parte).
» » DICKIE, Alg. Sutherl. 1, p. 142. (?)
» » GOBI, Algenfl. Weiss. Meer. p. 22.
» » KJELLM. Spetsb. Thall. 1, p. 3, Algenv. Murm. Meer, p. 7 et alibi.

Description of the species. External shape. When young, the plant forms an almost circular, thin crust on hard objects that lie free on the bottom, especially on shells of Balanidæ, and on muscles and stones. At this stage it resembles a *Lithothamnion polymorphum.* The crust is slightly rose-coloured, furnished at the edge with feebly marked, rounded lobes. At first it is smooth, but it soon gets feeble concentrical ridges especially towards the margin. It is closely attached to its substratum. Its surface is enlarged by marginal growth and at the same time its thickness increases considerably from the centre outwards. If the objects on which the plant has germinated, are of small size — to about 10 cm. in diameter — they are completely surrounded by the crust; if they are larger, this is the case only partially. As soon as it has attained a slight thickness, 2—3 mm., the crust produces more or less densely crowded, short, conical or wart-like, simple protuberances, on which as well as on the basal layer conceptacles of sporangia are developed in large numbers (pl. 3, fig. 1 and 2). Old specimens differ in shape according to the form of the object included. However, they approach generally the shape of a sphere or hemisphere, and have a considerable size and weight. Such a ball is often 15—20 cm. in diameter. On their lower side, which is turned towards the bottom, such individuals are often furnished with a large opening, through which the originally included object has fallen out (pl. 2, fig. 2). The thickness of the crustaceous basal portion varies much in different parts of the same full-grown specimen, from one half to one or two cm. (pl. 3, fig. 3).

The basal crust often puts forth clumsy protuberances or lobes, varying in circumference, thickness, and height, and bearing, like the rest of the crust, thinly scattered, regular, straight, simple, conical, blunt or cylindrically conical processes or branches, sometimes very low and wartlike, sometimes higher, 7—8 mm. long, even 5 mm. in diameter at the base. Older specimens have a greyish colour with a faint rose-red tint. The surface is never smooth, but finely rugged. The crust-like portion of older specimens is traversed with numerous, wider or finer passages made by worms, and is rich in cavities produced by boring-muscles.

Structure of the frond. The fracture of the plant is white, sprinkled with small, yellowish-brown spots. A section shows these spots to consist of conceptables of sporangia, that have been grown over. Such are to be found throughout the whole frond, in the branches as well as in the basal crust, at certain places very densely crowded. These fact point to the thickening meristema of the frond being superficial, overlapping the roofs of the conceptables. With regard to structure, this species agrees in the main with *L. soriferum.* However, on a cut parallel with the longitudinal axis of a branch, the cup-shaped layers of tissue appear less sharply marked. This results partly from the stratification being disturbed by the buried conceptacles, partly from the inner cells of each layer being less different in size from the outer ones, than is the case in *L. soriferum* (pl. 3, fig. 6, 9). The corners of the cell-rooms are rounded, their walls are thicker than in the last-mentioned species. The thickness of the cells in a median section varies between 6 and 10 μ., their length between 10 and 22 μ. In a cross cut of a branch the centre is seen to be occupied by a layer of 5—6-angular cells having a thick membrane with double contours (pl. 3, fig. 5, 8). This layer passes outwards without marked limit into a more or less mighty layer of tissue whose cells arrange themselves in rows more and more radially, and while maintaining their multangular periphery, extend themselves in the radial direction, the walls getting thinner. Beyond this layer there is a greater or smaller number of layers less sharply defined from one another, formed of transversely square or rectangular cells arranged in pretty regular lines radial as well as concentric. When the cells are rectangular, their longitudinal axis is often at right angles to the radius (pl. 3, fig. 7). The surface cells are isodiametric in the tangential direction, with rounded cell-rooms, 5—7 μ. in diameter. The thickness of the wall amounts to 2—4 μ. (pl. 3, fig. 10).

Organs of propagation. In this species I have seen only conceptacles of sporangia. These are very numerous, disseminated both over the processes in their whole length and on the crust between them without apparent order. They are small, 250—300 μ. in diameter, slightly elevated above the surface of the frond, with convex roof; fig. 4, pl. 3. The gelatiniferous canals of the roof, in cross section, are 5—6-angular, 7—10 μ. in diameter. Their orificial cells differ scarcely or not at all from the adjoining surface cells (fig. 11, pl. 3).

The sporangia are bisporic. This statement is founded on an examination of specimens from widely distant parts of the Arctic Sea. I have never, amongst the pretty numerous sporangia I have examined, found any containing more than 2 spores. As to shape and size they vary within wide limits. They are often pyriform or elongated-pyriform, sometimes slenderly spindleshaped-cylindrical, sometimes almost perfectly cylindrical. Some of those measured by me, were 80—90 μ. long, 60 μ. thick, others about 120 μ. long, 40 μ. thick, again others 135—140 μ. long, 50—60 μ. thick a. s. o. (fig. 12, 13, 14, pl. 3).

Remark on the synonomy. When I mentioned this species from the Arctic Sea for the first time, I gave it the name of *L. calcareum*, being induced thereto by HARVEY's description of *Melobesia calcarea* in Phyc. Brit. agreeing in certain respects with the specimens I brought home from Spitzbergen. KLEEN having seen these and having found a form of Lithothamnion occurring in Nordlanden to be identical with the form from Spitzbergen, followed my example and recorded this plant under the name of *L. calcareum.* Finding, however, on closer examination that the form from Spitzbergen could not be the English *Melobesia calcarea* and looking round for some known species with which the arctic form might be identified, it seemed to me that it might be referred to *L. fasciculatum*, under which name I adopted it in Spets. Thall. 1. I had then only little opportunity to occupy myself with the genus of *Lithothamnia*, imperfectly known at that time and very feebly represented in collections. Above all, I did not know in what degree these forms vary and what importance should be attached to external differences. All this made me unwilling to set down the plant from Spitzbergen as a separate species. In a treatise published shortly afterwards, J. E. ARESCHOUG, the monographer of the Corallineæ, recorded the plant under this name, and thus I was prevailed upon not to abandon my former view in my subsequent work on the algæ of the Murman Sea. Having since then had the opportunity of studying the arctic as well as the Scandinavian forms of this genus longer and more closely, and having paid special attention to the present form, I have found that it is ever like itself everywhere in the Arctic Sea and does not show any variation at all in the direction of *L. fasciculatum*, from which it is easily distinguished at first sight even by the external shape, besides differing from it by general and essential diversities of structure. The only species I know, to which it shows any greater affinity, is *L. intermedium* described below, which differs however rather considerably in external characteristics, besides bearing regularly tetrasporic sporangia. Thus the present alga seems to me to be a good species, and I have consequently had no hesitation in giving it a separate name. The genus *Lithothamnion* has been hitherto rather much put aside, on account of the difficulties connected with a nearer investigation and characterization of the forms; in consequence whereof certainly only some of the most remarkable of its numerous forms have been hitherto described and cleared up satisfactorily. After the publication of SOLMS-LAUBACH'S excellent work (Corall. Monogr.) and HAUCK'S comprehensive researches, the Mediterranean forms may be considered to be essentially cleared up. That much remains to be done with regard the northern species, is a fact of which I have clear evidence.

Hab. The present plant is a deep-water form. Most often and in the greatest number it is met with at a depth of 10—20 fathoms. It thrives best on a bottom consisting of shingle, gravel, and shells, and is found on open shores as well as in sheltered bays. It is gregarious. On the coasts of Spitzbergen and Novaya Zemlya it covers the bottom in deep layers for several miles, and altogether determines the general aspect of the vegetation, wherever it occurs. In the formation of future strata of the earth's crust in these regions it must become of essential importance. On the coast of Norway I have met with it only occasionally, amongst other Lithothamnia. At Spitzbergen it bears ripe sporangia both in summer, in July, and in mid-winter, in the months of November and December. On the coast of Norway and the west coast of Novaya Zemlya, specimens with such organs have been collected in summer, in July and August. I have never seen specimens bearing carpospores.

Geogr. Distr. The species is dispersed over the greater part of the Arctic Sea. Only from the Kara and Siberian Seas it is not known. It attains its most vigorous development, as far as I know, at Spitzbergen and on the west coast of Novaya Zemlya, where it occurs also in the greatest numbers. The northernmost place where it has been found, is Treurenberg Bay on the north coast of Spitzbergen Lat. N. 79° 56'.

Localities: The Norwegian Polar Sea: Nordlanden at several places in deep water; Finmarken rather scarce and local, as at Maasö, Gjesvær, and in Magerö Sound.

The Greenland Sea: common and plentiful on the west and north coasts of Spitzbergen.

The Murman Sea: the coast of Russian Lapland and the west coast of Novaya Zemlya, in the latter locality common and abundant.

The American Arctic Sea: Probably the plant reported from here by DICKIE (Alg. Sutherl. 1, p. 142) under the name of *L. fasciculatum*, is the present species.

Baffin Bay: I have seen specimens from the west coast of Greenland, collected there by Prof. TH. M. FRIES.

Lithothamnion intermedium nob.

L. fronde subglobosa, dilute rosea, scabriuscula, diametro circa 7 cm., parte centrali solida, plus minus distincte lobata, ramos vel breves, verrucæformes vel longiores usque 4—5 mm. altos, basi 2 mm. crassos vel simplices, conico-cylindricos, apicibus obtusis, vel infra apicem uno alterove ramulo brevissimo, verrucæformi præditos undique emittente; conceptaculis sporangiferis demum innatis, minutis, convexiusculis at parum prominentibus, creberrimis, nullo ordine in tota fronde sparsis; sporangiis quaternas sporas foventibus, 130—150 μ. longis, circa 40 μ. crassis; Tab. 4.

Description of the species. External shape. The plant forms spherical or spheroidical balls, which, when older, are about 7 cm. in diameter (fig. 1). In several of the specimens examined by me there was nothing included in the interior of these globular masses; others encompassed small stones. The frond is constituted of a thick central contained portion intersected with cavities and canals, and projecting in more or less distinct, simple or divided, clumsy, thick lobes. These carry partly simple, short, and wartlike, partly longer and branch-like processes. The latter reach a length of 4—5 mm., with a diameter of about 2 mm. at the base. They are sometimes simple, conically cylindrical, obtuse, sometimes furnished with one or two, generally short, wartlike side-branches below the tip. The plant is more or less deeply rose-coloured. Its surface is uneven, on account of local, scaly thickenings of the surface. As in the preceding species, the central mass is rich in holes after boring-muscles, and in passages produced by worms.

Structure of the frond. The fracture is white or faintly rose-coloured with rare, small, yellowish-brown dots — the grown-in conceptacles of sporangia. In longitudinal section the processes show distinct cup-shaped layers, although these are not so regular as for inst. in *L. soriferum*. The inner cells of these layers, in longitudinal section, are rectangular with rather thin walls, the outer ones have more rounded cell-rooms and somewhat thicker walls. The diameter of the former amounts to 7 μ., their greatest length to about 15 μ. fig. 8. A cross cut of a process shows essentially the same structure as in the preceding species, differing only by the cells of the outer concentric layers having thinner walls and a greater length in the direction of the radius (fig. 4, 6, 7). The surface cells of the frond are isodiametrical in a tangential section, 4—6-angular, about 10 μ. in diameter, with walls 2,5 μ. thick (fig. 9).

Organs of propagation. The conceptacles of sporangia are spread over the whole frond and become grown over in the present species as in the preceding. Although there are certainly found older grown-in organs of this kind nearer the centre of the processes, most of them are peripherical. The superficial ones are but little prominent above the surface of the frond, small, with slightly convex roof (fig. 3). This is traversed with numerous gelatiniferous canals. In none of the specimens examined I have seen these reach to the surface of the roof, and from that cause do not know how their orificial cells may be constructed. The sporangia are tetrasporic, spindleshaped-cylindrical or claviform, 130—150 μ. long, about 40 μ. thick (fig. 10).

Relation of the present species to others. The present species is probably most closely related to *L. glaciale*. By its general habit and its grown-in sporangia, it reminds one most of this. In other respects, however, its structure approaches more nearly that of *L. soriferum* and the species resembling it. It differs from *L. glaciale*, besides in structure, even by the lesser thickness and more cylindrical shape of the processes and by its tetrasporic sporangia. With *L. soriferum* or *L. Ungeri* it cannot be confounded, the habit being different and the sporangia becoming grown over. The name is meant to denote that in character it stands between *L. glaciale* and *L. soriferum* and the species most closely allied to the latter.

Habitat. The specimens I have collected myself, were taken in 5—10 fathoms water on stony bottom, in a pretty well-sheltered locality. Here scattered individuals were found. On the coast of Norway it has ripe tetrasporangia in the month of June.

Geogr. Distr. It is known hitherto only from the Atlantic region of the Polar Sea. The most northerly locality where it has been found, is Karlsö at Lat. N. 70°.

Localities: The Norwegian Polar Sea: Tromsö amt, for inst. at Tromsö (Foslie) and Karlsö; Finmarken, at Vadsö (Foslie).

Lithothamnion flavescens nob.

L. fronde crustacea, arcte adnata; crusta tenuiore, vix 1 mm. crassa, e roseo flavescente, scabriuscula, limbo lævi, subnitido, obsolete concentrice striato, margine subundulato, e cellulis majoribus formata; conceptaculis sporocarpiferis et sporangiferis in eodem specimine sparsis, illis depresso-conicis, apice perforatis, his demum innatis, creberrimis, magnis, diametro 700 μ. hemisphæricis, prominentibus; sporangiis quaternas sporas foventibus, sporis maturis, 190—220 μ. longis, 50—100 μ. crassis. Tab. 6, fig. 1—7.

Description of the species. Habit. The plant forms incrustations on other Lithothamnia, for inst. *L. glaciale, L. compactum*, and on shells of Balanidæ. The crust is closely adherent to the substratum, thin, scarcely one mm. thick, always uneven when older, finely rugged and squamellate on the greater part of its surface. The greater or smaller unevenness of the surface is caused by the substratum, to which it clings closely. However the brim is smooth, feebly shining, with few indistinct concentric stripes; the edge is uneven, shallowly undulating. When younger the crust has a faint rosy colour, which passes afterwards into faint brownish-yellow, which colour increases when the plant dies, and appears particularly strong in the fracture (fig. 1).

Structure of the frond. The lower co-axil system is feebly developed, its anticlinals converge gradually toward the matrix. In fragments of the frond, which have a thickness of 0,3 mm., it takes up about 25 μ. The cells are elongated, about twice as long as broad, rectangular or rhomboidical in radial section (fig. 3—4). In the upper thickening-layer of the frond the cells, on a radial section, are four-angular, squarish or rectangular, their longitudinal axis in the latter case running sometimes in the direction of the radius, sometimes in that of the tangent. The outermost cells in particular are in the latter position. Their thickness amounts to 10—13 μ.; their length does not exceed 15 μ. The cell-rooms are rounded, the walls about 2,5 μ. thick. The surface cells are nearly isodiametrical in tangential direction, with rounded or rounded-angular cell-rooms, 5—8 μ. in diameter; the thickness of the dissepiment amounts to 4—5 μ. (fig. 5).

Organs of propagation. Conceptacles of sporocarps and sporangia are to be found on the same individual. The former are conical, low, with a canal at the tip, almost as wide at the base as the conceptacles of the sporangia. That part of them which rises above the surface of the frond, becomes easily detached and falls away at last, whereupon a cup-shaped scar with somewhat elevated edges appears on the surface of the frond. This hollow is gradually filled with new-formed tissue so as to efface the scar. These local new formations contribute to the unevenness of the frond. I have never seen conceptacles of sporocarps with distinct spores. The sporocarpal bed is plane, and the sporigen cell-rows are developed peripherically on it.

The conceptacles of the sporangia become finally immersed. They are numerous, scattered, large, about 700 μ. in diameter, much elevated hemispherically. The roof, whose thickness amounts to 125 μ. when the sporangia are ripe, is intersected with numerous, 80—90, transversely six-angular canals, whose orifices are surrounded with a ring of cells differing in shape and size from the other surface cells (fig. 2, 3, 6). The sporangia are tetrasporic, cylindrical, cylindrically spindleshaped, or slightly claviform, large, 190—220 μ. long, 50—100 μ. thick (fig. 7).

The relation of the present species to others. In sterile condition and superficially considered, the present species may be easily confounded with other crustaceous Lithothamnia. However, it is sharply distinct from these by its large, strongly prominent conceptacles of sporangia and its coarse structure.

Habitat. It grows scattered, in company with other Lithothamnia, at a depth of 5—10 fathoms on stony and gravelly bottom, on open coasts as well as in sheltered places. In June it bears ripe sporangia, on the coast of Norway at the beginning of the month, on the west coast of Novaya Zemlya at its end. The formation of carpospores appears to set in earlier.

Geogr. Distr. This alga belongs to the Atlantic as well as the arctic region of the Polar Sea. Its northernmost known place of growth is Karmakul Bay on the west coast of Novaya Zemlya, about 72° 30′ N. Lat.

Localities: The Norwegian Polar Sea: Tromsö amt at Karlsö; Finmarken on the south coast of Magerö, everywhere local and scarce.

The Murman Sea: Karmakul Bay, scanty and local.

Lithothamnion foecundum nob.

L. fronde crustacea, initio arcte adnata, demum soluta, circa 2 mm. crassa, in statu juvenili lævissima, nitida, ætate provectiore conceptaculis sporangiferis inæquali, dilute rosea, limbo albido, margine undulato-lobato, e cellulis majoribus constructa; conceptaculis sporangiferis immersis, tecto margine elevato circumdato, demum innatis, depresso-globosis, numerosissimis; sporangiis quaternas sporas foventibus, 120—185 μ. longis, 45—65 μ. crassis. Tab. 5, fig. 11—19.

Syn. Lithothamnion polymorphum KJELLM. Kariska hafvets Algv. p. 15.

Description of the species. Habit. The plant covers stones and other hard objects in the shape of a crust. At first it is fastened closely and firmly to its substratum, but when older is easily separated from it. The form of the crust depends on that of the substratum, and does not exhibit any tendency to become circular, as in the preceding species. The central parts thicken more rapidly and considerably than the peripherical. New crusts may be produced upon others, whereby may be formed crust-layers severel mm. thick. The margin of the crust is shallowly crenate with rounded lobes. The nature of the surface is determined by that of the substratum; if this is smooth, the crust is also smooth and shining when young. Older crusts always become uneven and finely rugged on the surface, by growing over and covering up small extraneous objects, and especially on account of the peculiar shape of the conceptacles. Fresh fractures of older crusts are of a pure white colour, in younger crusts they are white with a faint rose-coloured tinge, at least outwards. The surface is faintly rose-coloured, the brim whitish (fig. 11).

Structure of the frond. A basal, co-axil layer is almost always distinctly and vigorously developed, with pretty strongly incurved, anticlinal cell-rows, whose cells are about twice as long as thick (fig. 13, 14). The cells of the upper thickening-layer, which on a radial section are arranged in distinct rows, that are slightly curved only nearest to the median, but otherwise straight, are square or rectangular, with the height greatest, 7—9 μ. thick, and even 15 μ. long, with thick walls and the corners of the cell-rooms rounded (f. 15). The surface cells are angular, 7—10 μ. in diameter, with the diaphragms about 3 μ. thick (fig. 16).

Organs of propagation. The conceptacles of sporangia are very numerous, densely crowded both in the internal and external portions of the frond. Hence the name of the species is derived. They are not, or scarcely not, raised above the surface of the frond, but they are sharply marked on it by their circular or oblong roofs being surrounded with a strongly prominent, annular border (fig. 12, 17). The roof is almost plane or slightly concave, traversed with comparatively few — I have numbered about 40 — transversely six-sided gelatiniferous canals, whose orifices are surrounded with a ring of cells different from the other cortical cells of the roof (fig. 18).

The sporangia are tetrasporic, slightly club-shaped or cylindrically spindle-shaped, from 120 to 185 μ., generally about 150 μ. long and 45 μ. thick (fig. 19). Sporocarps unknown.

Habitat. It occurs on rather open coasts within the sublitoral zone in 5—15 fathoms water, and seems to grow scattered in small number. Specimens taken at the end af August had most of their sporangia void. The development of spores takes place probably before that time.

Geogr. Distr. It belongs to the arctic region of the Polar Sea. Here it has apparently a pretty wide range. The northernmost place where it is known with certainty to grow, is Actinia Bay, about 76° N. Lat.

Localities: The Kara Sea: local and scarce at several places, as at Uddebay at 76° 8' N. Lat., 90° 25' O. Long., Actinia Bay.

Baffin Bay. A Lithothamnion brought home by TH. M. FRIES from the west coast of Greenland seems to me to belong to this species.

Lithothamnion compactum nob.

L. fronde crustacea, initio arcte adnata, demum crustis numerosis superimpositis formatis usque 2 cm. crassa, e matrice soluta; crusta primaria valida, circa 5 mm. crassa, subnitida, in statu juvenili striis brevioribus, densis, radiatim et concentrice dispositis, nudo oculo inconspicuis, in statu sporangifero foveolis, minutissimis, creberrimis inæquali, dilute vinoso-purpurea, flavescente vel albescente, e cellulis minutis constructa; conceptaculis sporangiferis immersis, demum innatis, numerosis, circumscriptione globosis vel depresso-globosis; sporangiis? Tab. 6, fig. 8—12.

Syn. Lithothamnion polymorphum KJELLM. Algenv. Murm. Meer. p. 8.

Description of the species. Habit. If the plant is allowed to develop freely on a plane surface, it forms an almost circular crust, whose extent depends on that of the substratum and whose thickness is comparatively considerable, amounting sometimes to 5 mm. in profile. It increases pretty uniformly in thickness, and thus the peripherical portions are not much thinner than the central ones. When young and sterile the crust appears perfectly smooth to the naked eye, as if it were polished. Magnifying shows, however, the surface to be uneven in consequence of very fine striæ partly radiated partly concentric. Older specimens with conceptacles of sporangia have their surface densely furnished with very small, point-like cavities, imperceptible to the naked eye, whose bottoms are formed by the roofs of the conceptacles. These roofs being dissolved in older, dead crusts, the surface is covered with distinct little holes. When young and living, it is feebly wine-coloured, and this is also the colour of the surface in older, living specimens. The fracture of these is however chalky-white with a faint tint of yellow. The young plant adheres closely and firmly to its substratum. Upon the young primary crust new crusts are formed, one upon the other, so closely united to one another, that the limit between them is very difficult to detect on a section. These crust-layers often attain 2 cm. in thickness, and when older detach themselves easily from the objects, other Lithothamnia and stones, over which they have spread (fig. 8).

Structure of the frond. The lower, co-axil system of the frond is scarcely perceptible on a radial section. Like the boundary-layer of the upper system, it has a faint yellowish colour differing from that of the other frond. It consists of rather elongated cells (fig. 10). In the upper thickening-layer of the frond the cells are seen on a radial section to be arranged in straight, well marked rows, square or rectangular, with their greatest extent in the vertical direction of the frond, no more than 10 μ. long and only about 5 μ. thick. The corners of the cell-rooms are scarcely rounded (fig. 10, 11). The surface cells are isodiametrical in a tangential direction with almost circular cell-rooms. Their diameter amounts to 5 μ., the thickness of the dissepiments to scarcely 2 μ. (fig. 12).

Organs of propagation. The conceptacles of the sporangia are always immersed, never rising above the surface. They are to be detected externally only by means of the small cavities that are to be seen above them in the surface of the crust. Afterwards they grow down into the frond. They are numerous, rather small, spherical or flattened-spherical. The texture of the roof seems to be easily dissolvable; in dead specimens it is destroyed, in consequence of which the frond is covered with a number of small holes perceptible to the naked eye. I have not seen specimens with sporocarps, nor any with mature sporangia. Thus I know nothing about the shape and size of these.

Relation to other species. Amongst the species of Lithothamnion that I know, the present species exhibits the greatest resemblance to *Lithophyllum incrustans* PHIL. ARESCH. Cp. Solms-Laubach, Corall. Monog. p. 16. In structure, however, it differs essentially from this.

Habitat. It clothes rocky ledges to a large extent in the upper part of the sublitoral zone, and stones and Lithothamnia in the lower part of the same region. I have found it down to a depth of 15 fathoms on stony bottom. It seems to prefer sheltered places. I do not know at what season it bears ripe spores. The specimens examined by me, were collected at the end of June and during the latter part of July.

Geogr. Distrib. Hitherto known only from the arctic region of the Polar Sea. Its most northern place of growth is Karmakul Bay on the west coast of Novaya Zemlya, about 72° 30′ N. Lat.

Localities: The Murman Sea: in Karmakul Bay and Kostin Shar on the west coast of Novaya Zemlya, in both places rather plentiful, though local.

Lithothamnion polymorphum (L.) ARESCH.

In J. G. AG. Spec. Alg. 2, p. 524; Millepora polymorpha L. Syst. Nat. p. 1285; ex parte.
Descr. Lithothamnion polymorphum SOLMS-LAUBACH, Corall. Monogr. p. 16—17, sub Lithophyllo incrustante.
Fig. » » HAUCK, Meeresalg. t. 1, fig. 4.
Exsicc. » » ARESCH. Alg. Scand. exsicc. N:o 302.
Syn. Lithothamnion polymorphum KLEEN, Nordl. Alg. p. 11, non ARESCH. Obs. Phyc. 3, p. 5, quoad plantam Spetsbergensem, nec KJELLM. Algenv. Murm. Meer. p. 8, Kariska hafvets algv. p. 15.
» Melobesia polymorpha CROALL, Fl. Disc. p. 459.
» » » DICKIE, Alg. Sutherl. 1, p. 142; 2, p. 192. (?)

Remark on the synonomy. It has been the general practice of algologists to unite all or nearly all crust-like Lithothamnia into a single species, *L. polymorphum*. I have also been guilty of this fault. Having now had the advantage of examining more thoroughly a greater number of specimens, I am however perfectly convinced that such a proceeding is incorrect and that there are among the crustaceous Lithothamnia several well marked and easily characterized species. I have above brought into view some species from the Arctic Sea, and I hope soon to get an opportunity of elucidating the two or three species occurring on the west coast of Sweden, but referred by all Swedish algologists to the same species, *L. polymorphum*. On account of the uncritical treatment these plants have been subjected to, it is impossible, without having access to original specimens, to determine what an author has meant by a plant called by him *L. polymorphum*. A plant thus named has been reported from the American Arctic Sea, but as no specimens of it have been at my command, I refer it only with hesitation to the species of *L. polymorphum*, as I understand it here. This is also the case in some degree with respect to that Corallinea which is mentioned by CROALL in

Florula Discoana under the name of *Melobesia polymorpha*. But having convinced myself, through an examination of the collections of Corallineæ brought home by TH. M. FRIES from the west coast of Greenland, that *L. polymorphum* is really to be found in Baffin Bay, I think it possible indeed that CROALL's determination is quite correct. ARESCHOUG l. c. records *L. polymorphum* from Spitzbergen, though with reservation. It may be that it occurs really here. But as I have not seen it here myself and as it was not to be found among the large collections of Lithothamnia once made here by myself, and as moreover *L. glaciale* at certain stages of its development may exhibit great outward resemblance to *L. polymorphum*, I consider myself justified in excluding it, at least temporarily, from the Flora of Spitzbergen. That *Lithothamnion polymorphum* which I have myself reported from the Murman and the Kara Seas, belongs to other species, as I have already demonstrated.

Habitat. The plant is properly a litoral alga, living principally in rock-pools within the lower part of the litoral zone. However it descends also into the sublitoral zone and is met with even at such a considerable depth as 10—15 fathoms. It flourishes in exposed as well as in sheltered localities and is gregarious, although in the north it does not occur in any very great masses at the same place. I know only sterile specimens from the Polar Sea.

Geogr. Distrib. This species occurs with certainty in the Atlantic as well as the arctic region of the Polar Sea, although there are wide reaches of the latter, where it seems to be replaced by other species. Its maximum of frequences lies probably within the Atlantic region. The question of how far it goes northwards in Baffin Bay, is not yet cleared up. The northermost place where it is known with certainty to grow, is Gjesvær, about Lat. N. 71°.

Localities: The Norwegian Polar Sea: common and pretty plentiful both in the south at Nordlanden and in the north within Tromsö amt and Finmarken amt, as at the town of Tromsö, at Karlsö, Maasö, Gjesvær, the south coast of Magerö (ipse), Ingö, Honningsvaag and Berlevaag (FOSLIE).

The Siberian Sea: Pitlekay, scanty, local.

The American Arctic Sea: Erebus and Terror Bay, Union Bay, Beachey Island, Cape Spencer (?).

Baffin Bay: The west coast of Greenland according to specimens collected by TH. M. FRIES; Disco Isle (?).

Gen. **Lithophyllum** (PHIL.) ROSAN.

Melob. p. 79; Phil. Wiegm. Arch. 1, p. 385; lim. mut.

Lithophyllum Lenormandi (ARESCH.) ROSAN.

l. c. p. 85; Melobesia Lenormandi Aresch. in J. G. AG. Spec. Alg. 2, p. 514.

Descr. Lithophyllum Lenormandi ROSAN. l. c. p. 85.
Fig. » » » » t. 5, fig. 16 et 17; t. 6 fig. 1, 2, 3 et 5.
Exsicc. Melobesia » » HOHENACK. Alg. Mar. N:o 296.
Syn. Lithophyllum Lenormandi GOBI, Algenfl. Weiss. Meer. p. 21.
» Melobesia Lenormandi KLEEN, Nordl. Alg. p. 11.

Habitat. I have never myself met with this plant in the Polar Sea. According to existing statements, it occurs here, fastened to stones and shells, within the litoral as well as the sublitoral zone, in the latter in 5—6 fathoms water. KLEEN has found it with »sporfrukter» (conceptacles of sporangia?) in July and August.

Geogr. Distr. It belongs to the Atlantic region of the Polar Sea and to the neighbouring part of the arctic region. Its northern limit, as far as known at present, is on the coast of Russian Lapland.

Localities: The Norwegian Polar Sea: Nordlanden, common and abundant. The western *Murman* and the *White Seas*, common and abundant.

Lithophyllum Arcticum KJELLM.

Kariska hafvets algv. p. 16.

Descr. Lithophyllum arcticum KJELLM. l. c. p. 16.
Fig. » » » » t. 1, fig. 1—13.
Syn. Melobesia lichenoides DICKIE, Alg. Sutherl. 1, p. 142 (?).

Habitat. It grows gregarious on stony bottom in 5—10 fathoms water, on pretty open coasts, attached to stones and Lithothamnia. Specimens with mature tetrasporangia have been taken in the Kara Sea towards the end of August.

Geogr. Distrib. It is known with certainty only from the Kara Sea at 74° 5′ N. Lat. I think it is probable however that DICKIE's *Melobesia lichonoides* from Baffin Bay is identical with the present species.

Localities: The Kara Sea: Uddebay, plentiful.

Baffin Bay: Fiskernes, Hunde Islands, and Cape Adair (?).

Gen. Melobesia (LAMOUR.) ROSAN.

Melob. p. 53; LAMOUR. BULL. SOC. PHIL. 1812, sec. ROSAN. l. c. p. 60.

Melobesia membranacea LAMOUR.

Hist. Polyp. p. 515.

Descr. Melobesia membranacea ROSAN. Melob. p. 66.
Fig. » » » » t. 2, fig. 13—16 och t. 3, fig. 1.
Syn. Melobesia membranacea KLEEN, Nordl. Alg. p. 11.

Habitat. This species has been found in the Polar Sea, within the upper part of the sublitoral zone, fastened to *Fucus vesiculosus* f. *vadorum;* it has been found here in August with reproductive organs.

Geogr. Distrib. It is known only from the southern part of the Polar Sea, on the coast of Norway.

Locality: The Norwegian Polar Sea: Nordlanden at Fleinvær.

Melobesia macrocarpa ROSAN.

Melob. p. 74.

Descr. Melobesia macrocarpa ROSAN, l. c.
Fig. » » » » t. 4, fig. 2—8 et 11—20.

Syn. Nelobesia macrocarpa KLEEN, Nordl. Alg. p. 11.

Habitat. It has been found in the Polar Sea growing on *Laminaria digitata* within the upper part of the sublitoral zone. Here it has ripe sporangia in July and August.

Geogr. Distrib. It is known only from the Atlantic region of the Polar Sea, on the coast of Norway.

Locality: The Norwegian Polar Sea: Nordlanden.

Melobesia Lejolisii ROSAN.

Melob. p. 62.

Descr. Melobesia Lejolisii ROSAN. l. c.
Fig. » » » » tab. 1, fig. 1—12.

Syn. Melobesia spec. KJELLM. Spetsb. Thall. 1, p. 4. (?)

Habitat. It occurs fastened to *Ptilota plumosa* (and *P. pectinata?*) and accordingly belongs to the sublitoral zone. I have seen specimens with conceptacles of sporangia from Greenland, but I have no information as to the time of the year when they were taken.

Geogr. Distrib. It is known with certainty from the arctic region of the Polar Sea, and it may be supposed with pretty great certainty to occur also within the Atlantic region of it. Besides this, its distribution is uncertain. I have seen specimens from Greenland belonging surely to this species, and this may possibly be the case also with those sterile specimens which I found once on the north coast of Spitzbergen.

Localities: The Greenland Sea: the northern coast of Spitzbergen (?).

Baffin Bay: on the west coast of Greenland.

Fam. RHODOMELACEÆ J. G. AG.

Symb. p. 23; Spec. Alg. 2, p. 787.

Gen. **Odonthalia** LYNGB.

Hydr. Dan. p. 9.

Odonthalia dentata (L.) LYNGB.

l. c. Fucus dentatus L. MANT, p. 35.

Descr. Odonthalia dentata J. G. AG. Spec. Alg. 2, p. 899.
Fig. » » HARV. Phyc. Brit. t. 34.
Exsicc. » » ARESCH. Alg. Scand. exsicc. N:o 56.

Syn. Atomaria dentata RUPR. Alg. Och. p. 209.
» Fucus dentatus GUNN. Fl. Norv. 2, p. 91.
» Odonthalia dentata J. G. AG. Spetsb. Alg. Progr. p. 3, Bidr. p. 11; Till. p. 28.
» » » ARESCH. Phyc. Scand. p. 261.
» » » DICKIE, Alg. Cumberl. p. 238.
» » » EATON, List. p. 44.
» » » GOBI, Algenfl. Weiss. Meer. p. 23.
» » » HARV. Fl. West-Eskim. p. 49.
» » » KJELLM. Spetsb. Thall. 1, p. 5; Algenv. Murm. Meer. p. 9; Kariska hafvets Algv. p. 19.
» » » KLEEN, Nordl. Alg. p. 12.
» » » NYL. et SÆL. Herb. Fenn. p. 73.
» » » POST et RUPR. Ill. Alg. p. II.
» Rhodomela dentata LINDBL. Bot. Not. p. 157.
» » » SCHRENK, Ural. Reise 2, p. 547.

Remark on the forms of the species. Two forms have been distinguished, one with broader, the other with narrower frond, the latter one named f. *angusta* by HARVEY (Fl. West-Eskim.). They pass however so gradually into each other, that no limit can be drawn between them.

Habitat. The present species always grows within the sublitoral zone. In the Norwegian Polar Sea it has been found by KLEEN on deep stony and shelly bottom. I have taken it here myself sometimes luxuriantly developed on gravelly bottom in 10—15 fathoms water, sometimes poorly developed on so-called dead bottom at a depth of 5—6 fathoms, in the former case together with several purely arctic algæ within the formation I have above called the arctic. Within the arctic region of the Polar Sea it belongs chiefly to the formation of *Laminariaceæ*. It is properly a pelagic species, but is nevertheless to be found also in the interior of deep bays, although it is more rare and less richly developed here. Almost without exception it grows scattered. On the west coast of Sweden and the coasts of Great Britain it bears spores in winter. In the Polar Sea I have not met with any specimens with reproductive organs during that time, but I have found individuals with tetraspores in summer, in August. RUPRECHT mentions having collected such specimens in the month of June at Triostrowa (RUPR. Alg. Och. p. 212). In some specimens taken at Finmarken in the interior of Altenfjord at the end of August, there are to be seen the beginnings of sporocarps. Hence it seems as if the present species should develop its organs of propagation at another season in the Polar Sea than farther to the south.

Geogr. Distrib. The plant is circumpolar. I have found it most common and luxuriant at Gjesvær in the Norwegian Polar Sea and in the eastern part of the Murman Sea. Its northernmost locality is Treurenberg Bay on the north coast of Spitzbergen 79° 56′ N. Lat.

Localities: The Norwegian Polar Sea: Nordlanden, scanty, local; Finmarken at several places, but local and not plentiful, as at Maasö, Gjesvær, Talvik; Vardö (GUNNERUS).

The Greenland Sea: scarce and local along the west and north coasts of Spitzbergen; Beeren Eiland.

The Murman Sea: commonly spread, as on the Murman coast, Cisuralian Samoyede-land, the west coast of Nowaya Zemlya rather abundant, Kolgujew Isle, the main land at Jugor Shar.

The White Sea: common and abundant according to GOBI.

The Kara Sea: scanty in Uddebay on the east coast of northern Novaya Zemlya.

The Siberian Sea: scarce at Irkaypi.

The American Arctic Sea: Western Eskimaux-land; Hudson Strait.

Baffin Bay: Cumberland Sound, rather rare.

Gen. **Rhodomela** (AG.) J. G. AG.

Sp. Alg. 2, p. 874; AG. Spec. Alg. 1, p. 368; ex parte.

Rhodomela lycopodioides (L.) AG.

l. c. p. 377; Fucus lycopodioides L. Syst. Nat. 2, p. 717.

f. *typica*

α. *compacta* nob.

Descr.	Rhodomela	lycopodioides	J. G. AG. Spec. Alg. 2, p. 885.
Fig.	»	»	HARV. Phyc. Brit. t. 50.
Exsicc.	»	»	ARESCH. Alg. Scand. exsicc. N:o 3.

β. *laxa* nob.

f. fronde quam in priore laxius ramosa, axi principali et inferne et superne ramos longiores emittente, ramulis laxius dispositis. Tab. 9, fig. 1.

γ. *tenera* nob.

f. fronde 15—30 cm. alta, tenera et flaccida, axi primario 5—6 cm. longo, residuis ramorum dejectorum et ramis brevibus, rigidis, cylindricis, basi plus minus attenuatis, adpressis, densissime vestito, ramosque emittente nonnullos longiores, flaccidos, systemata ramorum breviora, oblongo-lanceolata, laxe disposita et ramulos simplices pauciores apice et basi attenuatos gerentes. Tab. 9, fig. 2.

Description. The form γ *tenera* becomes nearly a foot long, and is flaccid and slender. The frond has a short, generally 5—6 cm. long, main axis which is densely beset with short, 4—5 mm. long, rigid, appressed secondary axes, some of which are plainly remains of branches, while others are side-axes that have stopped in their growth (leaves). The latter have generally a cylindrical or almost spindle-shaped form, tapering commonly somewhat towards the base. From the remains of the branches new side-axes may be developed. Besides, there issue from the short main axis, which has ceased to grow longer, one or more, but commonly only a little number, of long, flaccid, slender, repeatedly racemose branch-systemes, which are linear-lanceolate in circumference, and have a distinct main axis, beset with short, scarcely 2 cm. long, thin branchsystems of the second order, oblong-lanceolate in periphery, and few, scattered, simple branches, often incurved in the shape of a sickle and attenuated towards the base and the tip. In other specimens the main stem puts forth some few branches dissolved at the top and resembling the main stem. In this case it is these that give rise to the long, flaccid branch-systems.

f. *cladostephus* J. G. AG. (KJELLM.)

Spetsb. Thall. 1, p. 8; Rhodomela cladostephus J. G. AG. Spetsb. Alg. Till. p. 48.

α. densa nob.

Descr. Aphanarthron cladostephus J. G. AG. Spetsb. Alg. Bidr. p. 8—9.
Fig. » » » » » t. 2.

β. distans nob.

f. quam prior laxius ramosa, ramulis longioribus, magis distantibus.

f. *setacea* nob.

f. fronde usque 15 cm. alta, fusco-purpurea, siccata subfusca; axi primario plus minus distincto, setaceo, ramos breves nonnullos, basi subattenuatos, subfalcatos et longiores vel simplices vel apice parce vel decomposito subcorymboso-ramulosos ramulis longioribus emittente. Tab. 9, fig. 3.

Description. As far as I know, this plant attains no more than 15 cm. in height, most often less. Its main axis is scarcely thicker than a bristle, of a dark red-brown colour, which in drying grows brownish with a darkish tint. It is brittle, and the younger branches become flattened in drying. It is nearly corymbose in circumference. The main axis, issuing from a callus radicalis, can sometimes be traced throughout the whole frond, sometimes only a bit upwards, the latter condition of the plant being caused by one branch-system or some branch-systems being developed above their branching-point as strongly as the main axis. The elements of ramification are 1:o) short branches more or less sickle-shapedly incurved, tapering towards the base and the tip; these are few in number; 2:o) simple, slenderly spindleshaped-cylindrical branches; 3:o) long branch-systems whose main axis is most often unbranched in the greater part of its length, bearing only towards the top some short, racemosely arranged, simple or scantily branched secondary axes, and 4:o) long, profusely branched, fastigiatedly developed branch-systems, composed of elements 2 and 3. These elements are more or less combined into denser or thinner, fastigiatedly developed branch-systems of higher orders. Reproductive organs are unknown, but it seems as if the axes of the last order and the last order but one should produce tetrasporangia near their tops. In structure the present form differs but little, at least when older, from the typical *Rh. lycopodioides.*

f. *flagellaris* nob.

f. quam prior parcius ramosa, ramis longioribus. Tab. 10, fig. 1—2.

Description. The plant reminds one much of the preceding one and, no doubt, stands near this. It is less tinged with brown and retains its red colour in drying. The mode of ramification is the same as in the preceding form, the difference consisting in the branching of f. *flagellaris* being less decompound than in f. *setacea.* The peculiar aspect of the present form, is principally effected by the long branches of the last order. In structure it approaches nearly to other forms, especially when these are younger (fig. 1—2).

f. *tenuissima* RUPR. (nob.)

Fuscaria tenuissima RUPR. Alg. Och. p. 221.

α. prolifera nob.

f. fronde prioris anni a ramis persistentibus prolifera.

Descr. Fuscaria tenuissima RUPR. l. c.
Fig. » » » tab. 10.

β. glacialis nob.

Descr. Rhodomela tenuissima KJELLM. Spetsb. Thall. 1, p. 6.
Fig. » » » » » tab. 4, fig. 1—2.
Exsicc. » » » in ARESCH. Alg. Scand. exsicc. N:o 402.

Syn. Aphanarthron cladostephus J. G. AG. Spetsb. Alg. Bidr. p. 9.
Fucus lycopodioides GUNN. Fl. Norv. 2, p. 80.
» » Wg. Fl. Lapp. p. 505.
» subfuscus » » » » » Cfr. ARESCH. Obs. Phyc. 3, p. 7.
Fuscaria tenuissima RUPR. Alg. Och. p. 221.
Rhodomela cladostephus J. G. AG. Spetsb. Alg. Till. p. 48.
» lycopodioides J. G. AG. Grönl. Alg. p. 111.
» » ARESCH. Phyc. Scand. p. 262.
» » KJELLM. Spetsb. Thall. 1, p. 8; Algenv. Murm. Meer. p. 10
» » KLEEN, Nordl. Alg. p. 12.
» » Nyl. et Sæl. Herb. Fenn. p. 74.
» subfusca (var?) J. G. AG. Spetsb. Alg. Progr. p. 3; Bidr. p. 11.
» » DICKIE, Alg. Cumberl. p. 238.
» » GOBI, Algenfl. Weiss. Meer. p. 24.
» » KJELLM. Spetsb. Thall. 1, p. 5.
» » KLEEN, Nordl. Alg. p. 12.
» » POST. et RUPR. Ill. Alg. p. II.
» tenuissima KJELLM. Vinteralg. p. 64; Spetsb. Thall. 1, p. 6; Algenv. Murm. Meer. p. 10 et Kariska hafvets Algv. p. 19.

Remark on the species. GOBI, in his account of the Flora of the White Sea, has defended the opinion entertained also by other algologists, that *Rhodomela lycopodioides* is not specifically distinct from *Rh. subfusca* (WOODW.) AG. In support of this opinion he alleges, firstly that these two forms of Rhodomela, taken for different species, agree perfectly in anatomical structure and differ only by outward characteristics, amongst which is a somewhat different variety of colour, secondly that there are to be found transitions between them, as has been pointed out already by HARVEY, and lastly that *Rh. subfusca*, whose distribution is more southerly, changes gradually in habit as it advances towards the north, passing into that *Rh. lycopodioides* which is characteristic of higher latitudes. Amongst the transitional forms that *Rhodomela* ought to be numbered according to GOBI, which has been named *Fuscaria tenuissima* by RUPRECHT and whose claim to be regarded as a separate species I have tried before to justify.

I cannot but accede to this view so far on the one side, as to allow the claim of *Rh. tenuissima* to be considered as a separate species to fall. My observations and studies of its forms as occurring in the south-eastern part of the Siberian Sea, compel me to admit that this plant, so sharply distinguished in its purely arctic form from *Rh. lycopodioides*, is yet so nearly related to it that no definite limits can be drawn between them. Besides, I must concede, on the other side, that the uniting of *Rh. lycopodioides* and *Rh. subfusca* insisted on by GOBI, however unnatural it may appear at first, has nevertheless a great probability, considering the number of diverse forms under which *Rh. lycopodioides* appears, several among which come extremely near *Rh. subfusca* (WOODW.) AG. However, in this admission I must make a decided restriction. By the excellent works of J. E. ARESCHOUG Swedish algologists have been made well acquainted with the fact that there are to be found on the west coast of Sweden three forms of *Rhodomela*, one characterized by the just mentioned algologist as *forma extratæniensis et normalis*, the second as *forma intratæniensis præcedentis magnitudinis et crassitiei*, the third as *forma gracilis*. ARESCHOUG regards all three as forms of *Rhodomela subfusca* (WOODW.) (see ARESCH. Obs. Phyc. 3, p. 6). All have been distributed in magnificent specimens in Alg. Scand. exsicc. under N:o 57, 58, 303, Ser. 2 and N:o 54 Ser. 1. I have had an opportunity of more closely examining living specimens of the two first of these in different stages of development and at different times of the year, and I have found them to disagree so essentially in habit, morphological development, anatomical structure, and biological conditions, that I must account them different species, if indeed any species of *Rhodomela* are to be distinguished at all. Only one of these two can be regarded as nearly related to *Rh. lycopodioides;* the other is surely sharply distinct from it. In order to be able to expose this question more fully, I think best to give already here a description of these two Swedish species of *Rhodomela*, although this does not belong strictly to the immediate subject of this work. I shall begin with the above-mentioned *forma intratæniensis*. As far as I can see, it has not been described or figured under any specific name, at least not so as to be recognizable. I propose to name it *Rh. virgata*.

Rhodomela Virgata novum nomen.

Tab. nostra 7. Exsicc. ARESCH. Alg. Scand. exsicc. N:o 303.

Description. A spring plant of the first year, according to a specimen from Bohuslän taken in the middle of May; fig. 1. About 20 cm. high; when dry, with flat main stem and flat primary branches, which do not become black; of a red-brown colour. The hold-fast is a *callus radicalis*. The frond is distinctly repeatedly racemosely branched. A main axis is distinguishable throughout the whole frond; it attains its greatest thickness at the middle, tapering rather swiftly towards the summit, gradually towards the base, abruptly only in the vicinity of the hold-fast. Throughout its whole length, it bears branch-systems, diminishing upwards in length and strength. The lower ones have a lanceolate, the upper ones an ovato-triangular periphery. The main axes of the lower larger branch-systems are thickest at the middle and taper strongly towards the tip as well as towards the base. At the point where they branch off from the main stem, their thickness is considerably smaller than that of the stem. The lower lateral axes of the first order carry few and very finely decompound, short, branch-systems of the second order, which are ovate or ovato-triangular in circumference, of pretty equal size, very thin at the lower part of the side axis, somewhat denser upwards, although even here they are thin. With these branch-systems of the second order those systems of the first order, that issue nearer the summit of the main stem, agree. Branches of a higher order than the third are rare. The branches of the last order are fine as a hair, and there is a great difference in thickness between branches of the first and the second order.

The summer and autumn plant of the first year presents the aspect shown by fig. 2. This is produced by the upper branch-systems and all lower branches of a higher order than the first in the spring plant, having fallen off either completely or so that only the lowest basal parts remain. The frond accordingly consists here of the main axis of the frond and the lower side-axes of the first order, all of which have increased in solidity and thickness.

Older sterile specimens in spring. The plant reaches a considerable size, at least 1—2 feet high, very bushy. In ramification, older individuals agree with a plant of the first year's growth, differing from it only by the ramification being more decompound. At the commencement of the new season, the organs of reproduction being developed, new branch-systems spring out from the perennial portions, sometimes resembling the primary branch-systems of the young plant, sometimes the whole plant of the first year, and sometimes being even more decompound than this.

Older individuals in autumn resemble younger individuals at the same time of the year, differing only by being larger, more robust, and more decompound.

Individual with antheridia, sporocarpia and tetrasporangia. I have found such individuals only during winter, in the months of December and January. They agree with the autumn plants, except in the surviving parts being more or less densely covered with short, decompound branch-systems, sometimes single, sometimes gathered into thin tufts, issuing without visible order, ovate in circumference, with scarcely distinguishable main axis; fig. 3. When grown to the length of 2—3 mm., these already bear ripe antheridia or sporocarpia in various stages of development, from recently etablished until almost ripe; fig. 4—5. Certain tetrasporangic branch-systems or stands of tetrasporangia resemble the stands of antheridia and sporocarpia, and carry ripe tetrasporangia, even when only about 2 mm. long; fig. 6. Others are longer, less metamorphosed, with certain axes sterile, growing and branching, while the other side-axes bear a few tetrasporangia. I do not know for certain, which is the ultimate fate of these metamorphosed branch-systems thrown out by the autumn plant for the development of the organs of reproduction; but I have reason to believe that they are dissolved or fall off, when their functions are accomplished. However, one finds now and then, although rarely, at the end of spring or the commencement of summer, specimens that differ in habit from the common spring plants by those portions, which have persisted through the winter, bearing bushy, long- and richly branched, not distinctly racemose branch-systems, whose axes throughout their whole length are beset with thinly scattered, very short, blunt, sometimes slightly club-shaped processes. These seem to point to these branch-systems being stands of sporocarpia grown out, possibly such whose carpogon has not, from some cause or other, come to a normal development. It is also possible, that in certain cases the branch-systems destined to the formation of tetrasporangia may, after some few reproductive cells are produced, develop themselves for vegetative purposes.

The structure of the frond. Fig. 7 is a part of a cross-section, fig. 8 of a longitudinal section, of the main axis of the frond, near the base, of an older specimen. These sections show that the largest part of the frond consists of a parenchymatous tissue, whose thin-walled cells, that are destitute of or very poor in endocrom, diminish gradually in size from within outwards, where they are surrounded by a cortical layer of small cells rich in endocrom, which layer is sharply defined from the inner mass of the tissue. The pericentral siphons as well as the central siphon are of inconsiderable width.

The organs of propagation. The sporocarpia are ovato-urceolate, with a short neck. The spores are pyriform, about 100 μ. long and 50 μ. thick. The tetrasporangia are large. The antheridia are slenderly cylindro-conical, greatly variable in size, but in general about 100 μ. in diameter near the base.

The other species of *Rhodomela* from Bohuslän, the *Rh. subfusca f. extratæniensis vel normalis* of J. E. ARESCHOUG is clearly identical with *Fucus subfuscus* WOODW., TURN., *Stackh.* and *Fl. Dan.*, and accordingly ought to be called *Rhodomela subfusca* (WOODW.) AG. It is certainly a well-known species,. But in order to point out its discrepancies from the preceding one, I give figures on plate 8 referring to it, together with a description of specimens from Bohuslän.

A first year's plant, sterile, from a specimen from Bohuslän, taken in the month of December; fig 1.

The plant becomes altogether black in drying and adheres firmly to the paper. The axes retain their terete form, or are at least only almost imperceptibly compressed. The hold-fast is a callus, from which issue oftentimes several systems of axes, in general unequally developed. My description refers to an one-stemmed, thinly branched specimen. A main axis is traceable only for some distance upwards in the frond. As elements of ramification we may consider, I think, 1) short cylindrically subulate branches, about 5—6 mm. in length; 2) fastigiate branch-systems of the same length with the preceding, whose primary axis bears a few side-axes only near the top; 3) branch-systems, 5—8 cm. long, generally slenderly lanceolate, whose main axis bears side-axes of the two preceding types. These elements are combined in a more or less distinctly racemose manner. The racemose arrangement often becomes, however, difficult to follow, because the main stem, as well as the primary axes in the branch-system of the first order, becomes sooner or later untraceable, a branch-system of the next higher order being developed of equal strength with the general axis above the point where the branch-system springs out. The difference in thickness between axes of next different orders in not considerable. The varieties I have observed with regard to the above-mentioned ramifications, are limited to the elements 1 being sometimes more numerous, sometimes less numerous than elements 2, and to elements 3 being now and then shorter than I have stated above.

In summer the plant always presents an aspect of this kind. J. E. ARESCHOUG in Alg. Scand. Exsicc. Ser. 2. N:o 57 has distributed such specimens, collected in August. Only exceptionally one finds some specimens of this kind during winter.

The plant in winter habit. During winter and the earlier part of spring the plant has the appearance shown by fig. 2 and produced by all the elements of ramification being more or less completely dissolved. The elements 1 and 2 are most strongly resorbed. Specimens of this kind are very common on the coast af Bohuslän in winter during the months of December and January.

The plant in spring habit. Fig. 3. Cp. ARESCH. Alg. Scand. exsicc. Ser. 1. N:o 54. From the portions that have persisted through the winter, branch-systems, sometimes scattered, sometimes somewhat tufted, are developed, which produce sporocarpia and tetrasporangia. These systems are decompound, with a corymbose development, and in this species attain a more considerable size before the ripeness of the spores, than in the preceding one. I do not know any antheridia in this species. I have taken specimens with ripe sporocarpia in May, with ripe tetraspores in April.

The structure of the frond. The figures 4 and 5, both representing sections of the lower part of the frond, show that outside the siphons there begins a mighty layer of large-celled parenchyma, sharply defined without against a small-celled layer of tissue, that is also mighty and passes without marked limit into the cortical layer. All cell-walls are thick. The large-celled parenchyma is destitute of or poor in endochrome, the small-celled is rich in endochrome.

It is evident, that of these two species *Rh. virgata* has nothing to do with *Rh. lycopodioides*. *Rh. subfusca*, on the contrary, presents so great a resemblance to certain forms of this species, especially f. *typica β. laxa*, that it may be questioned whether they are indeed specifically distinct. Both have very often been confounded with each other. All the specimens of the so-called *Rh. subfusca*, brought home by KLEEN from Nordlanden and come under my notice, are undoubtedly forms of *Rh. lycopodioides;* and that plant from the coasts of Spitzbergen, which I have mentioned under the name of *Rh. subfusca*, I must now allow to be a form of *Rh. lycopodioides*. Many instances of that kind might be quoted. On that account, one might be inclined, like GOBI, to unite these two *Rhodomelæ* and to regard *Rh. subfusca* as a southern form of the other. But, on the other hand, it is remarkable, that both the forms occur quite characteristical on the coasts of England, and that on the coast of Sweden *Rhodomela subfusca*, in whatever localities it may grow, whether near the surface or in deep water, is constantly alike in form, and, above all, never appears here in any densely branched *compacta*- or *densa*-form; whereas *Rh. lycopodioides* on the coast of Norway, when growing between tide-marks, exhibits regularly the form *typica compacta*, but in other cases assumes readily the aspect of f. *typica β laxa*, which proves that these two species or forms vary in a different manner. I must, moreover, call attention to a difference between them, which, as far as my researches go, has shown itself to be universal and constant. *Rh. lycopodioides*, in whichever of its numerous and extremely variable forms it may occur, always bears on its more robust axes short, slightly bow-

or sickle-shaped, erect ramuli, which are thickest a little below the middle and attenuated towards the base. They occur without any apparent order, sometimes very numerous, sometimes very few, and are, as far as I have been able to see, a sort of adventive branches. A good figure is to be found in J. G. AG. Spetsb. Alg. Bidr., pl. 2, fig. 2. In *Rhodomela subfusca* there are certainly branches, that remind one of these, viz. those above indicated as elements 1 of the ramification, but these are subulate or cylindrically subulate, issuing from broad bases and always developed in strictly acropetal order. I have never seen any formations resembling those of *Rh. lycopodioides* on the considerable number of *Rh. subfusca* from Bohuslän, that I have examined. These circumstances seem to me to imply, that *Rh. lycopodioides* and *Rh. subfusca* are two distinct, although feebly differentiated species, which have possibly once sprung from one type, but afterwards developed differently.

GOBI's opinion that different species should be perceptibly unlike in anatomical structure can scarcely be regarded as defensible. If such a condition should be carried out in algology, a considerable number of species, constant, easily recognizable, and regarded as good, must be suppressed and subsumed under others in long series. External morphological diversity ought certainly even here to be considered valid as a character of species. Small anatomical diversities are indeed to be found even between *Rh. lycopodioides* and *Rh. subfusca*, but the anatomical structure of both species being essentially different in different, older or younger, portions of the frond, and the different forms of what is undoubtedly *Rh. lycopodioides* being also somewhat different from one another in this respect, it is necessary to examine a great many specimens of different ages and places of growth, in order to be able to ascertain what is essential or unessential. I have not had an opportunity to undertake such an examination, and I am thus obliged to confine myself to stating that, with regard to structure, *Rh. subfusca* and *Rh. lycopodioides* are very similar to each other, and differ essentially from *Rh. virgata.*

Remark on the form. As appears from the list of synonyms, I have united in one species all that *Rhodomela* which has been reported from the Arctic Sea. I have thus regarded as variations of the same type a great many forms that, at first sight and in their extremes, differ most considerably from the typical form. As far as I can judge, there is ta present no other course left. The principal forms, that I have tried to discern, do not stand isolated, but are combined with one another by more or less numerous intermediate forms. The two forms which I have called f. *flagellaris* and f. *tenuissima β glacialis* are those most unlike the typical one. To the former one of these I have not before been able to assign a place, but have mentioned it as an alga *incertæ sedis* in my description of the marine algæ of Spitzbergen; cf. KJELLM. Spetsb. Thall. I. p. 33. Having since that time become acquainted with the form named above f. *setacea*, which, as is shown by the figures given, stands undoubtedly near f. *flagellaris*, I do not hesitate to regard it as a peculiarly developed *Rh. lycopodioides*. That f. *setacea* belongs to the series of forms of *Rh. lycopodioides*, is shown by a comparison of figures 1 and 3 on plate 9.

I formerly regarded *Rh. lycopodioides* f. *tenuissima β glacialis* as a good species, but, having found its subform *prolifera*, which merges in *Rh. lycopioides* f. *typica β laxa*, I can no longer maintain this opinion. It is a high-arctic form, adapted to the peculiar physical circumstances of the Polar Sea. My opinion of f. *cladostephus α densa* I have stated Spetsb. Thall. 1. p. 8, and I have found no reason to abandon it. But later observations have convinced me, that that plant from Spitzbergen which I once determined as *Rh. subfusca* must be regarded as a form of *Rh. lycopodioides*. It is allied with its f. *cladostephus α densa*, just as f. *laxa* of the typical form of the species is allied with the subform *β compacta*.

That these two forms are very nearly related, will be doubted by no one that has had an opportunity of seeing a greater number of specimens of both of them. They are connected by numerous intermediate forms. The elegant subform *γ tenera* is assuredly another variety, of the same value as these. It is a *Rh. lycopodioides* f. *typica*, that has grown in brackisk or almost fresh water.

Hab. It is a priori probable, that a plant appearing in so many different shapes will present many diversities even in its habitat. This is indeed the case. Still it is remarkable, that one form at least, although exposed to very different external conditions, retains its characteristic appearance in a most noteworthy degree. I have already mentioned that this is the case with f. *typica α congesta*. There are specimens of it, grown within the sublitoral zone on the west coast of Novaya Zemlya, that cannot be distinguished from specimens grown on rocks between tide-marks on the coast of Norway. It is also noticeable, that this form and the very nearly related *Rh. lycopodioides* f. *cladostephus α densa* are often more luxuriant than the typical form on the north coast of Norway. This is rather reduced in size, and generally keeps about the length of 5—10 cm., scarcely ever reaching 15 cm. On the west coast of Norway it becomes at least 25 cm. long, on the coast of Britain still larger, upwards of 2 feet, according to HARVEY. In order to give a clear idea of the habitat of this plant, it will be most fit to treat of each form separately.

Rh. lycopodioides f. *typica*. In the Norwegian Polar Sea, where this is the most common form, it occurs almost always within the litoral zone, partly on rocks between tide-marks, partly in tide-pools. In the former case, it always appears as *α compacta*, in the latter as well as when descending rarely into the upper part of the sublitoral zone, it sometimes, though far from always, has the aspect of *β laxa*. When growing in places where the saltness of the water is little, it invests itself with the peculiar habit of *γ tenera*. On the west coast of Greenland it is now litoral, now sublitoral. In the arctic region proper of the Polar Sea, the plant is always sublitoral and belongs here to the formation of Laminariaceæ. It is properly a pelagic form; however it enters also into deep bays, although, according to my experience, it is far more rare here than on unsheltered coasts. In the Norwegian Polar Sea it is gregarious, but not in the Arctic Sea proper. I have not seen specimens with sporocarpia from the Arctic Sea. Individuals bearing tetrasporangia have been found on the coast of Norway in the months of June, July, and August; on the west coast of Novaya Zemlya I have found such specimens in July.

Rh. lycopodioides f. *cladostephus* resembles the preceding form with respect to its habitat, when it occurs within the arctic region of the Polar Sea.

Rh. lycopodioides f. *setacea* I have found only in one single place. It grew attached to stones, in the interior of the deep Altenfjord, within the litoral zone at the mouth of a stream. When gathered at the end of the summer, it was sterile.

Rh. lycopodioides f. *flagellaris* has been taken, in the latter part of July, within the sublitoral zone, on shingly bottom on the north coast of Spitzbergen, in the interior of a bay. It was then sterile.

Rh. lycopodioides f. *tenuissima*. This form, the most common in the Arctic Sea proper, always grows within the sublitoral zone, from near its upper limit to about 6—8 fathoms, attached to small stones, shells, or large algæ. It is not seldom gregarious and very large masses of individuals occur within small areas. It grows on open as well as sheltered coasts. On the north coast of Spitzbergen it is found throughout the whole winter and is in development during the whole year, although only with the month of March a stronger and livelier formation of new vegetative parts sets in. On the north-east coast of Siberia it has a period of rest during some part of the year. This seems to be broken towards the end of June, as I infer from the fact that in a great number of specimens which I had the advantage of examining daily from the 7:th to the 14:th of July, the older portions, that had persisted through the winter, were clothed with new branch-systems in the first stage of development. I do not know at what time the period of rest commences. Specimens taken in the middle of September had already assumed their winter habit. At Spitzbergen the plant probably bears its organs of reproduction, sporocarpia and tetrasporangia, during the whole year. I have found specimens furnished with either of these organs, in all the months of the year except May. They are most richly developed during the latter part of July, during August, November and the commencement of December; however, specimens with copious sporocarpia and tetrasporangia were found even in January. On the west and east coasts of Novaya Zemlya I have collected specimens with sporocarpia at the end of June and the beginning of August, and specimens richly furnished with tetrasporangia in the middle of July. Some specimens taken in the eastern part of the Siberian Sea in the earlier half of July had no tetrasporangia on the new branch-systems sprung from the winter plant.

Geogr. Distr. The species, as understood in the above-mentioned comprehensive sense, is probably circumpolar. But it is not as yet known from the American Arctic Sea. The most northern place where it has been found, is Treurenberg Bay on the north coast of Spitzbergen near Lat. 80° N. Within the Atlantic region of the Arctic Sea the typical form predominates, within the arctic region proper f. *tenuissima*, which is to be regarded as one of the most characteristic algæ of this region.

Localities: The Norwegian Polar Sea: at Nordlanden common and abundant (f. *typica*); at Finmarken common, but often in rather small numbers, as at Maasö, Oxfjord, Talvik; at some places abundant, for inst. at Gjesvær and on the south coast of Magerö (f. *typica*); at Talvik also f. *setacea* was found.

The Greenland Sea: f. *cladostephus* scarce and local, f. *tenuissima* common and abundant on the north and west coasts of Spitzbergen; f. *flagellaris* at Treurenberg bay.

The Murman Sea: f. *typica*, on the coast of Russian Lapland and the west coast of Novaya Zemlya, at the latter place local and scarce; f. *cladostephus*, on the west coast of Novaya Zemlya, more common and abundant than the preceding form; f. *tenuissima*, on the coast of Cisuralian Samoyede-land, on the west cost of Novaya Zemlya and Waygats, here common and abundant.

The Kara Sea, on the east coast of Novaya Zemlya, on the northern coast of Siberia at Cape Palander and Aktinia bay, everywhere scarce.

The Siberian Sea: f. *tenuissima*, at Irkaypi scarce, at and about Pitlekay common and abundant.

Baffin Bay: f. *typica*, at Cumberland Sound common; on the west coast of Greenland in several places, as at Lichtenau, Julianeshaab, Godthaab, Sukkertoppen, Holstenborg, Godhavn and Rittenbenk.

Rhodomela larix (TURN.) AG.

Spec. Alg. 1, p. 376: Fucus larix TURN. Hist. Fuc. 4, p. 23.

Descr. Rhodomela larix J. G. AG. Spec. Alg. 2, p. 886.

Fig. Fucus larix TURN. l. c. t. 207.

Syn. Rhodomela larix HARV. Fl. West.-Esk. p. 49.

Hab. Nothing is known about the habitat of this species in the Arctic Sea.

Geogr. Distr. I have not myself had the opportunity of seeing this species in the Arctic Sea. It seems to be of very local occurrence here and to belong only to that part of it, which lies to the north-east of Behring Strait, having probably immigrated there from the Behring Sea, where it occurs abundantly even far towards the north.

Locality. According to SEEMANN (l. c.) it has been found in that part of the American Arctic Sea, which stretches along the north coast of western Eskimaux-land.

Gen. **Polysiphonia** GREV.

Fl. Edinb. p. 308, sec. J. G. AG. Spec. Alg. 2, p. 900.

Polysiphonia parasitica (HUDS.) GREV.

Fl. Edinb. p. 309; Conferva parasitica HUDS. Fl. Angl. p. 604.

f. *typica*.

Descr. Polysiphonia parasitica J. G. AG. Spec. Alg. 2, p. 930.

Fig. » » HARV. Phyc. Brit. t. 147.

Exsicc. » » CROUAN, Alg. Finist. N:o 315.

Syn. Polysiphonia parasitica KLEEN, Nordl. Alg. p. 14.

Habitat. It grows attached to shells in several fathoms water together with *Plocamium coccineum*. Only found sterile in the Polar Sea.

Geogr. Distrib. It belongs to the Atlantic region of the Polar Sea.

Locality. It has been taken by KLEEN in the southern part of the Norwegian Polar Sea at the Givær isles in Nordlanden.

Polysiphonia urceolata (LIGHTF.) GREV.

Fl. Edinb. p. 309. Conferva urceolata LIGHTF. in Dillw. Intr. p. 82.

f. *typica*.

Descr. Polysiphonia urceolata α urceolata J. G. AG. Spec. Alg. 2, p. 970.
Fig. » » HARV. Phyc. Brit. t. 167.
Exsicc. » » ARESCH. Alg. Scand. exsicc. N:r 68.

f. *roseola* AG. (J. G. AG.)

l. c. p. 971; Hutchinsia roseola AG. Spec. Alg. 2, p. 92.

Descr. Polysiphonia urceolata ε roseola J. G. AG. l. c.
Fig. » formosa HARV. Phyc. Brit. t. 168.
Exsicc. » roseola ARESCH. Alg. Scand. exsicc. N:o 69.

Syn. Conferva stricta WG. Fl. Lapp. p. 512.
Polysiphonia pulvinata GOBI, Algenfl. Weiss. Meer. p. 25, excl. syn.
» roseola POST. et RUPR. Il. Alg. p. II, see. GOBI, l. c.
» » Nyl. et Sæl. Herb. Fenn. p. 74.
» urceolata CROALL, Fl. Disc. p. 459; ex parte.
» » DICKIE, Alg. Sutherl. 2, p. 191.
» » GOBI, Algenfl. Weiss. Meer, d. 26.
» » KLEEN, Nordl. Alg. p. 13.

Remark on the definition of the form. According to my experience, there are to be found in the Polar Sea only two forms of this species, of which the one is identical with that distributed by ARESCHOUG in Alg. Scand. exsicc. N:o 68 under the name of *P. urceolata*, the other with the plant called *P. roseola*. KLEEN mentions certainly that there occur in the Norwegian Polar Sea two other forms, f. *patens* and f. *formosa*, but there are not, in the collections rich in specimens of *P. urceolata* which he has brought home from there, to be found any specimens that I think can be referred to the varieties so called. Some specimens show indeed differences from one another, but these are not so distinctly marked as to make it possible to draw any definite limit. There are some specimens differing in many respects from the typical *P. urceolata*, but the peculiar development of these ought rather to be regarded as a monstrosity than as a difference of type. These are densely tufted, more robust than the typical specimens; the upper main branches carry dense, corymbose clusters of branches, whose secondary branches are short, robust, curved backwards or angularly, closely compact and connected with one another by peculiar fastening-organs. These are sometimes longer sometimes shorter, hyaline, unicellular, with very thick walls, and terminate in a fastening-disk with crenate margin.

Remark on the synonymy. I have referred to *P. urceolata* the plant recorded by GOBI in his Algenflora des weissen Meeres under the name of *P. pulvinata* J. G. AG. Spec. Alg. p. 957 and ARESCH. Alg. Scand. Exsicc. N:o 67. J. G. AGARDH l. c. has already identified the plant called by him *P. pulvinata* AG. with that mentioned by ARESCHOUG in Phyc. Scand. p. 270—280 under the name of *P. pulvinata* ROTH, and I suppose it is in consequence of this that this plant is stated to grow in *Sinus Codanus*.

Gobi on this point follows the example of J. G. Agardh. However, these two plants cannot possibly be identified, because they differ essentially in structure from each other.

Polysiphonia pulvinata J. G. Ag. and Gobi, according to both authors, has 4-siphonic articles, whereas Areschoug's *P. pulvinata* in Phyc. Scand. is 6-siphonic. He says expressly »Interstitia sub microscopio visu tristriata» and the specimens distributed by him in Alg. Scand. exsicc. Ser. 1. N:o 60, to which he refers, possess indeed 6-siphonic articles. The same plant has been distributed by him afterwards in the second series of this work of exsiccatæ N:o 67. This is quoted by Gobi as identical with his *P. pulvinata* from the White Sea. All the specimens of that *P. pulvinata* of Areschoug, which I have had an opportunity of examining, have 6 pericentral siphons, differing thereby from the *P. pulvinata* Gobi found in the White Sea. That this structure is an essential characteristic of *P. pulvinata* Aresch., is evident by this author's detailed description of it in Obs. Phyc. 3. p. 7—8, where it stands under the name of *P. hemisphærica* Aresch., Syn. *P. pulvinata* Aresch. Phyc. Scand. p. 57, Alg. Scand. exsicc. Ed. I. N:o 60 and Ed. II. N:o 67. I dare not allege with certainty that *P. pulvinata* J. G. Ag. does not occur on the coast of Scandinavia. Areschoug neither records it in Phyc. Scand. nor did he mention it as Scandinavian in his public lectures on the algæ of Scandinavia delivered some years ago. I have myself never seen any plant, neither at Bohuslän nor on the coast of Norway, that might be identified with *P. pulvinata* J. G. Ag. But on the other hand I have found at several times on the west coast of Scandinavia a *Polysiphonia* much resembling in habit *P. pulvinata* i. e. *P. hemisphærica* Aresch. Like this, it forms very dense, nearly hemispherical tufts, which assume a brownish colour in drying. Like this, it possesses a dense plexus radicalis, formed of the prostrate, intertwisted, lower parts of the frondal axes, which throw out short, hyaline rhizines furnished at the top with a crenate, scutiform fastening-disk. It is, however, always 4-siphonic and passes by plainly intermediate forms into the typical *P. urceolata*. I think it is a *P. urceolata* of this kind from the Polar Sea that Gobi has seen and determined as *P. pulvinata* J. G. Ag. This seems to be indicated, besides by Gobi's decided statement as to its having 4 pericentral siphons, by the fact of its constituting »*ziemlich dichte Büschel*» — *P. pulvinata* J. G. Ag. is densely tufted — and of such an experienced algologist as Ruprecht having called it *P. roseola* Ag. Cp. Gobi l. c. p. 26, note. *P. stricta* Croall vide *P. arctica*.

Habitat. The present plant is properly and usually litoral in the Polar Sea, but it occurs also within the sublitoral zone, even descending to its lower limit. I have taken it in Finmarken in 15—20 fathoms water, but it was usually met with in the lower part of the litoral zone. It is fastened sometimes to other algæ sometimes to stones, and seems to prefer an exposed coast. For although it penetrates also into the interior of deep bays, it does not there, according to my experience, develop to the same luxuriancy as in exposed localities. It grows scattered, though sometimes in rather large numbers. According to Kleen it bears sporocarps and tetrasporangia during the whole summer in the southern part of the Norwegian Polar Sea. On the coast of Finmarken I have found specimens with such organs at the end of July and the beginning of August.

Geogr. Distrib. This species belongs to both regions of Polar Sea, the Atlantic as well as the arctic, but in the latter it has only a limited distribution. It goes far northwards in Baffin Bay, where it is said to be found at Lat. N. 73° 20'. It reaches its maximum of frequency in the Norwegian Polar Sea. The most common form is f. *typica;* f. *roseola* is known to me only from the southern part of the Norwegian Polar Sea.

Localities: The Norwegian Polar Sea: at Nordlanden common and abundant, at Finmarken common and abundant on open shores, at Maasö, Gjesvær and Öxfjord, at Talvik and in Magerö sound.

The Murman Sea: on the coast of Russian Lapland and of Cisuralian Samoyede-land.

The White Sea: scarce (?).

Baffin Bay: probably pretty plentiful on the west coast of Greenland, as at Julianeshaab, Ameralik, Godthaab and at Lat. N. 73° 20', Long. W. 57° 20', Egedesminde (?).

Polysiphonia Brodiæi (DILLW.) GREV.

in HOOK. Brit. Fl. 2, p. 328; Conferva Brodiæi DILLW. Brit. Conf. t. 107.

f. *Kützingii* nob.

f. parvula, circa 6 cm. alta, cæspitosa, densa.

Fig. Polysiphonia Brodiæi Kütz. Tab. Phyc. 14, t. 1.

f. *Agardhii* nob.

Descr. Hutchinsia Brodiæi α AG. Spec. Alg. 2, p. 63.

Exsicc. Polysiphonia penicillata ARESCH. Alg. Scand. exsicc. N:o 64.

f. *Lyngbyei* nob.

Descr. Hutchinsia Brodiæi Lyngb. Hydr. Dan. p. 109.

Fig. » » » » » t. 33.

α. *laxa* nob.

forma sequente robustior, penicillis ramulorum paucioribus, permagnis, distantibus.

β. *confluens* nob.

forma penicillis ramulorum creberrimis, apicem axis primarii versus valde approximatis, confluentibus.

Syn. Polysiphonia Brodiæi KLEEN, Nordl. Alg. p. 13.

Remark on the definition on the form. In KLEEN's collections, that are rich in specimens of this species from the Norwegian Polar Sea, there are to be found three distinctly marked forms, besides several intermediate ones. Unfortunately this algologist has given no information as to their occurrence and habitat. The form denominated by me f. *Kützingii*, much resembles in habit a low *Rhodomela lycopodioides* f. *laxa*. It agrees in ramification with the plant figured l. c. by KÜTZING under the name of *P. Brodiæi*, with which however the diagnosis given by him in Spec. Alg. of an homonymous alga accords but little. In the last-quoted passage he means apparently the same form that AGARDH describes in Spec. Alg. under the name of *Hutchinsia penicillata.*

Forma Kützingii appears to have grown in narrow, densely overgrown rock-pools within the litoral zone, rich in Mytilus edulis and shells. Whether AGARDH by his *Hutchinsia Brodiæi typica* has really meant the beautiful alga distributed by ARESCHOUG in Alg. Scand. exsicc. N:o 64 under the name of *Polysiphonia penicillata*, is a question I must leave undecided at present. However, his diagnosis accords well with this form. Several specimens from the Polar Sea belong evidently to *P. penicillata* ARESCH. That this cannot well be regarded as one of the other forms in a young condition, appears to be proved by its being profusely fructiferous.

The form named by me f. *Lyngbyei α laxa*, which is most probably to be considered as the typical *P. Brodiæi*, is well represented by the quoted figure in LYNGBYE. Near it stands a more robust form with dense clusters of branches that are confluent upwards. I have found this form on the west coast of Sweden in open places exposed to a heavy surge.

Habitat. According to KLEEN, this alga occurs in the southern part of the Norwegian Polar Sea on exposed coasts in rock-pools within the litoral zone, but it was not to be found in the interior of the large bay, Saltenfjord, investigated by him. It seems thus to be a pelagic form. It possibly grows here, as farther to the south, pretty gregarious in large masses. KLEEN says nothing on this point. On the coast of Nordlanden it bears sporangia in July and August.

Geogr. Distrib. It is known only from the Atlantic region of the Polar Sea, and exclusively from its southern part.

Locality: *The Norwegian Polar Sea*: at Nordlanden pretty common.

Polysiphonia fibrillosa (DILLW.) GREV.

in HOOK. Brit. Fl. 2, p. 334; Conferva fibrillosa DILLW. Brit. Conf. p. 86.

Descr. Polysiphonia fibrillosa J. G. AG. Spec. Alg. 2, p. 991.

Fig. „ lasiotricha Kütz. Phyc. gener. t. 49.

Syn. Polysiphonia violacea KLEEN, Nordl. Alg. p. 13.

Remark on the determination of the species. The plant recorded by KLEEN in his work on the algæ of Nordlanden under the name of *P. violacea*, judging by the specimens in his herbarium, is not that species, but *P. fibrillosa*. It differs certainly by the richer branching, less strongly developed cortical layer, and greater flaccidity, from English specimens of this plant, but still it agrees with these in the main. It comes nearest to the *P. laciotricha* figured by KÜTZING l. c., which according to J. G. AGARDH ought to be identified with *P. fibrillosa*. The difference exhibited by the specimens from Nordlanden as compared with the English, depends probably on the former having grown in deep water, while on the coasts of England the plant is litoral.

Habitat. It has been found in the Polar Sea in several fathoms water, attached to shells, or dead parts of *Fucus*, or to *Desmarestia aculeata*, and bearing tetraspores in the month of August.

Geogr. Distrib. It is known only from the southern part of the Atlantic region of the Polar Sea.

Localities: *The Norwegian Polar Sea*: Nordlanden. All the localities reported by KLEEN for *P. violacea*, viz. Röst, Givær, and Fleinvær, are probably to be referred to this species. At least his collections contain no *P. violacea*.

Polysiphonia Schübelerii FOSLIE.

Arct. Havalg. p. 3.

Descr. Polysiphonia Schübelerii FOSLIE, l. c.
Fig. » » » t. 1, fig. 1—3.

Remark on the species. The present species approaches very nearly to *P. fibrillosa*, as its author remarks himself, and can hardly be specifically distinguished from it. Through the kindness of Mr FOSLIE I have had the advantage of seeing several specimens. These differ rather much in habit from *P. fibrillosa*, especially by their very robust mainstems and lateral axes of the first order, which condition however may be partly caused by hard pressing in preserving. Until the plant has been more closely studied in nature, I think fit however to maintain it as a species.

Habitat. It grows on sandy bottom mixed with stones, in 2—4 fathoms water, fastened to small stones and shells, together with *Ceramium* and *Punctaria*. It bears tetrasporangia in summer.

Geogr. Distrib. It is known only from the Atlantic region of the Polar Sea.

Locality: *The Norwegian Polar Sea*: Finmarken in Porsanger fjord.

Polysiphonia elongata (HUDS.) HARV.

in HOOK. Brit. Fl. p. 333; Conferva elongata HUDS. Fl. Angl. p. 599.

f. *Lyngbyei* J. G. AG.

Spec. Alg. 2, p. 1004.

Descr. Polysiphonia elongata f. Lyngbyei J. G. AG. l. c.
Fig. Ceramium brachygonium LYNGB. Hydr. Dan. t. 36.
Exsicc. Polysiphonia elongata ARESCH. Alg. Scand. exsicc. N:o 60.
Syn. Polysiphonia elongata J. G. AG. Spetsb. Alg. Bidr. p. 11.
» » KLEEN, Nordl. Alg. p. 12.

Habitat. This plant occurs in the Norwegian Polar Sea on shelly and dead bottom within the sublitoral zone in 8—15 fathoms water, bearing sporocarps and tetrasporangia in July and August. On the west coast of Sweden I have found it with such organs at this season, but also earlier, in May and June. Judging by the specimens in KLEEN's herbarium, the plant on the coast of Nordlanden enters into its period of rest towards the end of August.

Geogr. Distrib. There exists a statement by J. G. AGARDH that the present species should have been brought home from Spitzbergen by the expedition of TORELL 1861. I have never seen it there myself nor anywhere else in the Arctic Sea. In the southern part of the Norwegian Polar Sea it has been found by KLEEN.

Localities: *The Norwegian Polar Sea*: Nordlanden at Givær isles and Fleinvær isles. *The Greenland Sea*: the coast of Spitzbergen, the particular place not being stated.

Polysiphonia fastigiata (ROTH) GREV.

Fl. Edinb. p. 308; Ceramium fastigiatum ROTH. Fl. Germ. 3, p. 463,

Descr. Polysiphonia fastigiata J. G. AG. Spec. Alg. 2, p. 1029.
Fig. » » HARV. Phyc. Brit. t. 299.
Exsicc. » » ARESCH. Alg. Scand. exsicc. N:o 4.

Syn. Conferva polymorpha GUNN. Fl. Norv. 2, p. 92; fide. syn.
» » WG. Fl. Lapp. p. 511.
Hutchinsia fastigiata LYNGB. Hydr. DAN. p. 108.
Polysiphonia » J. G. AG. Spetsb. Alg. Bidr. p. 11.
» » ARESCH. Phyc. Shand. p. 278.
» » KJELLM. Spetsb. Thall, 1, p. 9.

Habitat. This plant belongs to the litoral zone and appears to attach itself almost exclusively to *Ozothallia nodosa.* When growing, as happens sometimes, on fragments of this species torn off and carried into deep (10—15 fathoms) water, it becomes less densely branched, finer, longer, and less corymbose, the axes at the same time tapering more strongly towards the top. Such specimens, differing rather much in habit from the typical form, I have found at Maasö in Finmarken. The present species is pelagic, avoiding at least deep bays, and somewhat gregarious. According to KLEEN, it bears sporocarps and tetrasporangia during all the summer in the southern part of the Norwegian Polar Sea. On the coast of Finmarken it occurred with sporocarps during the months of August, September and October, with tetrasporangia in August.

Geogr. Distrib. This species has its maximum of frequency in the Norwegian Polar Sea. LYNGBYE has reported it from Baffin Bay. It seems uncertain as yet whether it occurs really also at other places in the Arctic Sea. It is certainly considered as native in the Greenland Sea on the coast of Spitzbergen. I have indeed found it here myself, though never attached, but only washed ashore, having probably drifted there from the south. This was possibly the case also with those specimens which were brought home from the same region by Torell's expedition 1861. In the White Sea it has not been noticed at all. For the time being, Gjesvær immediately north of 71° N. Lat. must be regarded as the northernmost place where it has been with certainty found to grow.

Localities: *The Norwegian Polar Sea*: Nordlanden, Tromsö amt at Tromsö, Renö, and Karlsö; Finmarken at Maasö, Gjesvær, Öxfjord, and the southern coast of Magerö, everywhere common and plentiful.

The Greenland Sea: the coast of Spitzbergen (?).

Baffin Bay: the west coast of Greenland.

Polysiphonia arctica J. G. AG.

Spec. Alg. 2. p. 1034.

Descr. Polysiphonia arctica J. G. AG. l. c. et GOBI, Algenfl. Weiss. Meer. p. 26.
Exsicc. » » KJELLM. in ARESCH. Alg. Scand. exsicc. N:o 403.

Syn. Conferva nigra R. Br. in Scoresby, Account. 1. App. 5(?)
Hutchinsia badia POST et RUPR. Ill. Alg. p. II. Cfr. GOBI, l. c. p. 27.
» stricta LINDBL. Bot. Not. p. 158.

Polysiphonia arctica J. G. AG. Spetsb. Alg. Progr. p. 3; Bidr. p. 11.
» » DICKIE, Alg. Cumberl. p. 238.
» » EATON, List. p. 44.
» » GOBI, l. c. p. 26-
» » KJELLM. Winteralgv. p. 64. Spetsb. Thall. 1, p. 9; Algenv. Murm. Meer. p. 11; Kariska hafvets Algv: p. 19.
» stricta ZELLER, Zweite d. Polarf. p. 85; fide spec.
» urceolata CROALL, Fl. Disc. p. 459 saltem ex parte.
» » WITTR. in Heugl. Reise 3, p. 284; fide spec.

Remark on the synonymy. It is probably impossible to decide at present what R. BROWN meant by his *Conferva nigra* mentioned in Appendix V to SCORESBY's Voyage. I think it highly probable, however, that this is the *P. arctica* which is common on the coast of Spitzbergen and becomes very black in drying. Of *P. stricta* CROALL some specimens — those with five siphons — belong probably to *P. arctica*, others, especially those from Egedesminde, to *P. urceolata*. There are most probably no other species than these to choose between. Cp. KJELLM. Spetsb. Thall. 1, p. 9 and GOBI, Algenfl. weiss. Meer., p. 27.

Habitat. It grows on exposed as well as sheltered coasts, fastened partly to other algæ partly to stones, within the sublitoral zone, generally in the upper part of it in 1—10 fathoms water, sometimes in its lower part together with several deep-water forms, or even in the uppermost part of the litoral zone. On the arctic coast of Norway I have met with it only in the lower part of the sublitoral zone at a depth of 10—20 fathoms together with several other species common and widely spread in the Arctic Sea. Even in the White Sea it appears to occur most often in deeper water, at 10—12 fathoms, sometimes at a less depth 3—8 fathoms, but even then together with several purely arctic forms, as *Odonthalia dentata, Delesseria sinuosa, Ptilota pectinata*, and *Phyllophora interrupta*. It belongs chiefly to the formation of *Laminariaceæ*, and grows here scattered, never gregarious in greater masses. On the north coast of Spitzbergen it persists through the winter and develops during the whole dark and cold season, although slowly. Specimens with young vegetative organs in a state of development are however continually found. In April their development becomes more vigorous, it reaches its maximum of energy in the middle of May and continues during the summer months.

Although I have had the opportunity to examine a great many specimens of this alga at all seasons, I have but very rarely met with any furnished with organs of propagation. Only once, in the month of August 1872, I have found a specimen with young sporocarps, and in July of the same year another specimen with young formations that were propably the beginning of antheridia. I found specimens with tetrasporangia on the 8:th and 21:st of November, on the 19:th and 20:th of December 1872, and the 18:th of January 1873. That the species at Spitzbergen produces tetrasporangia also in summer, is proved by J. G. AGARDH describing these organs from specimens brought home from there by VAHL and by TORELL's expedition who stayed on the coasts of Spitzbergen only during the summer.

Geogr. Distrib. This species has its maximum of frequency in the Greenland Sea on the coasts of Spitzbergen. It attains its maximum of luxuriancy on the north coast of Norway where it forms rich, dense tufts more than 20 cm. in length. It is reported from all parts of the Arctic Sea except the American Arctic Sea. But it is probably to be found even here, and may thus be numbered among the circumpolar species. The northernmost place where it is known with certainty, is the North Cape of Spitzbergen Lat. N. 80° 31'.

Localities: The Norwegian Polar Sea: Finmarken at Maasö and Gjesvær, scarce and local.

The Greenland Sea: common and very plentiful on the north and west coasts of Spitzbergen; known also from Storfjord to the east of western Spitzbergen and from Sabine Isle on the east coast of Greenland.

The Murman Sea: the coast of Russian Lapland, Kolgujew Isle, the west coast of Novaya Zemlya and Waygats, in the latter place common but less abundant.

The White Sea: common and plentiful.

The Kara Sea: Uddebay on the east coast of Novaya Zemlya Lat. N. 76° 18' Long. O. 92° 20', Cape Palander, and Actinia Bay, everywhere scanty, though at several place pretty widely dispersed.

The Siberian Sea: the north coast of Tshuktshland, scanty and local.

Baffin Bay: Cumberland Sound, pretty common, at several places on the south and west coasts of Greenland, as Nenese, Tessarmiut Bay, Godhavn, Jacobshavn (?), Disco Isle.

Polysiphonia atrorubescens (Dillw.) Grev.

Fl. Edinb. p. 308. Conferva atrorubescens Dillw. Brit. Conf. t. 70.

Syn. Polysiphonia atrorubescens J. G. Ag. Spetsb. Alg. Till. p. 48.

The habitat of this species in the Arctic Sea is unknown to me.

Geogr. Distrib. and *Localities.* Cp. J. E. Ag. l. c.

Polysiphonia byssoides (Good. et Woodw.) Grev.

Fl. Edinb. p. 309. Fucus byssoides Good. et Woodw. Linn. Trans. 3, p. 229.

Descr. Polysiphonia byssoides J. G. Ag. Spec. Alg. 2, p. 1042.

Fig. » » Harv. Phyc. Brit. t. 284.

Exsicc. » » Areschi. Alg. Scand. exsicc. N:o 66.

Syn. Polysiphonia byssoides Kleen, Nordl. Alg. p. 14.

Habitat. It grows, according to Kleen, in Nordlanden on shelly bottom in deep water, and has been found here with sporocarps in July and August.

Geogr. Distrib. It is known only from the southern part of the Atlantic region of the Polar Sea.

Localities: The Norwegian Polar Sea: at Fleinvær and Givær Isles in Nordlanden, which are accordingly the northernmost places of growth hitherto known of the species.

Polysiphonia nigrescens (HUDS.) HARV.

Brit. Fl. 3, p. 332. Conferva nigrescens HUDS. Engl. Bot. t. 1717.

f. *pectinata* AG.

Hutchinsia nigrescens β pectinata AG. Syst. Alg. p. 151.

Descr. Polysiphonia nigrescens α pectinata J. G. AG. Spec. Alg. 2, p. 1058.
Fig. Conferva nigrescens Engl. Bot. t. 1717.
Exsicc. Polysiphonia Brodiæi ARESCH. Alg. Scand. exsicc. N:o 63 et 152.
» nigrescens » » » » N:o 62 et 304.

f. *protensa* J. G. AG.

Spec. Alg. 2, p. 1058.

***β gracilis* nob.**

f. setacea, circa 10 cm. alta, fragilis, dilute violacea, fastigiato-ramosa.

Syn. Conferva atrorubens WG. Fl. Lapp. p. 511; fide herb.
Polysiphonia nigrescens ARESCH. Phyc. Scand. p. 271.
» » DICKIE, Alg. Sutherl. 1, p. 142.
» » GOBI, Algenfl. Weiss. Meer. p. 29.
» » Nyl. et Sæl. Herb. Fenn. p. 74.
» » KLEEN, Nordl. Alg. p. 13.
Rhodomela gracilis » » » » 12.

Remark on the synonomy. In his work on the algæ of Nordlanden, KLEEN records *Rhodomela gracilis* as found in a little lake with almost fresh water called Kosmovandet. There is indeed to be found in his collections a plant taken at this place, which approaches very nearly *Rhodomela gracilis* in habit, and there is no other alga of his that could by any possibility be called *Rh. gracilis*. This plant, which is accordingly, as far as I can see, precisely that which KLEEN has meant by his *Rh. gracilis*, is no *Rhodomela*, however, but a peculiar form of *Polysiphonia nigrescens* that approaches most nearly J. G. AGARDH'S f. *protensa*, although differing from it rather considerably. I have called it above *Rh. nigrescens* f. *protensa β gracilis*.

Habitat. The common form, f. *pectinata*, occurs on the coast of Norway in rock-pools within the litoral zone, in the White Sea in the sublitoral zone down to a depth of 18 fathoms, generally on sandy and pebbly bottom. According to DICKIE the plant has been brought up from a depth of 40—50 fathoms in Baffin Bay. Cp. above p. 11. It flourishes on exposed as well as sheltered parts of the coast, growing scattered. It has been found with tetrasporangia on the coast of Norway in July and August, in the White Sea at the middle of July. The form *protensa β gracilis* is a brack-water form, as has been mentioned above. It is known only in sterile condition.

Geogr. Distrib. The species belongs indeed to the Atlantic as well as the arctic region of the Polar Sea, but it has its maximum of frequency within the former region, being but little distributed in the latter. It has assuredly immigrated from the south into the Arctic Sea proper. Its northernmost certain locality is the coast of Russian Lapland. According to DICKIE it has been collected in Baffin Bay much farther to the north, viz. in Whale Sound at 77° N. Lat., but here it was found washed ashore.

The other locality in Baffin Bay from which it is reported, Hunde Islands, is situate at about the same latitude as Russian Lapland and moreover is not quite certain.

Localities: The Norwegian Polar Sea: Nordlanden commonly spread, but scanty; Finmarken »frequenter» according to Wahlenberg; I have not myself been able to find it on this coast.

The Murman Sea: the coast of Russian Lapland, the western part of the coast of Cisuralian Samoyede-land, Kolgujew Island.

The White Sea, common and plentiful.

Baffin Bay: the west coast of Greenland, Hunde Islands (?), Whale sound, washed ashore. In the collections of the Copenhagen Museum I have seen a small fragment of the species »e Groenlandia» without no defined locality being stated.

Fam. SPONGIOCARPEÆ (GREV.)

sec. J. G. AG. Epigr. Alg. p. 628.

Gen. Polyides AG.

Spec. Alg. 1, p. 390.

Polyides rotundus (GMEL.) GREV.

Alg. Brit. p. 70, sec. J. G. AG. Spec. Alg. 2, p. 721. Fucus rotundus GMEL. Hist. Fuc. p. 110.

f. *typica*.

Descr. Polyides lumbricalis J. G. AG. Spec. Alg. 2, p. 721.
Fig. » rotundus HARV. Phyc. Brit. t. 95.
Exsicc. » » ARESCH. Alg. Scand. exsicc. N:o 252.
Syn. Polyides lumbricalis KLEEN, Nordl. Alg. p. 15.
» rotundus GOBI, Algenfl. Weiss. Meer. p. 32.
» Nyl. et Sæl. Herb. Fenn. p. 74.

f. *fastigiata* TURN.

Hist. Fuc. 1, p. 9.

Descr. Fucus rotundus γ fastigiatus TURN. l. c.
Syn. Polyides rotundus KJELLM. Algenv. Murm. Meer. p. 14.

Remark on the determination of the species. In my paper on the algæ of the Murman Sea, I referred an alga taken at Matotshin Shar and at Besimannaja Bay on the east coast of Novaya Zemlya to *Polyides rotundus*. I have subjected this alga to a renewed examination and succeeded in finding tetrasporangia in some of those specimens which were most similar in habit to *Furcellaria fastigiata*. These tetrasporangia prove distinctly that the plant is a *Polyides*.

Habitat. On the coast of the Norwegian Polar Sea this alga is litoral, occurring on exposed shores, chiefly in rock-pools between tide-marks. On the west coast of Novaya Zemlya it is sublitoral. Here as elsewhere it grows scattered. KLEEN has found it with sporocarps at Nordlanden at the end of July. At the same season I have gathered specimens with young sporocarps and ripe tetrasporangia at Novaya Zemlya. On the west coast of Sweden the proper time for the development of the propagative organs seems to be in winter, in the months of December and January; however ARESCHOUG states that he has found here individuals bearing sporocarps also in August and September. Cp. ARESCH. Phyc. Scand. p. 309.

Geogr. Distrib. This species is known from the Atlantic as well as the arctic region of the Polar Sea. It is but little spread in the latter and probably is a species immigrated from the south. Its northernmost locality is Matotshin Shar on the west coast of Novaya Zemlya Lat. N. 73° 15', where it occurs in the dwarfed form *fastigiata.*

Localities: The Norwegian Polar Sea: Nordlanden, f. *typica*, common, abundant; Finmarken, f. *fastigiata*, scarce, local at Maasö, Gjesvær and Öxfjord.

The Murman Sea: the coast of Cisuralian Samoyede-land, the west coast of Novaya Zemlya and Waygats, pretty common and plentiful.

The White Sea: scarce (?).

Fam. WRANGELIACEÆ (J. G. AG.) HAUCK.

Meeresalg. p. 14; J. G. AG. Spec. Alg. 2, p. 701; lim. mut.

Gen. **Spermothamnion** ARESCH.

Phyc. Scand. p. 334.

Spermothamnion Turneri (MERT.) ARESCH.

l. c. p. 335. Ceramium Turneri Mert. in ROTH, Cat. Bot. 3, p. 127.

Descr. Spermothamnion Turneri ARESCH. l. c.
Fig. » roseolum PRINGSH. Morph. Meeresalg. t. 4—6.
Exsicc. » » ARESCH. Alg. Scand. exsicc. N:o 83.

Habitat. It usually occurs within the litoral zone, attached to other algæ. However it has also been found on stones at Nordlanden. In the Polar Sea it grows scattered, chiefly on exposed coasts, bearing tetrasporangia and sporocarps in summer. On the west coast of Sweden I have found it with plenty of tetrasporangia even in winter, at the end of December.

Geogr. Distrib. It is known only from the Atlantic region of the Polar Sea. Its most northerly place of growth is Öxfjord in Finmarken at the mouth of Altenfjord, about Lat. N. 70°. Its maximum of frequency is in the southern part of the Norwegian Polar Sea.

Localities: The Norwegian Polar Sea: Nordlanden common and abundant; Finmarken: Öxfjord, local, rare.

Gen. **Chantransia** (DC.) FRIES.

Syst. Veg. p. 338; DC. Fl. Fr. 2, p. 49; lim. mut.

Chantransia efflorescens (J. G. AG.) KJELLM.

Spetsb. Thall. 1, p. 4. Callithamnion efflorescens J. G. AG. Spec. Alg. 2, p. 15.

f. *tenuis*.

f. laxe cæspitosa, qvam forma typica in Sinu Codano proveniente multo tenuior et flaccidior; articulis axis principalis 5 μ. diametro non attingentibus, (in f. typica 6—8 μ.)

Fig. Chantransia efflorescens f. tenuis tab. nostra 12, fig. 1—2.

Exsicc. Cfr. Trentepohlia Daviesii var. α ARESCH. Alg. Scand. exsicc. N:o 16.

Syn. Chantransia Daviesii GOBI, Algenfl. Weiss. Meer, p. 50.

» efflorescens KJELLM. Spetsb. Thall. 1, p. 14; Algenv. Murm. Meer. p. 14; Kariska hafvets Alg. p. 20.

Remark on the determination of the species. The *Chantransia efflorescens* found by me at several places in the Polar Sea, differs from that occurring on the coast of Bohuslän and distributed by ARESCHOUG l. c. by being scarcely half as thick, and more flaccid, and by forming thinner tufts than the Bohuslän form. I have assumed it to be the same form as that reported by GOBI from the White Sea, and propose to give it the name of *tenuis*.

Habitat. This alga is sublitoral everywhere in the Polar Sea, and grows attached to various other algæ, as *Lithothamnion soriferum, Odonthalia dentata, Polysiphonia arctica, Delesseria Bærii, Chætomorpha melagonium* a. o. It has not been met with in the interior of deep bays. It lives scattered. It has been found with reproductive organs at Spitzbergen in July and August, on the west coast of Novaya Zemlya in July and on the eastern coast at the end of August. On the coast of Sweden I have observed sporiferous specimens in August.

Geogr. Distrib. It belongs both to the Atlantic and the arctic region of the Polar Sea, and is pretty widely spread within the latter. I have found it to be scarce every where. Its maximum of frequency seems to be in the White Sea, Cp. GOBI l. c.

Localities: The Norwegian Polar Sea: Finmarken at Maasö, scarce and local.

The Greenland Sea: the north and north-west coasts of Spitzbergen, scarce and local.

The Murman Sea: the west coast of Novaya Zemlya, scarce and local.

The White Sea: common and abundant.

The Kara Sea: Uddebay on the east coast of Novaya Zemlya, scarce.

Chantransia Daviesii (DILLW.) THUR.

in Le Jol. List. Alg. Cherb. p. 106. Conferva Daviesii DILLW. Brit. Conf. Intr. p. 73.

Descr. Callithamnion Daviesii J. G. AG. Epicr. p. 8.

Fig. » » HARV, Phyc. Brit. t. 314.

Syn. Chantransia Daviesii KLEEN, Nordl. Alg. p. 16.

Habitat. This species grows scattered, fastened to litoral algæ, for inst. species of *Corallina* and *Cladophora* a. o. In the Polar Sea it bears spores at least during the months of July, August, and September.

Geogr. Distrib. It is known at present only from the Atlantic region of the Polar Sea, and has its maximum of frequency in the southern part of this region. Its northernmost locality is Öxfjord in Finmarken at the mouth of Altenfjord, about Lat. N. 70°.

Localities: The Norwegian Polar Sea: Nordlanden, common and plentiful; Finmarken at Öxfjord, scarce.

Chantransia virgatula (HARV.) THUR.

in Le Jol. List. Alg. Cherb. p. 106. Callithamnion virgatulum HARV. in Hook. Brit. Fl. 2, p. 349.

f. *Farlowii* nob.

Descr. Trentepohlia virgatula FARL. New Engl. Alg. p. 109.
Fig. » » » t. 10. fig. 3.

Remark on the determination of the species. In the description of the marine species of the genus *Chantransia* there prevails a great confusion. As almost every author differs from the others in his views about the different species, the synonymy has become entangled in the highest degree. In the Norwegian Polar Sea there are to be found, besides the two species mentioned above, two other species of *Chantransia*, the one of which is surely identical with that named by FARLOW *virgatula* HARV. and figured l. c., while the other coincides with that which ARESCHOUG has distributed in his work of exsiccatæ under the name of *Trentepohlia secundata* LYNGB. FARLOW quotes *Trentepohlia virgatula* HARV. Phyc. Brit. pl. 313 as being identical with his species. I cannot but doubt his being fully justified in doing so, in case one may suppose that the figures by HARVEY and FARLOW are both of them true to life. For the two figures are very different from each other, so as to make the impression that the two authors have meant specifically distinct algæ by one and the same name. However, it is possible that *Ch. virgatula* varies much, and that the plant figured by FARLOW is connected by intermediate forms with the species originally described by HARVEY under the name of *Callithamnion virgatulum.* For the present I am obliged to assume this to be the case, and I accordingly denote the plant in question by the above combination of names.

Habitat. I have only succeeded to collect a very slight number of specimens of this plant. These grew at sheltered places of the coast in rock-pools within the litoral zone, attached to *Cladophora gracilis.* Those collected in September were sporiferous.

Geogr. Distrib. The present species is known only from the Atlantic region of the Polar Sea. Its most northerly locality is the same as that of the preceding one, viz. Öxfjord.

Localities: The Norwegian Polar Sea: Tromsö and Öxfjord, very rare at both places.

Chantransia secundata (LYNGB.) THUR.

in Le Jol. List. Alg. Cherb. p. 106. Callithamnion Daviesii β secundatum LYNGB. Hydr. Dan. p. 129.
Exsicc. Trentepohlia secundata ARESCH. Alg. Scand. exsicc. N:o 84.

Syn. Chantransia secundata KJELLM. Algenv. Murm. Meer. p. 15.
» » KLEEN, Nordl. Alg. p. 16.

Remark on the species. This plant, as I understand it here, is identical with that distributed by ARESCHOUG l. c. Neither LYNGBYE'S, nor J. G. AGARDH'S, nor even ARESCHOUG'S, descriptions and diagnoses of *Chantransia* (*Callithamnion*) *secundata* agree very well with that alga, and it is doubtful, I think, whether it is really identical with LYNGBYE'S *Callithamnion Daviesii β secundatum*. This is differently understood by different authors, the short description and incomplete figure furnished by LYNGBYE easily giving rise to different explanations [1]).

I shall give a description of the Scandinavian *Chantransia secundata* at another time, when I have had an opportunity of examining living individuals. Dried specimens are not well fit for a closer investigation.

Habitat. On the coast of Norway the present alga is litoral, growing chiefly on *Porphyra laciniata*. On the west coast of Novaya Zemlya I have found it within the sublitoral zone, attached to *Odonthalia dentata*. It has been collected with spores in the Polar Sea in July and August.

Geogr. Distrib. It belongs both to the Atlantic and the arctic region of the Polar Sea, but in the latter is only little spread. Nor does it seem to be more widely distributed or particularly abundant within the former region. Its most northerly locality is Rogatshew Bay on the west coast of Novaya Zemlya, Lat. N. 71° 23'.

Localities: The Norwegian Polar Sea: Nordlanden, pretty common.

The Murman Sea: Rogatshew Bay, rare.

Fam. DELESSERIACEÆ J. G. AG.

Epicr. p. 444. Cfr. Alg. med. p. 155.

Gen. **Delesseria** (LAMOUR.) J. G. AG.

Epicr. p. 477; LAMOUR. Ess. p. 122; ex parte.

Delesseria rostrata (LYNGB.) J. G. AG.

Spec. Alg. 2, p. 685. Gigartina purpurascens γ rostrata LYNGB. Hydr. Dan. p. 46.

Descr. Delesseria rostrata J. G. AG. Spec. Alg. 2, p. 685.
Fig. Gigartina purpurascens γ rostrata LYNGB. l. c. t. 12 B.
Syn. Gigartina Fabriciana LYNGB. l. c. p. 48, t. 11, D. Cfr. J. G. AG. l. c. p. 698.

Habitat: In the collection of Greenland algæ at the Copenhagen Museum there are some specimens of this alga with sporocarps and tetrasporangia. By the statement of VAHL on the label affixed to them, they have been collected in March and were found attached »ad saxa maritima». Besides this, I know nothing about the habitat of this species.

---

[1]) FARLOW'S *Trentepohlia virgatula var. secundata*, New. Engl. Alg., p. 109 is probably not *Callithamnion Daviesii β secundatum* LYNGB., but *C. luxurians* J. G. AG., KÜTZ., *Trentepohlia virgatula* ARESCH.

Geogr. Distrib. It is known only from Baffin Bay along the south and south-west coast of Greenland.

Localities: Baffin Bay: Julianeshaab Lat. N. 60° 35', the only locality recorded with certainty. In the collection from Greenland mentioned above, there is a specimen of the plant called by LYNGBYE *Gigartina Fabriciana*, which I believe I have been able to identify with *D. rostrata.* According to the label affixed, it has been taken at Nennese in Greenland by J. VAHL.

In a collection of algæ, brought together by the Moravian mission, which by the kind intercession of Prof. TH. M. FRIES I have had the advantage of examining, there were to be found some individuals of *D. rostrata:* but the locality of their growth was not recorded. It is certainly most probable that they had been collected on the west coast of Greenland, but it is possible that they had come from Labrador.

Delesseria Bærii RUPR.

Alg. Och. p. 239.

Descr. Delesseria Bærii J. G. AG. Spec. Alg. 2, p. 685.
» » KJELLM. Spetsb. Thall. 1, p. 12.

Syn. Delesseria Bærii J. G. AG. Spetsb. Alg. Progr. p. 3; Till. p. 11.
» » GOBI, Algenfl. Weiss. Meer. p. 31.
» » KJELLM. l. c.; Algenv. Murm. Meer. p. 13.
» » RUPR. l. c.
Fucus clavellosus SCORESBY, Account 1, p. 132 (?)
» forsan nova spec. prope alatum? R. Br. in SCORESBY, l. c. 1, App. 5.
Rhodymenia Bærii POST. et RUPR. Ill. Alg. p. II.
» » Nyl. et Sæl. Herb. Fenn. p. 74.

Remark on the synonymy. It can now no more be decided with certainty whether the two synonyms taken from the lists of algæ in Scoreby's Account are really to be referred to the present species. But it is not impossible, I think, that this may indeed be the case [1]).

Remark on the species. The group *Cryptoneura* in the genus *Delesseria* comprises, besides *D. Jürgensii*, which is surely independent, four other species so closely similar to one another, that many algologists would certainly feel inclined to abandon J. G. AGARDH's view and to contest their right of being regarded as separate species. These species are *D. Bærii*, *D. angustissima*, the *D. rostrata* mentioned before, and *D. corymbosa.* After having examined these species, which are very scarce in collections, as carefully as possible, I must certainly admit, on the one hand, that they approach very closely to one another, besides according nearly with *D. alata*, but on the other hand there are indeed to be observed differences between them, which appear constant, both in structure and in the branching of the frond and the position of the reproductive organs. On this ground I think it right to embrace the opinion of AGARDH that

[1]) I remark here that the editor of R. BROWN's Vermischte Schrifter, Dr. E. MEYER, does Scoresby an injustice in alleging that he has not declared that it is R. BROWN who has furnished the list of algæ inserted in Appendix V to SCORESBY's work. SCORESBY states this expressly in the text Vol. 1. p. 148.

the plants in question are so strongly differentiated from one another, that they ought to be regarded as separate species which have issued, probably not long ago, from one or two fundamental types.

Habitat. *D. Bærii* is in the Arctic Sea a sublitoral alga occurring within the formation of *Laminariaceæ* and apparently preferring exposed coasts to the interior of deep bays and other sheltered places. It is usually fastened to other algæ, especially to the rhizines of the *Laminariaceæ*. On the north coast of Spitzbergen I have found it in full development in the middle of February. Specimens with sporocarps were met with on the west coast of Novaya Zemlya in June and July, at Spitzbergen in August; specimens with tetrasporangia at both these places in the month of July. During the development of the spores part of the branches are dissolved, but at the end of that time a vigorous development of new vegetative parts sets in, by which a great number of the new branch-systems are produced in the axils.

Geogr. Distrib. This species is known only from the arctic region of the Polar Sea, and in this region it has a pretty narrow range. For it has been found only in the eastern part of the Greenland Sea and in the Murman and the White Seas. Its maximum of frequency seems to be in the White Sea and in the eastern part of the Murman Sea. The most northerly place where it has been met with, is Musselbay on the north coast of Spitzbergen, Lat. N. 79° 53'.

Localities: The Greenland Sea: the north and west coasts of Spitzbergen, pretty common, but not abundant.

The Murman Sea: at several places on the coast of Russian Lapland and Cisuralian Samoyede-land, Kolgujew Isle, the west coast of Novaya Zemlya from Matotshin Shar to N. Gusinnoi Cape, on the latter coast more plentiful and luxuriant than at Spitzbergen.

The White Sea: pretty common and abundant.

Delesseria corymbosa J. G. Ag.

Spec. Alg. 2, p. 684.

Descr. Delesseria corymbosa J. G. Ag. l. c.
Fig. » » tab. nostra 10, fig. 3.
Syn. Delesseria angustissima Croall, Fl. Disc. p. 459(?),

Habitat. It appears to belong to the sublitoral zone and to the formation of *Laminariaceæ*. In support of this view I can quote the inscription of a label by J. Vahl »In stipibus Laminariæ saccharinæ».

Geogr. Distrib. It is known only from Baffin Bay. Only one *Locality* is quite sure, namely Godthaab on the west coast of Greenland. If however, as seems most probable, Croall's *D. angustissima* is the present species, it is reported also from Jacobshavn, which is, in that case, the most northerly locality where it is at present known to grow, about Lat. N. 69° 15'.

Delesseria angustissima (TURN.) GRIFF.

in HARV. Phyc. Brit. t. 83. Fucus alatus γ angustissimus TURN. Hist. Fuc. 3, p. 60.

Descr. Delesseria angustissima J. G. AG. Spec. Alg. 2, p. 686.
Fig. » » HARV. Phyc. Brit. l. c.

Syn. Delesseria alata var. angustissima KLEEN, Nordl. Alg. p. 14.

Remark on the determination of the species. I have never seen any specimens of this alga from the Polar Sea collected by others, nor have I ever met with it there myself. It is only on the authority of KLEEN that I give it a place in the present work. KLEEN'S collections contain no specimens of this species, but in his list of the algæ of Nordlanden he declares decidedly that *D. angustissima* HARV. Phyc. Brit. pl. 83 was found there common. I have no reason to suppose that KLEEN'S statement should not be founded on accurate comparisons, and the alga growing at Scotland and the Orkney isles, its occurrence at Nordlanden is highly probable.

Habitat. It is litoral, attached to stones beneath *Fucaceæ.*

Geogr. Distrib. It is known only from the Atlantic region of the Polar Sea, in the southern part of which region it is commonly spread, according to KLEEN.

Locality: the Norwegian Polar Sea: Nordlanden common and plentiful.

Delesseria alata (HUDS.) LAMOUR.

Ess. p. 124. Fucus alatus HUDS. Fl. Angl. p. 578.

Descr. Delesseria alata J. G. AG. Epicr. p. 483.
Fig. » » HARV. Phyc. Brit. t. 247.
Exsicc. » » ARESCH. Alg. Scand. exsicc. N:o 75.

Syn. Delesseria alata ARESCH. Phyc. Scand. p. 292.
» » KLEEN, Nordl. Alg. p. 14.
Fucus alatus GUNN. Fl. Norv. 2, p. 91.
» » WG. Fl. Lapp. p. 492.

Habitat. In the southern part of the Norwegian Polar Sea at Nordlanden this species grows partly in rock-pools between tide-marks, partly within the upper part of the sublitoral zone, attached to stones or more usually to *Laminariaceæ.* Farther northwards on the coast of Finmarken it always keeps, according to my experience, in the litoral region, forming in exposed localities, together with *Ptilota elegans,* a dense mat on the steep, flat or somewhat hollow, outsides of stones or rocks covered by masses of *Ozothallia* and other *Fucaceæ.* Here it is dwarfed, seldom, if ever, reaching more than 3—5 cm. in height by 1—2 mm. in width, while even at Nordlanden it becomes over 8 cm. long and about 4 mm. broad. The specimens from Nordlanden found in KLEEN'S collections are all sterile, nor does he mention in his treatise on the marine Flora of Nordlanden that he has ever seen any specimen with any kind of propagative organs. Myself found only sterile individuals at Finmarken. On the west coast of Sweden the plant bears tetrasporangia during the winter months, December and January, and ARESCHOUG states it to be »mensibus Martii atque Aprilis in mari Bahusiensi fructificans». Accordingly, the plant here develops its organs of propagation in winter and spring. Probably it does so also farther to the north.

Geogr. Distrib. It is known only from the Atlantic region of the Polar Sea, reaching its maximum of frequency at Nordlanden. Its northernmost locality is Gjesvær about Lat. N. 71°.

Delesseria Montagnei novum nomen.

Delesseria denticulata Mont. Syll. p. 408. Cfr. Ann. d. Sc. 9, p. 62.

Descr. Hypoglossum denticulatum Kütz. Tab. Phyc. 16, p. 6.

Fig. » » » » » » t. 15.

Syn. Delesseria alata β angustifolia Lyngb. Hydr. Dan. p. 8.

Remark on the species. In the collections of Greenland algæ belonging to the Copenhagen Museum there is a pretty great quantity of specimens, evidently gathered at different occasions and by different persons, of a *Delesseria* bearing the name of *D. alata*. Some of these agree very well with the above-quoted figure in Kützing, others approach more nearly to *D. alata*. However, they all differ from the latter species by their more spreading branches which are never so obliquely cut out at the base and consequently more regularly linear or elongated-cuneiform than in *D. alata*. Besides, the branches, especially those of the last order, are always distinctly, sometimes densely, serrate. Thus there can be no doubt, I think, that the specimens from Greenland are not to be referred to *D. alata*, but to *D. denticulata* Mont., which must be regarded as a species distinct from *D. alata* and most nearly related to *D. spinulosa* Rupr. J. G. Ag. known from the Pacific, if it be not indeed quite identical with this. It is difficult to draw any definite limit between them. Montagne has himself declared that his species is identical with *D. spinulosa*. Ruprecht states that *D. spinulosa* is closely allied to *D. alata* f. *denticulata*, but differs from it by the narrowness of the branches of the frond which are crispy at the margin, by the more spreading lower secondary axes which spring out almost at right angles, by some difference in regard to the lateral nerves which I do not quite understand, and by the tetraspores being developed somewhat farther down from the tips of the axes. With regard to the breadth of the frond, the specimens from Greenland vary much, from 4 mm. to 1,5—1 mm. and even less, in case *D. alata β angustifolia* Lyngb., as is most probable, is a slender form of the present species. Broader specimens from Greenland often have a distinctly crisp margin. The branches are in general very much expanded, so that the lower ones, in several specimens of *D. denticulata*, form a right or nearly right angle to the main axis. In *D. denticulata* the development of tetraspores both begins and extends farther down than in *D. alata*. Cp. Ruprecht, Alg. Och. p. 244.

J. G. Agardh has some doubts about the identity of *D. spinulosa* and *D. denticulata*, chiefly because the latter, as figured by Kützing, wants the microscopic lateral nerves, which are to be seen in *D. spinulosa*. But it should be remarked that it is stated expressly in the diagnosis that accompanies the figure in Kützing l. c. p. 6, »segmentis .... a costa ad margines venis obliquis percursis», and that the lateral nerves in the specimens of *D. denticulata* from Greenland are always, especially in broader sterile individuals, distinctly apparent on microscopical examination, sometimes even visible to the naked eye. Thus there can hardly be drawn any limit, I think, between these two plants, as far as they are known at present. With regard to *D. alata β angustifolia* LYNGB., I am quite aware of the difficulty of determining, by means of a few dried specimens, to which species such a reduced form as this should be referred. It seems more probable, however, that it should belong to *D. denticulata* than to *D. alata*, because, as has been correctly stated already by LYNGBYE, it has dentate upper axes and spreading, linear branches, which are not at all or very little cut out obliquely at the base.

In order to distinguish the present species, *D. denticulata* MONT., from *D. denticulata* HARVEY I have changed the name of *denticulata* for that of *Montagnei*.

Habitat. On this point I know but little. According to what is stated on the labels affixed to part of the specimens I have examined, this species grows within the sublitoral zone, attached to *Laminariaceæ*. Several individuals bear reproductive organs, but I do not know at what season these individuals were collected.

Geogr. Distrib. It belongs to Baffin Bay, where it seems to be rather common. The northernmost place where it has been ascertained to grow, is Godhavn 69° 15' Lat. N.

Localities: Baffin Bay: on the west coast of Greenland, for instance at Tessarmiut Bay, Godthaab, and Godhavn.

Delesseria sinuosa (GOOD. et WOODW.) LAMOUR.

Ess. p. 124. Fucus sinuosus GOOD. et WOODW. Linn. Trans. 3, p. 111.

f. *typica.*

Descr. Delesseria sinuosa J. G. AG. Spec. Alg. 2, p. 691.
Fig. » » HARV. Phyc. Brit. t. 259.

f. *quercifolia* TURN.

Hist. Fuc. I, p. 74.

Descr. Fucus sinuosus γ quercifolius TURN. l. c.
Exsicc. Delesseria sinuosa ARESCH. Alg. Scand. exsicc. N:o 74.

f. *lingulata* AG.

Spec. Alg. 1, p. 175.

Descr. Delesseria sinuosa γ lingulata AG. l. c.
Fig. Phycodrys sinuosa KÜTZ., Tab. Phyc. 16, t. 20, fig. c—f.
Syn. Delesseria sinuosa ARESCH. Phyc. Scand. p. 291.
» » var. J. G. AG. Spetsb. Alg. Progr. p. 3; Bidr. p. 11.
» » DICKIE, Alg. Sutherl. 1, p. 142.
» » EATON, List, p. 44.
» » GOBI, Algenfl. Weiss. Meer. p. 30.
» » KJELLM. Vinteralg. p. 64; Spetsb. Thall. 1, p. 10; Algenv. Murm. Meer. p. 12; Kariska hafvets algv. p. 20.
» » KLEEN, Nordl. Alg. p. 14.
» » Nyl. et Sæl. Herb. Fenn. p. 74.
» » POST. et RUPR. Ill. Alg. p. II.
» » SCHRENK, Ural. Reise 2, p. 547.
» » WITTR. in Heugl. Reise 3, p. 284.

Syn. Fucus Quercus PALL. Sib. Reise 3, p. 34. (?)
» rubens GUNN. Fl. Norv. 2, p. 69.
» sinuatus R. Br. in SCORESBY, Account 1, App. 5.
» sinuosus SCORESBY, Account 2, p. 131.
» » γ. WG. Fl. Lapp. p. 491.
Phycodrys sinuosa ZELLER, Zweite d. Polarf. 2, p. 86.

Remark on the arctic forms of this species. *Delesseria sinuosa* is one of the commonest algæ of the Polar Sea and occurs in a great many varieties. These may however, I think, be arranged under the three above-mentioned forms, which were distinguished long ago. For I have convinced myself that the form I have formerly called f. *angusta*, is to be referred to f. *lingulata* AG. The most common of the forms is the *D. sinuosa* of the older authors. In certain parts of the Polar Sea it attains a considerable size and is surely to be reckoned among the most magnificent algæ of this sea. I have seen specimens more than 30 cm. long, with leaf-shaped branches of even 18 cm. in length, by 3—4 cm. in breadth. Another form which approaches most nearly, although it never quite coincides with, the f. *quercifolia* common in Skagerack, is less often to be observed. F. *lingulata* is more common, differing, when most sharply marked, most considerably from the other forms, with which it is however connected by intermediate conditions. Such a transition to f. *typica*, extremely pretty and characteristic, is the variety recorded by KLEEN from Nordlanden. Other intermediate forms, that I have taken at other places in the Arctic Sea, resemble so closely the figure KÜTZING gives of D. *Lyallii* in Tab. Phyc. 16, t. 14, that they cannot be distinguished in habit from this species. I have pretty often found individuals of f. *lingulata* with some, or most, or all branches of the last order or the last order but one filiform, upwards of 3—4 cm. long. These branches are sometimes terete in their whole length, sometimes flattened at the tip. Some specimens found at Spitzbergen lying loose on loamy bottom are especially remarkable in this respect. HARVEY mentions such specimens from the coasts of Great Britain, cp. l. c. Even the typical form has sometimes such branches of the last order, but these are shorter.

Habitat. This plant everywhere in the Polar Sea is sublitoral or elitoral. It occurs usually in 10—20 fathoms, sometimes in more shallow water, 1½—2 fathoms deep, or at very great depths. It has been dredged at Spitzbergen quite fresh in 85 fathoms. When growing in shallower water, it belongs to the formation of *Laminariaceæ*, in deeper places it is mostly found in company with *Odonthalia dentata*, *Polysiphonia arctica*, *Phyllophora interrupta*, *Ptilota pectinata* and some others. It prefers exposed coasts and a bottom of solid rock, but is also met with in the interior of deep bays and on bottom formed of pebbles, shells and *Lithothamnia*.

During the earlier part of the winter only older specimens were to be found on the north coast of Spitzbergen, but from the beginning of January young individuals became common. Although in the older ones the looser parts of the frond were more or less injured and destroyed, probably by animals, they were however in course of developing new parts, that were easily distinguished by their greater slenderness and their lighter and clearer colour. Older specimens bore tetrasporangia throughout the winter, during the months of November, December, January, February and March. Individuals with sporocarps were found most numerous in February, March, April and May. Older specimens with sporocarps have been met with in great number also in August on the coast of Spitzbergen. The development of tetraspores seems here to reach its maximum in November and December, and it was most probably in tetraspores from this season that the young individuals occurring during the latter part of the winter had their origin. However, tetrasporiferous individuals are found here also in summer, in June, August, and September. KLEEN appears to have met with only sterile specimens at Nordlanden. I have collected tetrasporiferous individuals on the coast of Finmarken in September, in the Siberian Sea in August, and specimens with tetrasporangia and sporocarps in the eastern part of the Murman Sea in the months of June and July. It would thus seem that this alga bears propagative organs of some kind or other all the year round. With regard to the present species on the coast of Scandinavia ARESCHOUG says l. c. »Martii et Aprilis mensibus fructificans», to which may be added that on the coast of Bohuslän I have found individuals with sporocarps and tetrasporangia at the end of December and plants in germination at the beginning of January.

Geogr. Distrib. This species occurs both in the Atlantic and the arctic region of the Polar Sea, being widely distributed in both. It appears to have its maximum of frequency in those parts of the Arctic Sea which extend to the west of the Atlantic. Its northernmost known locality is the North Cape of Spitzbergen Lat. N. 80° 31'.

Localities: *The Norwegian Polar Sea:* Nordlanden, very common and plentiful; Finmarken, common, but not very abundant, as at Maasö, Gjesvær, the south coast of Magerö, Öxfjord, and Talvik.

The Greenland Sea: the north and west coasts of Spitzbergen, common and abundant; the eastern coast of Greenland at several places.

The Murman Sea: on the coasts of Russian Lapland and Cisuralian Samoyedeland, Kolgujew Isle, the west coast of Novaya Zemlya and Waygats, from Matotshin Shar to Jugor Shar, common and abundant.

The White Sea: common and abundant.

The Kara Sea: Uddebay, Actinia Bay, scanty; Kara Bay.

The Siberian Sea: Cape Jakan and the mouth of Koljutshin Bay, pretty common, but not abundant.

The American Arctic Sea: the north coast of Western Eskimaux-land.

Baffin Bay: the west coast of Greenland at Godhavn and Dark Head. Here the species appears to be scarce.

Of the forms mentioned, f. *typica* is the most common and abundant. Nevertheless it does not, as far as I know, occur in the Siberian Sea, being supplied here by f. *lingulata*, which is known, besides, from the east coast of Greenland, the north and west coasts of Siberia, the west coast of Novaya Zemlya and Waygats, and the Norwegian Polar Sea. Forms approaching most nearly to f. *quercifolia* I have seen on the west coast of Spitzbergen and of Novaya Zemlya. WAHLENBERG reports this form from Nordlanden.

Gen. **Nitophyllum** (GREV.) J. G. AG.

Epicr. p. 446; GREV. Alg. Brit. p. 77; lim. mut.

Nitophyllum punctatum (STACKH.) GREV.

Alg. Brit. p. 79. Ulva punctata STACKH. Linn. Trans. 3, p. 236.

Descr. Nithophyllum punctatum J. G. AG. Spec. Alg. 2, p. 659.

Remark on the determination of the species. In the collection of algæ belonging to the Copenhagen Museum there are two specimens, sterile and fragmentary, of a *Nitophyllum*, taken, according to the label affixed by WORMSKIOLD, on the coast of Greenland. I have not been able to determine the species with certainty, but as the specimens resemble *N. punctatum* in structure and agree in habit with the plant figured by KÜTZING (Tab. Phyc. 16. t. 35) under the name of *Aglaophyllum delicatulum*, which is referred to *N. punctatum α ocellatum* by J. G. AGARDH (Epicr. p. 448), I believe I may denominate them *N. punctatum*.

Habitat unknown.

Geogr. Distrib. and *Localities.* Nothing more is known on these points than I have stated above.

Fam. HILDBRANDTIACEÆ HAUCK.

Meeresalg. p. 13.

Gen. **Hildbrantia** NARDO.

Isis 1834, p. 675.

Hildbrandtia rosea KÜTZ.

Phyc. gener. p. 384.

Descr. Hildbrantia rosea J. G. AG. Epicr. p. 379.
Fig. » » KÜTZ. Tab. Phyc. 19 t. 91.
Exsicc. » » ARESCH. Alg. Scand. exsicc. N:o 159.
Syn. Hildbrantia rosea GOBI, Algenfl. Weiss. Meer. p. 23.
» » KJELLM. Spetsb. Thall. 1, p. 4; Algenv. Murm. Meer. p. 8.
» » KLEEN, Nordl. Alg. p. 12.

Habitat. It covers smaller stones, sometimes alone, sometimes together with other algæ, usually *Lithoderma fatiscens.* On the coast of Norway it occurs within the litoral zone, in other parts of the Polar Sea I have found it within the sublitoral zone in 5—10 fathoms water, most often belonging to the formation of *Lithoderma.* It grows at exposed as well as sheltered parts of the coast. Specimens with spores have been taken in June in the Arctic Sea on the west coast of Novaya Zemlya, in the middle af August at Finmarken.

Geogr. Distrib. It is found in the Atlantic as well as the arctic region of the Polar Sea, but it is only little spread within the latter. Its maximum of frequency is in the southern part of the Norwegian Polar Sea. The most northern place where it has been met with is the north-west coast of Spitzbergen about Lat. N. 79° 45′.

Localities: The Norwegian Polar Sea: Nordlanden, common and abundant; Finmarken at Maasö, Gjesvær, and Talvik, rather local and scanty.

The Greenland Sea: the west coast of Spitzbergen, local and scarce.

The White Sea, rare (?).

The Murman Sea, local and scarce.

Fam. SQUAMARIACEÆ (Zanard.) Hauck.

Meeresalg. p. 13; Zanard. Synops. p. 133, sec; J. G. Ag. Spec. Alg. 2, p. 485; lim. mut.

Gen. **Peyssonnelia** Dcsne.

Pl. Arab. p. 168; Class. p. 114.

Peyssonnelia Dubyi Crouan.

Ann. d. Sc. p. 368.

Descr. Peyssonnelia Dubyi J. G. Ag. p. 384.
Fig. » » Harv. Phyc. Brit. t. 71.
Syn. Peyssonnelia Dubyi Dickie, Alg. Sutherl. 1, p. 142.
» » Kleen, Nordl. Alg. p. 15.

Habitat. It has been found in very deep water off the arctic coast of Norway. Here it is sterile during the latter part of the summer. On the coast of Bohuslän it develops reproductive organs in great numbers in winter. In Baffin Bay it has been taken, according to Dickie, in 12—15 fathoms water, attached to stones.

Geogr. Distrib. This alga is known from the southern part of the Norwegian Polar Sea and from Baffin Bay. Its most northern locality is Cape Adair on the east coast of Baffin Land about Lat. N. 71°.

Localities: The Norwegian Polar Sea: Nordlanden, scattered and rather scarce.

Baffin Bay: Cape Adair.

Gen. **Petrocelis** J. G. Ag.

Spec. Alg. 2, p. 489.

Petrocelis Middendorffi (Rupr.) nob.

Cruoria Middendorffi Rupr. Alg. Och. p. 329.

Descr. Cruoria Middendorffi Rupr. l. c. sub Cruoria pellita.
Fig. » pellita » » t. 18, fig. a—b.

Remark on the species. On the coast of Finmarken I have found a species of *Petrocelis*, agreeing very well with that from the Ochotsh Sea described and figured

l. c. by RUPRECHT. He compares it — and he is quite justified in doing so — with *Cruoria pellita* HARV. Phyc. Brit. pl. 117, but he says that the Ochotsh plant differs in some respects from the English one. This accords well with my own observations. The differences between the English-French *Petrocelis cruenta* and RUPRECHT's *Cruoria Middendorffii* are in my opinion so considerable, that they ought to be regarded as separate species. The alga found by me on the coast of Finmarken differs, just as the Ochotsh form, from the more southern *Petrocelis cruenta* by a more strongly developed basal layer, by the vertical cell-rows being often branched, and, above all, by the different shape and position of the tetrasporangia. In *P. cruenta* these are developed above the middle of the vertical cell-rows. Cp. LE JOLIS Liste Alg. Cherb. t. 3 and RUPRECHT l. c.

Habitat. The present species has been found in the Polar Sea growing on rocks and stones between tide-marks on exposed coasts. Here it has young tetrasporangia in course of development in September.

Geogr. Distrib. Within the Polar Sea it is known only from the north coast of Norway. Its only certain *Locality* is Öxfjord at the mouth of Altenfjord.

Petrocelis cruenta J. G. AG.

Spec. Alg. 2, p. 490.

Descr. Petrocelis cruenta J. G. Ag. l. c.
Fig. » » THUR. in Le Jol. Liste Alg. Cherb. t. 3, fig. 3—4.

Syn. Petrocelis cruenta KLEEN, Nordl. Alg. p. 14.

Remark on the determination of the species. In KLEEN's collections from Nordlanden there are some microscopical preparations of that alga in a sterile condition which, in his list of the algæ of Nordlanden, he names *Petrocelis cruenta.* Judging from these preparations, this plant seems really to be another species of *Petrocelis* than the *P. Middendorffii* occurring on the coast of Finmarken. But on the other hand I am unable to decide with certainty whether the plant from Nordlanden is identical with the *P. cruenta* occurring on the coasts of France and England, or with the *P. Ruprechtii* common on the western coast of Sweden. This cannot be decided by means of sterile specimens. Accordingly I think I ought to adhere to the determination made by KLEN.

Habitat. The present alga grows, according to KLEEN, on rocks between tide-marks on exposed coasts as well as in the interior of bays. Only sterile specimens are known from the Polar Sea; they were collected in summer.

Geogr. Distrib. It is known only from the southern part of the Atlantic region of the Polar Sea.

Localities: The Norwegian Polar Sea: Nordlanden, common and abundant.

Gen. **Cruora** (FR.) J. G. AG.

Sp. Alg. 2, p. 490; FR. FL. Scan. p. 316; lim. mut.

Cruoria pellita (LYNGB.) FR.

l. c. p. 317. Chætophora pellita LYNGB. Hydr. Dan. p. 193.
Descr. Cruoria pellita J. G. AG. l. c. p. 491.
Fig. » » THUR. in Le Jol. Liste Alg. Cherb. t. 4.
Exsicc. » » ARESCH. Alg. Scand. exsicc. N:o 309.

Habitat. The present species, like the preceding one, is a litoral alga, at least in the Norwegian Polar Sea, attached to stones. Only sterile specimens have been met with. These were collected in summer. The plant probably here, as on the west coast of Sweden, develops its reproductive organs in winter. About the habitat of the plant in Baffin Bay I know nothing.

Geogr. Distrib. It belongs both to the Atlantic and the arctic region of the Polar Sea, being only little spread in both. It cannot be determined, by means of existing statements, how far it goes northwards.

Localities: The Norwegian Polar Sea: Nordlanden, commonly dispersed, but scanty. *Baffin Bay:* on the west coast of Greenland, according to specimens in the Royal Swedish Museum, brought home by Prof. TH. M. FRIES.

Gen. **Hæmescharia** nob.

Frondes depresso-hemisphæricæ in crustam mucosam confluentes, duobus stratis contextæ, inferiore tenui filis decumbentibus, superiore filis verticalibus muco uberiore laxius conjunctis constante. Fila verticalia triplicis generis: 1:o longiora et tenuiora vegetativa, 2:o longiora et tenuiora trichogynas vulgo plures portantia, 3:o breviora et crassiora partes definitas frondis formantia, quorum articuli, foecundatione peracta, singuli sporam singulam generant.

Hæmescharia polygyna nob.

H. frondibus minutis, purpureo-sanguineis.
Tab. 11.

Description. The plant forms small, flattened-hemispherical, gelatinous masses, which are confluent so as to constitute a crust of about one centimeter in diameter. It consists of a horizontal, feebly developed basal layer composed of cellular filaments connected with one another by a gelatinous substance, and of a thickening layer, that issues from this basal layer and consists of vertical rows of cells held together by a profusion of gelatine and easily separated in pressing. These rows are generally simple and vary in different parts of the frond, being sometimes comparatively long (about 250 μ.) and slender (6—8 μ.), composed of numerous, 15—20, cylindrical cells, which are even twice as long as thick. I believe these cell-rows to be vegetative; fig. 1, 3. In other parts of the frond several of the vertical cell-rows exhibit the appearance shown by fig. 7 and 8, i. e. from the terminal cell or the articular cells there issues a hair-shaped organ, sometimes even two from the same articular cell. These organs resemble trichogynes so much, that I have explained them as such. Again, other parts of the frond, which may be of comparatively rather wide extent, are formed exclusively of far shorter, thicker, and more intensely coloured, vertical cell-rows; fig. 2. These rows are in general not even half as long as the vegetative ones, being generally about 100 μ. in length. The horizontal rows supporting them are also somewhat different from those which give rise to the rows of vegetative cells, their cells being longer and richer in endochrome. It cannot well be doubted that in these cells spores are engendered. The figures 4—6 show this decidedly. Accordingly, in my opinion, they are cystidia. I have not been able to determine the manner of their development. Two cases seem possible: either they are produced independently of any act of fructification and are subsequently developed into cystidia after the fecundating substance has been transferred to them from the trichogynes, or else by the fecundation of the trichogynes there are developed from the cell-rows by which these are supported horizontal rows of long cells rich in endochrome, which produce the rows of cystidia. I have not observed any organ that might be regarded as intended to transfer the fecundating substance from the trichogynes to the carpogons. It is possible, however, that the horizontal cell-rows are able to perform this function.

Mature spores are about 10 μ. in diameter. Besides these organs I have found other formations of the kind represented in fig. 9, the nature of which I cannot decide. They may possibly be young tetraspores that have not yet undergone partition.

I found this little plant occasionally in very little number on stones covered with *Lithoderma fatiscens*. The material at my disposition is slight, and, being dried, it is little fit for a closer examination. If the explanation I have proposed with regard to the different parts be correct, the present plant cannot be referred to any genus of the *Squamariaceæ* that is known to me.

Habitat. I have found this plant attached to *Lithoderma fatiscens* and to stones within the sublitoral zone on an exposed coast. When collected in the middle of September, it had ripe spores.

Geogr. Distrib. It is known at present only from one place in the Siberian Sea.

Localities: The Siberian Sea: Irkaypi, scarce.

Fam. RHODYMENIACEÆ (HARV.) J. G. AG.

Epicr. p. 307; HARV. Brit. Syn. p. VIII; lim. mut.

Gen. **Hydrolapathum** (STACKH.) J. G. AG.

Epicr. p. 369; STACKH. Tentamen. sec. RUPR. Alg. Och. p. 247; char. mut.

Hydrolapathum sanguineum (L.) STACKH.

l. c. Fucus sanguineus L. Mant. p. 136.

Descr. Hydrolapathum sanguineum J. G. AG. l. c. p. 370.

Fig. Delesseria sanguinea HARV. Phyc. Brit. t. 151.
Exsicc. » » ARESCH. Alg. Scand. exsicc. N:o 73.
Syn. Fucus sanguineus GUNN. Fl. Norv. 2, p. 91.
» » WG. Fl. Lapp. p. 491.
Delesseria sanguinea ARESCH. Phyc. Scand. p. 290.
Wormskioldia sanguinea KLEEN, Nordl. Alg. p. 16.

Habit. At Nordlanden this plant occurs on exposed coasts within the sublitoral zone, attached partly to various species of *Laminaria*, partly to stones or shells. It is known here only as sterile, probably on account of its being observed only during summer. Its propagative organs are probably developed in winter and early spring here as well as farther southwards. It grows scattered, in small numbers at each place.

Geogr. Distrib. It is known from the Norwegian Polar Sea and Baffin Bay. As no exact localities have been recorded by others and I have never met with it myself within the Polar Sea, I cannot fix its northern limit. I have not found it on the coast of Finmarken.

Localities: The Norwegian Polar Sea: Nordlanden, rather rare according to WAHLENBERG (l. c.), common according to KLEEN.

Baffin Bay: according to a small specimen in the collections of the Botanical Museum at Copenhagen. The exact locality is not stated.

Gen. **Rhodophyllis** KÜTZ.

Bot. Scit. 1847, p. 23.

Rhodophyllis dichotoma (LEPECH.) GOBI.

Algenfl. Weiss. Meer. p. 35. Fucus dichotomus LEPECH. Commen. Petrop. p. 479, t. 22.
Descr. Rhodophyllis veprecula J. G. AG. Epicr. p. 362.
Fig. » » KÜTZ. Tab. Phyc. 19, t. 52. Cfr. tab. nostra 12, fig. 3.
Exsicc. » » KJELLM. in ARESCH. Alg. Scand. exsicc. N:o 404.
Syn. Calliblepharis ciliata DICKIE, Alg. Sutherl. 1, p. 142.
» » KLEEN, Nordl. Alg. p. 14.
Chondrus crispus ε pumilus LYNGB. Hydr. Dan. p. 16.
Ciliaria fusca RUPR. Alg. Och. p. 251.
Fucus pumilus Fl. Dan. t. 1066.
Rhodophyllis dichotoma GOBI, Algenfl. Weiss. Meer. p. 35.
» veprecula J. G. AG. Spetsb. Alg. Bidr. p. 10.
» » CRALL, Fl. Disc. p. 459.
» » KJELLM. Spetsb. Thall. 1, p. 16; Algenv. Murm. Meer. p. 16.
Rhodymenia ciliata POST et RUPR. Ill. Alg. p. II.
» jubata Nyl. et Sæl. Herb. Fenn. p. 74.
» Sphærococcus ciliatus β fuscus LYNGB. Hydr. Dan. p. 13.

Habitat. This species occurs in the lower parts of the sublitoral zone at a depth of 5—20 fathoms. It prefers pebbly and stony or shelly bottom, but it is usually attached to the algæ together with which it grows, as *Laminariaceæ*, *Delesseria sinuosa* *Ptilota pectinata* a. o. It is most vigorously and typically developed on exposed coasts, but enters also into deep bays, where however not seldom it assumes an appearance very different from the pelagic form. I have figured such a specimen from the interior of Altenfjord on the coast of Finmarken tab. 12. Sometimes it is found gregarious in quite large masses, but in general it constitutes a subordinate element in certain formations, such as those of the *Laminariaceæ* and *Corallineæ*. At Nordlanden it bears sporocarps in July and August, on the west coast of Novaya Zemlya at the end of June and in July, at Spitzbergen in August. I have taken individuals with tetrasporangia at Novaya Zemlya in July.

Geogr. Distrib. It belongs to the Atlantic as well as the arctic region of the Polar Sea, being widely spread within the latter. However, it reaches its maximum of frequency in the northern part of the Norwegian Polar Sea and in the White Sea. The most northern locality where it is known with certainty to grow, is the group of islands on the north-western coast of Spitzbergen, Lat. 79° 50′ N.

Localities: The Norwegian Polar Sea: Nordlanden, pretty common and abundant; Finmarken, pretty common, abundant at Maasö, Gjesvær and in Magerö sound, scarce at Talvik.

The Greenland Sea: scarce and local on the coast of Spitzbergen.

The Murman Sea: Russian Lapland and the coast of Cisuralian Samoyede-land; the west coast of Novaya Zemlya, local but, pretty plentiful.

The White Sea: common and abundant.

Baffin Bay: in several places on the western coast of Greenland, as Tessarmiut, Lichtenau, Julianeshaab, Friderichshaab, Godthaab, Egedesminde, Whale Islands; probably pretty abundant everywhere.

Gen. **Euthora** J. G. Ag.

Alg. Liebm. p. 11; Spec. Alg. 2, p. 383.

Euthora cristata (L.) J. G. Ag.

l. c. Fucus cristatus Lin. in Turn. Hist. Fuc. 1, p. 48.

f. *typica*.

Descr. Euthora cristata J. G. Ag. Epicr. p. 360.
Fig. Rhodymenia cristata Harv. Phyc. Brit. t. 307.
Exsicc. Euthora cristata Aresch. Alg. Scand. exsicc. N:o 308.

f. *angustata* Lyngb.

Hydr. Dan. p. 13.

Descr. Sphærococcus cristatus β angustatus. Lyngb. l. c.
Syn. Euthora cristata J. G. Ag. Spetsb. Alg. Progr. p. 3; Bidr. p. 11; Grönl. Alg. p. 111.
» » Croall, Fl. Disc. p. 459.

Euthora cristata DICKIE, Alg. Sutherl. 1, p. 142; Alg. Cumberl. p. 238.
» » EATON, List, p. 44.
» » GOBI, Algenfl. Weiss. Meer. p. 33.
» » HARV. Ner. Am. 2, p. 150.
» » KJELLM. Vinteralg. p. 64; Spetsb. Thall. 1, p. 16; Algenv. Murm. Meer. p. 16; Kariska hafvets Algv. p. 20.
» » KLEEN, Nordl. Alg. p. 17.
Fucus coccineus β pusillus WG. Fl. Lapp. p. 500.
» cristatus SOMMERF. Suppl. p. 184.
» » gigartinus GUNN. Fl. Norv. 2, p. 106; fide syn.
Nereidea cristata RUPR. Alg. Och. p. 255 sub Nereidea fruticulosa.
Rhodymenia cristata ARESCH. Phyc. Scand. p. 299.
» » Nyl. et Sæl. Herb. Fenn. p. 74.
» » POST. et RUPR. Ill. Alg. p. II.

Habitat. In the Arctic Sea this alga is generally sublitoral or elitoral. According to KLEEN it sometimes occurs at Nordlanden in rock-pools within the litoral zone. It prefers an open coast, though it may enter also into deep bays. It is usually attached to larger algæ, as *Laminaria digitata, Ptilota pectinata, Delesseria sinuosa* and *Lithothamnia,* more seldom to stones or shells. In most places it grows scattered, with isolated specimens. In the western part of the Kara Sea f. *angustata* was found gregarious in rather great masses. At Spitzbergen it continues to develop all the year round. I have found it at Musselbay growing and bearing sporocarps during the winter, although it attains here its highest development during the summer and the autumn. It has been found with sporocarps at Nordlanden during the whole summer, at Finmarken in August and September, at Spitzbergen in July and in January, on the west coast of Novaya Zemlya in July; with tetrasporangia at Nordlanden during the whole summer, at Finmarken in September, at Spitzbergen in July, in the White Sea at the end of June and the beginning of July. It seems thus as if this species could develop sporocarps in the Arctic Sea all the year round, but tetrasporangia chiefly during the summer.

Geogr. Distrib. It belongs to the Atlantic as well as the arctic region, and is widely distributed in the latter. However it is not known as yet from the Siberian Sea. I have found it most abundant in the eastern part of the Kara Sea. Its northernmost locality is Treurenberg Bay on the north coast of Spitzbergen, Lat. 79° 56′ N.

Localities: The Norwegian Polar Sea: Nordlanden, common, though not abundant; Finmarken scattered, but pretty plentiful, at Maasö, Gjesvær, the south coast of Magerö, Öxfjord and Talvik.

The Greenland Sea: at several places on the west and north coasts of Spitzbergen, but everywhere scanty.

The Murman Sea: the coast of Russian Lapland and Cisuralian Samoyede-land, Kolgujev Isle, the west coast of Novaya Zemlya and Waygats, at the last-mentioned places scattered and rare.

The White Sea: pretty common.

The Kara Sea: in Uddebay on the east coast of Novaya Zemlya, abundant.

The American Arctic Sea: according to HARV. Ner. Am. l. c.

Baffin Bay: Cumberland Sound; the west coast of Greenland at Godthaab, Holstenborg, Hunde Islands, Jakobshavn and Disco Island (?).

Of the two forms the more common one is f. *angusta*, as far as my observations go.

Gen. **Plocamium** (LAMOUR.) LYNGB.

Hydr. Dan. p. 39; LAMOUR. Ess. p. 137; lim. mut.

Plocamium coccineum (HUDS.) LYNGB.

l. c. Fucus coccineus HUDS. Fl. Angl. p. 586.

f. *typica.*

Descr. Plocamium coccineum J. G. AG. Epicr. p. 339.
Fig. » » HARV. Phyc. Brit. t. 44, fig. 1—8.
Exsicc. » » ARESCH. Alg. Scand. exsicc. N:o 203.

f. *uncinata* AG.

Delesseria Plocamium d uncinata AG. Spec. Alg. 1, p. 181.
Descr. Plocamium coccineum var. uncinatum J. G. AG. Epicr. p. 339.
Fig. » » β uncinata HARV. Phyc. Brit. t. 44, fig. 9.
Syn. Plocamium coccineum KLEEN, Nordl. Alg. p. 17.

Habitat. According to KLEEN it grows on deep stony and shelly bottom, and bears sporocarps and tetrasporangia in July and August at Nordlanden.

Geogr. Distrib. It is known only from the Atlantic region of the Polar Sea.

Localities: The Norwegian Polar Sea: Nordlanden, the most northern locality of the plant. It is generally pretty scarce here, being reported common by KLEEN only from one place, viz. Röst. Both the typical form and f. *uncinata* are to be found in KLEEN's collections.

Gen. **Rhodymenia** (GREV.) J. G. AG.

Alg. Liebm. p. 15; GREV. Alg. Brit. p. 84; char. mut.

Rhodymenia palmata (L.) GREV.

l. c. p. 93. Fucus palmatus L. Spec. Pl. 2, p. 1162.

f. *typica.*

α. *nuda* nob.

f. fronde a stipite cuneatim dilatata, dichotoma, subpalmata, margine nudo.

Descr. Fucus palmatus Turn. Hist. Fuc. 2, p. 114.
Fig. » » » » » t. 115, fig. a.

β. marginifera HARV.

Rhodymenia palmata β marginifera HARV. Phyc. Brit. t. 217.

f. antecedenti similis, at fronde e margine parce prolificante, prolificationibus oblongis.

Cfr. Rhodymenia palmata ARESCH. Alg. Scand. exsicc. N:o 154, quæ est forma inter hanc et f. proliferam intermedia.

f. *sarniensis* MERT. (GREV.)

Alg. Brit. p. 93. Fucus sarniensis MERT. in ROTH. Cat. Bot. 3, p. 103.

α. latiuscula nob.

Descr. Rhodymenia palmata β sarniensis J. G. AG. Spec. Alg. 2, p. 377.
Fig. Sphærococcus sarniensis KÜTZ. Tab. Phyc. 18, t. 88.

β. tenuissima TURN.

Descr. Fucus sarniensis β tenuissimus TURN. Hist. Fuc. 1, p. 96.
Exsicc. Rhodymenia palmata var. sobolifera ARESCH. Alg. Scand. exsicc. N:o 155.

f. *prolifera* KÜTZ.

Descr. Sphærococcus palmatus γ prolifer KÜTZ. Spec. Alg. p. 781.

α. purpurea nob.

f. fronde intense purpurea, prolificationibus vulgo oblongis, latitudine 2—3 cm. et ultra metientibus, raro iterum prolificantibus.

β. pallida nob.

f antecedenti similis at tota planta colore pallido, prolificationibus in lilacinum vergentibus.

Descr. Rhodymenia palmata ARESCH. Phyc. Scand. p. 298.
Exsicc. » » Alg. Scand. exsicc. N:o 9.

f. *angustifolia* nob.

f. fronde intense purpurea, prolificationibus oblongo-lanceolatis, angustis centimetrum latitudine vix excedentibus, vulgo iterum prolificantibus.

Syn. Fucus caprinus Fl. Dan. t. 1128.
» ovinus GUNN. Fl. Norv. 2, p. 96.
» palmatus WG. Fl. Lapp. p. 497.
Halymenia palmata LINDBL. Bot. Not. p. 157.
» » POST. et RUPR. Ill. Alg. p. 11.
» » SCHRENK, Ural. Reise, p. 547.
Rhodymenia palmata J. G. AG. Spetsb. Alg. Progr. p. 3; Bidr. p. 11; Till. p. 28; Grönl. Alg. p. 111.
» » ARESCH. Phyc. Scand. p. 298.
» » CROALL, Fl. Disc. p. 460.
» » DICKIE, Alg. Cumberl. p. 238.
» » EATON, List, p. 44.
» » GOBI, Algenfl. Weiss. Meer. p. 33.
» » KJELLM. Vinteralgv. p. 64; Spetsb. Thall. 1, p. 15; Algenv. Murm. Meer. p. 15
» » KLEEN, Nordl. Alg. p. 16.
» » Nyl. et Sæl. Herb. Fenn. p. 74.
Ulva caprina GUNN. Fl. Norv. 2, p. 127.
» delicatula » » » » p. 134.
» » Fl. Dan. t. 1190.
» palmata SOMMERF. Spitsb. Fl. p. 232.

Remark on the determination and the synonymy of this species. In the above attempt to group the arctic forms of this many-shaped species the forms constitute two series. The former of these, comprising f. *typica* and *sarniensis*, is characterized by the majority of the secondary axes of the frond being formed by repeatedly subdichotomous branching, the latter, comprehending f. *prolifera* and f. *angustifolia*, by all, or most, secondary axes being so-called prolifications. The specimens of the typical form, most common in the arctic parts of the Polar Sea, are very magnificent, large, and high-coloured. They are undoubtedly the largest *Florideæ* of this region. The typical form occurs more rarely in the habit represented by pl. 217 in Harv. Phyc. Brit. and named by HARVEY f. *marginifera*. It is easily seen, however, that the diagnosis, borrowed from Stackh. Ner. Brit. (p. 54), which accompanies the figure in HARVEY does not accord with the plant delineated The forms *sarniensis* and *prolifera* are sometimes rather difficult to distinguish from each other. The form distributed by ARESCHOUG in his work of Scandinavian exsiccates under the name of f. *sobolifera* approaches more nearly to f. *sarniensis*, especially to *β tenuissima*. I have assumed that KÜTZING by his *Sphærococcus palmatus γ prolifer* means the arctic broad-fronded, high-coloured form (*α purpurea* nob.). But I am not certain of it. This form is nearly related to what I have called f. *prolifera β pallida*, which is common in Kattegat and Skagerack and is possibly to be found also in the southern part of the Norwegian Polar Sea. However, *β pallida* differs so much in colour as to deserve special mention. Another form, very beautiful, which connects the two series with each other, but accords most closely with f. *prolifera*, is f. *angustifolia*, a prolificating form with narrow frond, that attains sometimes half a foot in length and becomes much branched by repeated prolifications. Sometimes it is very little, 1—2 inches long, with very small prolifications. It is probably a stunted form either of f. *angustifolia* or of f. *prolifera* that J. G. AGARDH has named f. *microphylla* in his list of the algæ brought home from Spitzbergen by the Swedish expedition of 1868 and distributed by him.

Habitat. In the Greenland Sea and the eastern part of the Murman Sea as well as on the west coast of Sweden the present species is sublitoral, in the Norwegian Polar Sea it is litoral. I do not know how it is in the other parts of the Arctic Sea. In the first-mentioned seas it grows on stony bottom at a depth varying between 3, seldom less, and 15 fathoms, sometimes gregarious, sometimes scattered together with *Fucoideæ*. On the coast of Norway it keeps chiefly to the lower part of the zone of the *Fucaceæ*, usually covering stones and rocks in large, dense masses together with *Fucaceæ*. At more free places where it is less covered and oppressed by other algæ, it appears here in its typical form or as a large-sized f. *prolifera α purpurea*. When growing, on the contrary, among dense masses of *Fucaceæ*, by which it is covered at low-tide, it assumes the habit charakteristic of f. *sarniensis*, *angustifolia*, or dwarfed *prolifera*. It occurs chiefly on exposed coasts, but enters also into deep bays, even where the water is comparatively little salt during some parts of the day. Cp. KLEEN, Nordl. Alg. p. 17. On the north coast of Spitzbergen the present plant is developing all the year round, even in winter. During the whole time from the beginning of November to May, accordingly during the whole dark and cold season, I have found here partly young plants, from such ones as were in germination to individuals of about one inch in length, partly older plants, some of which were in full vigour, developing new shoots, while others were more or less corroded and dissolved. From the beginning of December to the middle of May young individuals were common; most of them were sterile, others developing tetrasporangia. The older specimens during this period were almost always furnished with tetrasporangia. As far as I could see, the present alga on the coast of Spitzbergen is only annual, but during its life produces tetrasporangia at least twice, once when young, the other time immediately before dying. It is possible, however, that it develops such organs several times, for on the north and north-west coasts of Spitzbergen it bears ripe tetrasporangia also during the summer and the earlier part of the autumn, in July and October, especially in July, when it reaches its highest vegetative development. On the coast af Finmarken I have found only sterile specimens. At Nordlanden it seems also to be sterile in summer. KLEEN does not mention having found any individuals with reproductive organs during his summer visits there, nor are any to be found in his collections. The season for developing the tetrasporangia is probably here as farther southwards the winter. On the west coast of Novaya Zemlya I have collected specimens richly furnished with tetrasporangia in June and July.

Geogr. Distrib. The species appears to be limited to those parts of the Polar Sea which are situate north of the Atlantic. Although certainly attaining its maximum of frequency in the Norwegian Polar Sea, it is commonly spread and often met with in very large numbers even in the arctic parts of the Polar Sea. The northernmost place where it has been found as yet is Musselbay on the north coast of Spitzbergen Lat. 79° 53' N.

Localities: *The Norwegian Polar Sea:* Nordlanden, common and abundant; Finmarken, common and abundant at Maasö, Gjesvær, the south coast of Magerö, Öxfjord, Talvik.

The Greenland Sea: the coast of Russian Lapland, the west coast of Novaya Zemlya and Waygats from Matotshin Shar to Jugor Shar, common and abundant at least at the latter place.

The White Sea: rare, according to GOBI l. c.

Baffin Bay: Cumberland Sound, the west coast of Greenland in several places, as Tessarmiut, Nanortalik, Kakortok, Sukkertoppen, and Jakobshavn.

All the forms of the present species are met with in the Norwegian Polar Sea; besides, f. *typica* is found in the Greenland Sea, the eastern Murman Sea, and Baffin Bay, f. *sarniensis*, or a form most nearly related to this, in the western Murman Sea.

Rhodymenia pertusa (POST. et RUPR.) J. G. AG.

Spec. Alg. 2, p. 376; Porphyra pertusa POST. et RUPR. Ill. Alg. p. 20.

Descr. Rhodymenia pertusa J. G. AG. Spec. Alg. 2, p. 376 et Epicr. p. 329.

Fig. Porphyra pertusa POST. et RUPR. l. c. t. 36.

Syn. Rhodymenia pertusa KJELLM. Spetsb. Thall. 1, p. 15.

Habitat. It grows within the sublitoral zone on exposed coasts, scattered, as an element of the formation of *Laminariaceæ.* I have found individuals with tetrasporangia in August on the north coast of Spitzbergen.

Geogr. Distrib. It is known only from the arctic region, where it is but little spread. Its maximum of frequency is probably in Baffin Bay. Its northernmost locality is Fairhaven on the north-west coast of Spitzbergen, Lat. N. 79° 49'.

Localities: *The Greenland Sea*: the northwest coast of Spitzbergen, scattered and scanty.

Baffin Bay: on the west coast of Greenland. Here it appears to be pretty common. However, I know with certainty only one locality where it is found, namely Godthaab.

Fam. CHAMPIACEÆ J. G. Ag.

Epicr. p. 290.

Gen. Chylocladia (Grev.) J. G. Ag.

l. c. p. 295; Grev. in Hook Brit. Fl. 2, p. 297.

Chylocladia clavellosa (Turn.) Grev.

l. c. Fucus clavellosus Turn. Linn. Trans. 6, p. 133.
Descr. Chylocladia clavellosa J. G. Ag. Epicr. p. 297.
Fig. Chrysymenia clavellosa Harv. Phyc. Brit. t. 114.
Exsicc. Chylocladia clavellosa Aresch. Alg. Scand. exsicc. N:o 72.
Syn. Lomentaria clavellosa Kleen, Nordl. Alg. p. 16.

Habitat. It grows according to Kleen in deep water on stony bottom, bearing sporocarps and tetrasporangia in July and August at Nordlanden.

Geogr. Distrib. It is known only from the southern part of the Atlantic region of the Polar Sea.

Localities: *The Norwegian Polar Sea:* Nordlanden, generally scarce.

Chylocladia articulata (Huds.) Grev.

in Hook. Brit. Fl. 298. Ulva articulata Huds. Fl. Angl. p. 569.
Descr. Chylocladia articulata J. G. Ag. Epicr. p. 301.
Fig. » » Harv. Phyc. Brit. t. 283.
Exsicc. Lomentaria articulata Aresch. Alg. Scand. exsicc. N:o 7.
Syn. Lomentaria articulata Kleen, Nordl. Alg. p. 16.

Habitat. At that place where it has been found in the Polar Sea, it is litoral growing beneath *Fucus* and *Ozothallia*, according to Kleen. Only small sterile specimens are known from the Polar Sea.

Geogr. Distrib. It has hitherto been recorded only from one locality in the southern part of the Norwegian Polar Sea.

Locality: *The Norwegian Polar Sea*: Nordlanden at Fleinvær Isles.

Fam. DUMONTIACEÆ J. G. AG.

Epicr. p. 249. Spec. Alg. p. 346; char. mut.

Gen. Sarcophyllis (KÜTZ.) J. G. AG.

Epicr. p: 263; KÜTZ. Phyc. gener. p. 401; char. emend.

Sarcophyllis edulis (STACKH.) J. G. AG.

l. c. p. 265. Fucus edulis STACKH. Ner. Brit. p. 57.

Descr. Sarcophyllis edulis J. G. AG. l. c.
Fig. Iridæa edulis HARV. Phyc. Brit. t. 97.
Exsicc. » » ARESCH. Alg. Scand. exsicc. N.o 78.
Syn. Iridæa edulis POST et RUPR. Ill. Alg. p. II.
Sarcophyllis edulis GOBI, Algenfl. Weiss. Meer. p. 39.
Schizymenia edulis KJELLM. Spetsb. Thall. 1, p. 23.
» » KLEEN, Nordl. Alg. p. 19.

Habitat. Growing scattered on stony bottom in the deeper parts of the sublitoral zone, in localities exposed to the open sea. Found with tetrasporangia in July at Nordlanden.

Geogr. Distrib. Though this species is found in the Atlantic as well as the arctic region of the Polar Sea, it is rare here and beyond its proper sphere. The most northern locality where it has been collected is Geese Islands in Icefjord on the west coast of Spitzbergen about Lat. N. 78° 30'.

Localities: *The Norwegian Polar Sea:* Nordlanden at Givær.

The Greenland Sea: at the place just mentioned on the west coast of Spitzbergen. One individual found.

The White Sea: Tri-Ostrowa. One specimen.

Sarcophyllis arctica KJELLM.

Algenv. Murm. Meer. p. 17.

Descr. Sarcophyllis arctica KJELLM. l. c.
» Kallymenia? integra » Spetsb. Tall. 1, p. 19.
Fig. » » » » » t. 1, fig. 8—9.
» Sarcophyllis arctica tab. nostra 14, fig. 1—3.
Syn. Sarcophyllis arctica GOBI, Algenfl. Weiss. Meer. p. 39, sub. S. eduli sec. spec. benevole communic.
Kallymenia? integra KJELLM. Kariska hafvets Algv. p. 21.
» Pennyi DICKIE, Alg. Cumberl. p. 238; saltem ex parte.

Habitat. This species is sublitoral, usually growing scattered within different formations of algæ. I have found it most numerous and luxuriant together with *Laminariaceæ* in the most easterly part of the Siberian Sea. It prefers exposed coasts, but enters also into deep bays. On the north coast of Spitzbergen I have found only young individuals in winter. On the coasts of Spitzbergen as in other parts of the Arctic Sea it reaches its highest development in summer and early autumn. At this time it possesses in great numbers those organs which are explained as procarps by

J. G. Agardh. From this fact we may conclude that late autumn is its proper period for producing carpospores. I have never had the opportunity of examining the plant at this time. In my collections from the west coast of Novaya Zemlya there is a specimen with almost mature sporocarps taken in July. These are figured in the plate quoted above. Tetrasporangia are unknown.

Geogr. Distrib. The present species is a purely arctic alga very widely distributed within the arctic region. It is not known from the northern Atlantic, but in the northern part of the Pacific I have collected specimens of it. With the exception of the American Arctic Sea, it is known to me from all larger stretches of the Arctic Sea that have been investigated. Its maximum of frequency is in the eastern part of the Siberian Sea. Its northernmost locality, as far as is ascertained at present, is Musselbay on the north coast of Spitzbergen Lat. N. 79° 53'.

Localities: The Greenland Sea: the north and west coasts of Spitzbergen, local and scarce.

The Murman Sea: Cape Kanin according to specimens in Ruprecht's herbarium [1]); the west coast of Novaya Zemlya and Waygats from Matotshin Shar to Jugor Shar, common and pretty abundant.

The Kara Sea: Uddebay on the east coast of Novaya Zemlya, scarce.

The Siberian Sea: Irkaypi, Koljutshin Isle, Pitlekay, Tjakpa; at Pitlekay plentiful.

Baffin Bay: Cumberland Sound. I have supposed that the alga described by Dickie from here under the name of *K. Pennyi* is *S. arctica*, at least partly. The description accords perfectly with this, but it is too incomplete to afford full certainty.

In the above-mentioned collection of algæ made by Moravian missionaries there were also found a couple of specimens of *Sarcophyllis arctica*. These had come either from the west coast of Greenland or from Labrador.

Gen. **Halosaccion** (Kütz.) Rupr.

Alg. Och. p. 292; Kütz. Phyc. gener. p. 439: char. mut.

Halosaccion ramentaceum (L.) J. G. Ag.

Spec. Alg. 2, p. 358. Fucus ramentaceus. L. Syst. Nat. 2, p. 718, sec. Turn. Hist. Fuc. 3, p. 33.

f. *robusta* nob.

Planta annotina solitaria, thalli parte cauloidea simplici, prolificationes simplices, raro furcatas, distantes, apicem versus crebriores, usque pedales, diametro 0,5—2 mm., tetrasporangiis evolutis dejiciendas emittente.

Fig. Halosaccion ramentaceum f. robusta tab. nostra 12, fig. 4 et tab. 13, fig. 1—2.

Syn. Fucus graminifolius Lepech. Comment. Petrop. p. 481.

[1]) I am under the greatest obligation to Dr Chr. Gobi for having had the advantage of examining these specimens.

f. *ramosa* nob.

Planta annotina solitaria, thalli parte cauloidea parce vage ramosa, axi primario ramisque prolificationes simplices, raro furcatas, distantes usque digitales et ultra, diametro 0,5—2 mm., tetrasporangiis maturis dejiciendas emittentibus.

Fig. Halosaccion ramentaceum f. ramosa tab. nostra 13, fig. 4.

α. major nob.

Prolificationes digitales et ultra, diametro circa 2 mm.

Syn. Fucus tubulosus LEPECH. Comment. Petrop. p. 476.

β. minor nob.

Prolificationes pollicares, diametro 0,5—1 mm.

f. *subsimplex* RUPR.

Alg. Och. p. 270.

Descr. Halosaccion soboliferum var. subsimplex RUPR. l. c.
Fig. » ramentaceum f. subsimplex tab. nostra 13, fig. 3.

f. *densa* nob.

Descr. Scytosiphon ramentaceus LYNGB. Hydr. Dan. p. 61.
Fig. Fucus ramentaceus TURN. Hist. Fuc. t. 149.
Exsicc. Halosaccion ramentaceum ARESCH. Alg. Scand. exsicc. N:o 205.
Syn. Dumontia Lepechini POST et RUPR. Ill. Alg. p. II.
» ramentacea ARESCH. Phyc. Scand. p. 313.
» » Nyl. et Sæl. Herb. Fenn. p. 74.
» sobolifera DICKIE, Alg. Sutherl. 2, p. 191.
Fucus barbatus GUNN. Fl. Norv. 2, p. 129.
» ramentaceus GUNN. Fl. Norv. p. 79.
» » WG. Fl. Lapp. p. 504.
Gracilaria confervoides J. G. AG. Spetsb. Alg. Bidr. p. 11, sec. spec.
» » β procerrima POST. et RUPR. Ill. Alg. p. II—III sec. RUPR. Alg. Och. p. 274.
Halosaccion fistulosum RUPR. Alg. Och. p. 273.
» ramentaceum J. G. AG. Spetsb. Alg. Progr. p. 3; Bidr. p. 11; Grönl. Alg. p. 111.
» » CROALL, Fl. Disc. p. 460.
» » DICKIE, Alg. Cumberl. p. 239; Alg. Sutherl. 1, p. 143.
» » GOBI, Algenfl. Weiss. Meer. p. 38.
» » HARV. Ner. Amer. 2, p. 194.
» » KJELLM. Vinteralgv. p. 64; Spetsb. Thall. 1, p. 17; Algenv. Meer. p. 17.
» » KLEEN, Nordl. Alg. p. 18.
» soboliferum RUPR. ALG. Och. p. 268.
Halymenia ramentacea LINDBL. Bot. Not. p. 157.
Scytosiphon ramentaceus LYNGB. Hydr. Dan. p. 61.
Soliera chordalis ASHM. Alg. HAYES, p. 96(?) Cfr. FARL. New Engl. Alg. p. 148.

Remark on the definition of the form. Halosaccion ramentaceum is a very multiform species in the Arctic Sea. The forms I have recorded and tried to characterize above are far from being the only ones that occur, but the others may be grouped round them. Though it may certainly seem as if the form named by me f. *robusta* and the most known form, f. *densa*, were so essentially different from each other that they ought to be regarded as distinct species, still they are connected by numerous intermediate forms.

I have based the determination of the forms on the first year's plant, because it is in the first year, i. e. at the first time when the plants bear the prolifications producing tetrasporangia, that the difference between them comes out most sharply. These prolifications having for the greatest part fallen off and new, shorter ones being developed from the remaining stumps, especially the forms *robusta* and *ramosa* become very like f. *densa* in habit.

Besides the branching and the size and shape of the prolifications, there are other differences in colour and consistency between these forms, but these characteristics are found to change in the same individual at different stages of its development. It may be stated in general that f. *densa* is more cartilaginous than the others and that the prolifications in f. *robusta* are almost membranaceous, so that in drying they fall together and become flat. Younger individuals are more intensely coloured than older ones and pelagic forms more so than such as have grown in sheltered places.

The forms *robusta* and *ramosa* are in all probability not before unknown, no more than f. *subsimplex* and f. *densa*. For it does not appear doubtful to me that f. *robusta* is LEPECHIN'S *Fucus graminifolius* figured in Comment. Petrop. pl. 23, and that f. *ramosa* is the same author's *Fucus tubulosus* given in pl. 20. GOBI is indeed of opinion that *F. graminifolius* is a *H. ramentaceum*, but on the other hand he assumes *Fucus tubulosus*, which J. G. AGARDH and RUPRECHT refer also to this species, to be *Dumontia filiformis*. GOBI rests this assumption on the resemblance that is to be seen between LEPECHIN'S figure of *Fucus tubulosus* and HARVEY'S figure of *Dumontia filiformis* in Phyc. Brit. It is easily perceived that the strength of this demonstration is considerably weakened by that figure of what is assuredly a specimen of *Halosaccion ramentaceum* from Spitzbergen which I have given in pl. 13, fig. 4, and which might almost be thought to be a copy of the above-mentioned figure by LEPECHIN. However, there being still some uncertainty with respect to LEPECHIN'S two species of *Fucus* and his names being moreover, in case these so-called *Fuci* are identical with the *Halosaccia* in question, unsuitable and misleading, because the leaves in neither of them are flat and resemble the leaves of grasses, but are tubulous in both, although the wall has a more solid structure in the one than in the other, I have thought fit to choose new names for the forms now distinguished, stating however expressly at the same time that I hold *Halosaccion ramentaceum* f. *robusta mihi* to be most probably identical with *Fucus graminifolius* LEPECH. and f. *ramosa mihi* with *Fucus tubulosus* LEPECH.

Habitat. The form *densa* is litoral in the Norwegian Polar Sea, but in the other parts of the Arctic Sea it is, like the other forms of the present species, sublitoral, as far as my experience goes. Forma *robusta* and f. *ramosa α major*, probably also f. *subsimplex*, may be regarded as chiefly pelagic; even f. *densa* is most richly developed on exposed coasts, although it enters also into deep bays; f. *ramosa β minor*, on the contrary, prefers sheltered places, being most typically developed in bays with a loose bottom consisting of pebbles and small stones. It keeps generally in shallow water, at a depth of 1—4 fathoms, whereas the larger and more vigorous pelagic forms are often met with at a depth of 6—10 fathoms. The litoral form, f. *densa*, occurs gregarious, the others grow scattered, often constituting an element of the the *Laminariaceæ*-formation.

On the north coast of Spitzbergen the plant is to be found during the whole winter, partly with the prolifications fallen off — this is often the case —, partly with prolifications in course of development from the remaining portions — this is the case especially in March —, partly with older prolifications remaining and furnished with tetrasporangia. However, it reaches its highest development at Spitzbergen in July and August and at this season is also most richly provided with ripe tetrasporangia. At Spitzbergen specimens with tetrasporangia have been collected in April, July, August, October, November and December, on the west coast of Novaya Zemlya in July, on the north coast of Norway in July and August, in the Siberian Sea in September, on the west coast of Greenland in February. With regard to those peculiar outgrowths resembling nemathecia which are sometimes to be found in this plant, and the nature of which I am as yet unable to explain, I refer to my paper on the *Florideæ* of Spitzbergen (Spetsb. Thall. I, p. 18).

Geogr. Distrib. This species is circumpolar. It has its maximum of frequency in the northern part of the Norwegian Polar Sea, the eastern part of the Greenland Sea and the eastern part of the Murman Sea. The collections of algæ from Greenland contain a considerable number of individuals of this species, so that it may be supposed to be common and abundant there. Already in the southern part of the Norwegian Polar Sea it is scarce. The most northerly place where it has been found is Low Island on the north coast of Spitzbergen, Lat. N. 80° 20'.

Localities: The Norwegian Polar Sea: Nordlanden, local and scarce; Tromsö amt at Renö; Finmarken at Maasö, Gjesvær, the south coast of Magerö, Öxfjord, Talvik, common and abundant; at Vardö and Vadsö according to GUNNERUS l. c.

The Greenland Sea: common and abundant along the west and north coasts of Spitzbergen.

The Murman Sea: the coast of Russian Lapland; the west coast of Novaya Zemlya and Waygats from Matotshin Shar to Jugor Shar, common and pretty abundant.

The White Sea: rather local and scarce.

The Kara Sea: Cape Palander and Actinia Bay, local and rather scarce.

The Siberian Sea: Irkaypi, Koljutshin Isle, and Pitlekay, scanty, but pretty common.

The American Arctic Sea. Cp. HARVEY, l. c.

Baffin Bay: Cumberland Sound, pretty abundant; the west coast of Greenland: Nanortalik, Smallesund, Neuherrnhut, Godthaab, Sukkertoppen, Holstenborg, Disco Island, Jakobshavn, Sakak, Rittenbenk, Whale Island (washed ashore), Lat. N. 73° 26' and (?) in Smith's Sound between Lat. N. 78 and 82°, if, as I suppose, the plant reported from there by ASHMEAD under the name of *Soliera chordalis* is some form or other of *Halosaccion ramentaceum*.

Halosaccion saccatum (LEPECH.)

Fucus saccatus LEPECH. Comment. Petrop, p. 478.

Descr. Fucus saccatus LEPECH. l. c.

Fig. » » » tab. 21.

Remark on the species. There is no doubt but that LEPECHIN'S *Fucus saccatus* is a species of *Halosaccion,* resembling those species of this genus which are to be found in the northern part of the Pacific. The occurrence of such a species on the arctic coast is not at all impossible and the statement about the locality seems to be reliable. Accordingly this species ought to be recorded now, I think, in the Flora of the Arctic Sea.

Habitat. LEPECHIN says: ad instar glomerum integros investit lapides.

Geogr. Distrib. and *Locality.* It is reported by LEPECHIN from the Tri-Ostrowa Islands in the White Sea.

Gen. Dumontia (LAMOUR.) J. G. AG.

Spec. Alg. 2, p. 348, LAMOUR. Ess. p. 133; char. mut.

Dumontia filiformis (Fl. Dan.) GREV.

Alg. Brit. p. 165. Ulva filiformis Fl. Dan. t. 1480.

Descr. Dumontia filiformis J. G. AG. Epicr. p. 257.

Fig. » » KÜTZ. Tab. Phyc. 16, t. 81.

Exsicc. » » ARESCH. Alg. Scand. exsicc. N:o 79 et 157.

Syn. Dumontia contorta RUPR. Alg. Och. p. 295.

» filiformis ARESCH, Phyc. Scand. p. 312.

» CROALL, Fl. Disc. p. 459.

» GOBI, Algenfl. Weiss. Meer. p. 37.

» Nyl. et Sæl. Herb. Mus. Fenn. p. 74.

Ulva filiformis SOMMERF. Suppl. p. 187.

» » WG. Fl. Lapp. p. 508.

Habitat. This is a litoral alga, usually occurring in rock-pools between tides. It prefers sheltered places and is generally found somewhat gregarious. KLEEN has found it with sporocarps and tetrasporangia in the southern part of the Norwegian Polar Sea in the month of June.

Geogr. Distrib. It belongs to the Atlantic as well as the arctic region of the Polar Sea, but is not much spread in the latter. Its maximum of frequency is in the most southerly parts of the Polar Sea. I did not see it anywhere at Finmarken, but this may possibly have been caused by my investigations being carried on there too late in the year. For it is not unlikely that the development of this alga, here as well farther southwards, is finished within the earlier part of the year. It has been found here by WAHLENBERG in Altenfjord, which is the northernmost known locality of the species.

Localities: The Norwegian Polar Sea: Nordlanden, common; Tromsö amt at the town of Tromsö, pretty common and abundant; Finmarken in Altenfjord.

The Murman Sea: the coast of Russian Lapland and Cisuralian Samoyede-land.

Baffin Bay: the west coast of Greenland at Jakobshavn and Rittenbenk.

Fam. FURCELLARIACEÆ J. G. AG.

Epicr. p. 240.

Gen. **Furcellaria** LAMOUR.

Ess. p. 45.

Furcellaria fastigiata (L.) LAMOUR.

l. c. p. 46. Fucus fastigiatus L. Spec. Pl. 2, p. 1162.

f. *typica.*

Descr. Furcellaria fastigiata J. G. AG. Epicr. p. 241.
Fig. » » HARV. Phyc. Brit. t. 94 et 357 A.
Exsicc. » » ARESCH. Alg. Scand. exsicc. N:o 256.

f. *tenuior* ARESCH.

Alg. Scand. exsicc. N:o 257.

Syn. Fastigiaria furcellata GOBI, Algenfl. Weiss. Meer. p. 43.
Fucus caprinus GUNN. Fl. Norv. 1, p. 96.
» furcellatus » » » 2, p. 78.
» lumbricalis β WG. Fl. Lapp. p. 503.
Furcellaria fastigiata J. G. AG. Spetsb. Alg. Bidr. p. 11.
» » KLEEN, Nordl. Alg. p. 19.
» » Nyl. et Sæl. Herb. Fenn. p. 74.
» » POST. et RUPR. Ill. Alg. p. II.

Habitat. I have failed myself to find this species in the Polar Sea. According to KLEEN and WAHLENBERG it is litoral or sublitoral in the southern part of the Norwegian Polar Sea. In the other parts of the Polar Sea it probably keeps within the sublitoral zone. KLEEN has found it with sporocarps at Nordlanden in June, which is a remarkable fact, because farther to the south the plant seem to fruit in winter.

Geogr. Distrib. It is recorded from the Atlantic as well as the arctic region of the Polar Sea. Its maximum of frequency is in the southern part of the former; in the latter it is certainly rare. The northernmost place where it has been found is on the coast of Spitzbergen.

Localities: The Norwegian Polar Sea: Nordlanden, common and abundant; the north coast of Norway, BERGGREN, according to specimens in the herbarium of the Swedish Royal Museum.

The Greenland Sea: the coast of Spitzbergen; the exact place is not stated.

The Murman Sea: Kolgujew Isle and the coast of Novaya Zemlya.

The White Sea: Cp. GOBI l. c. p. 12 and 43.

Baffin Bay: the west coast of Greenland at Neuherrnhut.

At Nordlanden the species occurs in its typical form, at other places, as far as I have been able to ascertain, in a finer and more slender form, which agrees with that distributed by ARESCHOUG l. c. as *F. fastigiata* f. *tenuior*.

Fam. GIGARTINACEÆ (KÜTZ.) J. G. AG.

Epicr. p. 173; KÜTZ. Phyc. gener. 389; char. mut.

Gen. **Cystoclonium** KÜTZ.

Phyc. gener. p. 404.

Cystoclonium purpurascens (HUDS.) KÜTZ.

l. c. Fucus purpurascens HUDS. Fl. Angl. p. 589.

f. *typica*.

Descr. Cystoclonium purpurascens J. G. AG. Epicr. p. 239.
Fig. Hypnea purpurascens HARV. Phyc. Brit. tab. 116.
» Cystoclonium purpurascens KÜTZ. Tab. Phyc. 18, tab. 15.
Exsicc. » » ARESCH. Alg. Scand. exsicc. N:o 76.

f. *dendroidea* nob.

F. pumila, 7—10 cm. alta, thalli axi primario crassiusculo, inferne subnudo, supra medium decomposito-ramoso, ramis fasciculum densum, circumscriptione semicircularem vel sustriangularem formantibus. Polysiphoniam elongatam var microdendron habitu revocat. In formam typicam aperte abiens.

Syn. Cystoclonium purpurascens GOBI, Algenfl. Weiss. Meer. p. 40.
» » KLEEN, Nordl. Alg. p. 18.
» » ZELLER, Zweite d. Polarf. p. 85.
Fucus confervoides WG. Fl. Lapp. p. 504.
Gigartina purpurascens Nyl. et Sæl. Herb. Fenn. p. 74.
Hypnea purpurascens CROALL, Fl. Disc. p. 459.
Soliera chordalis ASHM. Alg. Hayes p. 64? Cfr. FARL. New Engl. Alg. p. 148.

Habitat. I know nothing about the habitat of the present species within the arctic region of the Polar Sea. In the Norwegian Polar Sea it is litoral. The typical form lives usually within the lowest part of the litoral zone, attached to other algæ or to steep rocks. In the upper part of this zone it is found chiefly in deep rock-pools between tide-marks. The form *dendroidea* I have met with only in the upper part of the litoral zone. It assumes here its most peculiar appearance, when growing on the walls of shallow rock-pools where *Mytilus edulis* lives in great numbers. The species flourishes both on exposed coasts and in sheltered places. It has been found with tetrasporangia at Nordlanden in summer, with sporocarps in August.

Geogr. Distrib. This species occurs in both divisions of the Polar Sea, but it is scarce in the arctic. It reaches its maximum of frequency in the southern part of the

Norwegian Polar Sea. The northernmost place where it has been found is North Cape on the north coast of Norway. If the supposition of FARLOW be right that ASCHMEAD'S *Soliera chordalis* is the present species, it would, on the west coast of Greenland, go up into Smith Sound to Lat. N. 78°—82°.

Localities: *The Norwegian Polar Sea*: Nordlanden, common and plentiful; Finmarken: Öxfjord, Gjesvær, and Altenfjord.

The Murman Sea: the coasts of Russian Lapland and Cisuralian Samoyede-land; Kolgujew Isle.

The White Sea: pretty common, but probably scanty.

Baffin Bay: the west coast of Greenland, at Neuherrnhut, Julianeshaab, and Egedesminde. Smith Sound?

The form *dendroidea* has been found at Gjesvær in the Norwegian Polar Sea.

Gen. **Callophyllis** KÜTZ.

Phyc. gener. p. 400.

Callophyllis laciniata (HUDS.) KÜTZ.

l. c. p. 401. Fucus laciniatus HUDS. Fl. Angl. p. 579.

Locality: DICKIE in Alg. Suth. 1, p. 143 states *C. laciniata* KÜTZ. to have been found floating in the sea and washed ashore at Whale Islands in Baffin Bay.

Gen. **Kallymenia** J. G. AG.

Alg. Med. p. 98; Epicr. p. 219.

Kallymenia rosacea J. G. AG.

Epicr. p. 220. Halymenia rosacea Spetsb. Alg. Till. p. 45.

Descr. Halymenia rosacea J. G. AG. Epicr. l. c.

Syn. Halymenia rosacea KJELLM. Spetsb. Thall. 1, p. 24; ex. parte.

Habitat. I have stated, in the passage quoted above, that I have found some young individuals of this species attached. I am now obliged to recall this statement, because I have convinced myself that those young individuals must be referred to another species of *Kallymenia* occurring in the Norwegian Polar Sea. Accordingly *Kallymenia rosacea* has been found at present only lying loose at the bottom within the sublitoral zone in 5—10 fathoms water, sometimes in larger masses together with other algæ: *Delesseria sinuosa*, *Phyllophora interrupta* a. o. It has been met with both on exposed coasts and in the interior of bays. Sporocarpiferous specimens have been collected in summer.

Geogr. Distrib. It is known only from the eastern part of the Greenland Sea. Its northernmost known locality is Treurenberg Bay on the north coast of Spitzbergen Lat. N. 79° 66'.

Localities: The Greenland Sea: at two or three places on the north and west coasts of Spitzbergen.

Kallymenia septemtrionalis nob.

Fronde callo radicali minuto affixa, stipite brevissimo, eramoso, lamina vulgo usque a primo initio orbiculari vel late reniformi, longitudine circa 6 cm. attingente subcarnoso-membranacea, crassitudine 100 μ, in statu adultiore parce vage lobata, lobis variæ formæ at semper basi latis, numqvam foliola propria stipite suffulta constituentibus, margine integerrimo vel leviter crenulato; colore planta Sarcophyllidi eduli simillima, sanguineo-purpurea. Tab. 14, fig. 4—6.

Syn. Halymenia rosacea KJELLM. Spetsb. Thall. 1, p. 24; ex parte.
Kallymenia reniformis CROALL, Fl. Disc. p. 459 (?).
» » KJELLM. Algenv. Murm. Meer. p. 20.
» » KLEEN, Nordl. Alg. p. 18.

Remark on the species. It seems to me probable in the highest degree that the species of *Kallymenia* occurring in the Polar Sea, called *K. reniformis* by KLEEN and formerly even by myself, is not identical with the species of this name occurring farther to the south, but rather a northern species differing from it in the shape of the frond, in colour and in consistency.

The arctic form does not apparently attain any considerable size. The largest specimen I have seen is one taken by KLEEN at Nordlanden and figured by me in pl. 14 fig. 5. It is 6 cm. in length by 7 cm. in breadth. Individuals almost as large as this have been collected by me on the west coast of Novaya Zemlya and at Spitzbergen. These appear to be full grown. The stipe is always very short, sometimes imperceptible. The lamina, even when very young, that is to say, when it has a length of 1—2 mm., is circular or reniform, rarely broadly ob-ovate; fig. 4. It becomes lobate only after having attained a more considerable size. The lobes are of different shape and size, but yet always have a broad base, being never sharply divided from the main lamina, as is the case in *Kallymenia reniformis*. It is distinguished from this species at first sight by means of its colour, which is the same as in *Sarcophyllis edulis*, in young individuals possibly somewhat darker than in this alga. This tone of colour is generally designed as »sanguineo-purpureus». According to J. G. AGARDH the thallus of *K. reniformis* is »pulchre coccineus», which seems to me to be the colour also of the figure of the species in HARVEY's Phyc. Brit., although it is said in the description to have the same colour as *Sarcophyllis edulis* i. e. blood-red. Even in respect to consistency the northern species appears to differ from the more southern, the former being membranaceous, slightly fleshy. *K. reniformis* is stated to be »gelatinoso-carnosa» (J. G. AG. Spec. Alg.), »gelatinoso-membranacea» (J. G. AG. Epicr.), »thickisk-membranaceous» (HARV. Phyc. Brit.) etc.; accordingly *K. reniformis* is in general thicker and more fleshy than *K. septemtrionalis*. In this species the central layer of the frond is feebly developed. The cells are few, the mass of gelatinous substance is considerable, the middle layer is distinct, more strongly developed in younger than in older individuals. The cortical layer consists of small cells the greatest extent of which is at right angles to the surface of the frond; fig. 6.

Reproductive organs are unknown.

I have supposed CROALL's *K. reniformis* to be the present species, but I have not seen any specimens of it.

Habitat. This plant is sublitoral, attached to other algæ, as *Ptilota pectinata* and *Lithothamnia*, or to small stones and old shells. It grows scattered on exposed coasts. The species probably develops its reproductive organs later in the year than it has been observed hitherto. Most of the specimens found by KLEEN at Nordlanden in July were young. At Finmarken at the beginning of August I met with only very small and young individuals. But in the arctic region I have collected some older and, I suppose, full-grown specimens in July.

Geogr. Distrib. It is known with certainty from the Norwegian Polar Sea, the eastern Greenland Sea, and the eastern Murman Sea. It has been found most common on the north coast of Norway. Its most northerly known locality is Treurenberg Bay on the north coast of Spitzbergen Lat. N. 79° 56′.

Localities: The Norwegian Polar Sea: Nordlanden, not rare; Finmarken, at Maasö, very plentiful, but local.

The Greenland Sea: the west coast of Spitzbergen at South Cape, washed ashore; the north coast at Musselbay and Treurenberg Bay, local, scarce.

The Murman Sea: Rogatshew Bay on the west coast of Novaya Zemlya, local, scarce.

Baffin Bay: without the locality being definitely stated, in case CROALL's *K. reniformis* belongs to this species. Cp. CROALL, l. c.

Kallymenia Pennyi HARV.

Ner. Am. 2, p. 172.

Descr. Kallymenia Pennyi J. G. AG. Epicr. p. 223.

Syn. Kallymenia Pennyi DICKIE, Alg. Walker. p. 86; Alg. Sutherl. 2, p. 192; Alg. Cumberl. p. 238 (?).
» » HARV. l. c.

Remark on the determination of the species. I have referred to this species two specimens of *Kallymenia* from Greenland, preserved in the herbarium of the Copenhagen Museum, taken by WORMSKIOLD and determined by MERTENS as *Fucus palmatus*. HARVEY's description of the present species accords well with them. They resemble much *K. reniformis*, but they are thinner than individuals of this species and their inner central cell-filaments are less dense. In colour they are pale red-brown. The one specimen is elongated in circumference, the other broadly reniform, the former is deeply divided, almost palmate, the latter irregularly lobed with slender or broad obovate, or linear, entire or subdivided lobes. The edge bears a few triangular teeth. The sporocarps are immersed in the upper part of the lobes near their surface, but not prominent above it, numerous, small, containing few spores with rounded angles. If my determination of these specimens is correct, *K. Pennyi* is nearly related on the one side with *K. reniformis*, on the other with *K. ornata* POST. et RUPR. It is distinguished from the former by its greater thinness, less dense central layer, thicker cortical layer, and more superficial sporocarps. The latter differs by its richer prolification and differently shaped sporocarps.

Habitat. The only fact known as to its habitat is that the plant has been taken in 15—20 fathoms water on a bottom of slate-shingle.

Geogr. Distrib. The species is known from the American province of the arctic region. Its northernmost known locality is Assistance Bay in arctic America, Lat. N. 74° 40'.

Localities: The American Arctic Sea: Port Kennedy and Assistance Bay.

Baffin Bay: Cumberland Sound (?), the west coast of Greenland.

Gen. **Phyllophora** (GREV.) J. G. AG.

Alg. Med. p. 93; GREV. Alg. Brit. p. 135; lim. mut.

Phyllophora Brodiæi (TURN.) J. G. AG.

l. c. Fucus Brodiæi Turn, Hist. Fuc. 2, p. 1.

Descr. Phyllophora Brodiæi J. G. AG. Epicr. p. 216.
Fig. » » HARV. Phyc. Brit. tab. 20, fig. 1.
Exsicc. » » ARESCH. Alg. Scand. exsicc. N:o 207.

Syn. Chondrus membranifolius POST. et RUPR. Ill. Alg. p. II. sec. GOBI, Algenfl. Weiss. Meer, p. 42—43.
» truncatus POST. et RUPR. Ill. Alg. p. II.
Coccotylus Brodiæi ZELLER, Zweite d. Polarf. 85.
Fucus truncatus PALL. Reise 3, p. 34. Cfr. GOBI, Algenfl. Weiss. Meer. p. 43 et AG. Spec. Alg. 1, p. 239.
Phyllophora Brodiæi (?) J. G. AG. Spetsb. Alg. Progr. p. 3; Bidr. p. 11.
» » GOBI, Algenfl. Weiss. Meer. p. 42.
» » KJELLM. Spetsb. Alg. Thall. 1, p. 23; Algenv. Murm. Meer. p. 21; Kariska hafvets Algv. p. 22.
» » Nyl. et Sæl. Herb. Fenn. p. 74.
Sphærococcus Brodiæi SCHÜBELER, in Heugl. Reise p. 317.

Remark on the species. Already in my account of the marine vegetation of Spitzbergen I have mentioned that I have found there a certain form of *Phyllophora* which may be referred with as good reason to *Ph. interrupta* as to *Ph. Brodiæi.* I have collected specimens of the same character also on the west coast of Novaya Zemlya, and GOBI has found others of the same habit amongst the collections of alga from the White Sea and the western Murman Sea examined by him. Even on the west coast of Norway I have seen a *Phyllophora* of rare occurrence, most individuals of which coincided with the *Ph. Brodiæi* occurring in Kattegat, while some resembled the arctic *P. interrupta* with regard to the shape of some of the branches. It cannot be doubted that these two species are very closely allied to each other. Probably one of them — in my opinion *Ph. Brodiæi* — has had its origin from the other. But as they are in general characteristically different from each other, it is more fit, I think, to treat them as different species than as forms of the same species.

Habitat. This alga in the Arctic Sea is sublitoral, sometimes elitoral, usually occurring among the formation of *Laminariaceæ*. It grows in general scattered, sometimes somewhat gregarious, attached to small stones and shells. Although preferring exposed coasts it does not limit itself exclusively to such localities. I have found it with young nemathecia at Finmarken in August, on the west coast of Waygats at the end of July.

Geogr. Distrib. According to my experience, this species has its maximum of frequency in the most south-eastern part of the Murman Sea. It is rare in the Atlantic region of the Polar Sea. Its northernmost known locality is on the north-west coast of Spitzbergen, Lat. N. 79° 45'.

Localities: The Norwegian Polar Sea: Finmarken: Gjesvær, Öxfjord, Talvik, local and scarce.

The Greenland Sea: the north-west coast of Spitzbergen, rare; the east coast of Greenland.

The Murman Sea: the coast of Cisuralian Samoyede-land, Kolgujew Island, the west coast of Novaya Zemlya from S. Gusinnoi Cape and the west coast of Waygats, at the latter place common and abundant; Jugor Shar.

The White Sea: at several places.

The Kara Sea: Uddebay on the east coast of northern Novaya Zemlya; Karabay.

Phyllophora interrupta (GREV.) J. G. AG.

Spetsb. Alg. Progr. p. 3; Sphærococcus interruptus GREV. Act. Leop. 14,2, p. 423.

Descr. Phyllophora interrupta J. G. AG. Epicr. p. 217.
» » KJELLM. Spetsb. Thall. 1. p. 21.

Fig. Sphærococcus interruptus Kütz. Tab. Phyc. 19, t. 20.

Exsicc. Phyllophora interrupta KJELLM. in ARESCH. Alg. Scand. exsicc. N:o 405.

Syn. Phyllophora Brodiæi HARV. Fl. West-Esk. p. 49. (?)
» » WITTR. in Heugl. Reise p. 284; sec. spec.
» interrupta J. G. AG. Spetsb. Alg. Progr. p. 3; Bidr. p. 11.
» » EATON, List, p. 44.
» » GOBI, Algenfl. Weiss. Meer. p. 41.
» » KJELLM. Vinteralgv. p. 64; Spetsb. Thall. 1, p. 20, Algenv. Murm. Meer. p. 20; Kariska hafvets Algv. 21.
Rhodymenia interrupta ASHM. Alg. HAYES, p. 96.
» » DICKIE, Alg. Walker, p. 86.
» » HARV. Ner. Am. 2, p. 149.

Habitat. A sub- or e-litoral species, growing scattered within the formation of *Corallineæ*, *Lithoderma*, and, not unfrequently, even of *Laminariaceæ*, generally on exposed coasts, but even in sheltered places. It has been mentioned before, that it is sometimes found lying loose on the bottom in great masses, cp. p. 17. On the north coast of Spitzbergen it continues to develop all the year round. I found plants in germination in January, and during all the winter specimens with young prolifications were common. Individuals bearing nemathecia were collected in November, December, January and March. The plant occurring here with such organs even at other season, I think it may be assumed to develop reproductive organs throughout the year. In the eastern Murman Sea, the Kara Sea, and the Siberian Sea I have seen individuals with nemathecia in the months of July, August, and September.

Geogr. Distrib. One of the commonest and most widely spread algæ of the Arctic Sea. It is circumpolar, with the exception of not being found within the Atlantic province. It has been met with in the greatest number at Spitzbergen and in the eastern part of the Kara Sea. The most northerly place where it is known with certainty to occur is Treurenberg Bay, Lat. N. 79° 56'. It may possibly go still farther northwards on the west coast of Greenland, as it has been reported from Smith Sound between Lat. N. 78 and 82.

Localities: *The Greenland Sea*: the coasts of Spitzbergen, common, sometimes plentiful.

The Murman Sea: the coast of Russian Lapland; Kolgujew Isle; the west coast of Novaya Zemlya and Waygats, common, but not plentiful.

The White Sea: common and abundant according to GOBI.

The Kara Sea: the east coast of Novaya Zemlya, Lat. N. 76° 8', Long. O. 90° 25', Lat. N. 76° 16', Long. O. 92° 20', Cape Palander, Actinia Bay, in the last two places common and abundant.

The Siberian Sea: to the west of Blishni Isle, Lat. N. 73° 40', Long. O. 140° 6', loose on the bottom; Irkaypi, common, but not abundant.

The American Arctic Sea: remarked at several places.

Baffin Bay: north of the 78:th latitude in Smith Sound.

Phyllophora membranifolia (GOOD. et WOODW.) J. G. AG.

Alg. Med. p. 93. Fucus membranifolius GOOD. et WOODW. Linn. Trans. 3, p. 120.

Descr. Phyllophora membranifolia J. G. AG. Epicr. p. 218.
Fig. » » HARV. Phyc. Brit. tab. 163.
Exsicc. » » ARESCH. Alg. Scand. exsicc. N:o 206.

Syn. Phyllophora membranifolia KLEEN, Nordl. Alg. p. 17.

Habitat. Litoral or sublitoral in the interior of deep bays in the Polar Sea, fastened to stones, muscles, or shells. It has been met with here only as sterile. It probably here, as farther to the south, bears reproductive organs in winter.

Geogr. Distrib. Known only from the southern part of the Atlantic region of the Polar Sea.

Localities: The Norwegian Polar Sea: Nordlanden in Skjærsta Bay.

Gen. Ahnfeltia (Fr.) J. G. Ag.

Alg. Liebm. p. 12; Fr. Fl. Scan. p. 309; spec. excl.

Ahnfeltia plicata (Huds.) Fr.

l. c. p. 310. Fucus plicatus Huds. Fl. Angl. p. 589.

Descr. Ahnfeltia plicata J. G. Ag. Epicr. p. 206.
Fig. Gymnogongrus plicatus Harv. Phyc. Brit. t. 288.
Exsicc. Ahnfeltia plicata Aresch. Alg. Scand. exsicc. N:o 77.

Syn. Ahnfeltia plicata Dickie, Alg. Cumberl. p. 238.
» » Gobi, Algenfl. Weiss. Mecr. p. 39
» » Harv. Fl. West. Esk. p. 50.
» » Kjellm. Spetsb. Thall. 1, p. 20.
» » Kleen, Nordl. Alg. p. 17.
Fucus albus Gunn. Fl. Norv. 2, p. 92.
» plicatus Wg. Fl. Lapp. p. 504.
Gigartina plicata Nyl. et Sæl. Herb. Fenn. p. 74.
» » Post et Rupr. Ill. Alg. p. II.
Gymnogongrus plicatus Dickie, Alg. Walker, p. 86.
» » Rupr. Alg. Och. p. 326.
Sphærococcus plicatus Schrenk, Ural. Reise. p. 547.

Habitat. In the Norwegian Polar Sea it is litoral, growing scattered in rock-pools between tide-marks; in the rest of the Polar Sea it is sublitoral, generally occurring in 1—5 fathoms water on stony or pebbly bottom together with the usual elements of the Laminariaceæ-formation. It is to be met with both on exposed coasts and in sheltered places, and is known only as sterile.

Geogr. Distrib. The plant is certainly circumpolar, but occurs nowhere in the Polar Sea in greater numbers. Its maximum of frequency is in the Norwegian Polar Sea. The northernmost locality where it has been found is Skansbay Lat. N. 78° 30'.

Locality: The Norwegian Polar Sea: Nordlanden, common; Finmarken: Altenfjord and elsewhere according to Wahlenberg.

The Greenland Sea: the west coast of Spitzbergen, local and scarce.

The Murman Sea: the coast of Russian Lapland; the coast of Cisuralian Samoyede-land; Kolgujew Isle, Jugor Shar.

The White Sea: one of the most common algæ.

The Siberian Sea: the mouth of Koljutshin Bay, local, but pretty abundant; the Tshutsh village Tjapka.

The American Arctic Sea: the north coast of western Eskimaux-land; Port Kennedy in the arctic American archipelago.

Baffin Bay: Cumberland Sound, plentiful.

Gen. **Gigartina** (LAMOUR.) J. G. AG.

Epicr. p. 189; LAMOUR. Ess. p. 134; char. mut.

Gigartina mamillosa (GOOD. et WOODW.) J. G. AG.

Alg. Med. p. 104. Fucus mamillosus GOOD. et WOODW. Linn. Trans. 3, p. 174.

Descr. Gigartina mamillosa J. G. AG. Epicr. p. 199.
Fig. » » HARV. Phyc. Brit. t. 199.
Exsicc. » » ARESCH. Alg. Scand. exsicc. N:o 10.
Syn. Fucus mamillosus SOMMERF. Suppl. p. 183.
» » WG. Fl. Lapp. p. 496.
Gigartina mamillosa KLEEN, Nordl. Alg. p. 18.
Rhodymenia mamillosa ARESCH. Phyc. Scand. p. 296.

Habitat. On the coast of Norway the plant is litoral, attached to rocks or stones above low-water mark, or growing in rock-pools between tide-marks. It belongs chiefly to the lower portion of the sublitoral zone and is almost exclusively pelagic. Most often it occurs gregarious in large masses. In the southern part of the Norwegian Polar Sea it has been found with sporocarps in July and August, in the northern part with young organs of that kind in the latter part of August.

Geogr. Distrib. Known from the Norwegian Polar Sea and Baffin Bay. It has probably its maximum of frequency within the former region. Its most northerly certain locality is Gjesvær near North Cape about Lat. N. 71°.

Localities: *The Norwegian Polar Sea*: Nordlanden, common and plentiful; Finmarken at Öxfjord, Maasö, Gjesvær, and the southern coast of Magerö, pretty common and abundant.

Baffin Bay: the west coast of Greenland, according to specimens in the herbarium of the Copenhagen Museum.

Gen. **Chondrus** (STACKH.) J. G. AG.

Spec. Alg. 2, p. 244; STACKH. Ner. Brit. sec. J. G. AG. l. c.; char. mut.

Chondrus crispus (L.) LYNGB.

Hydr. Dan. p. 15. Fucus crispus L. Mant. p. 134.

Descr. Chondrus crispus J. G. AG. Epicr. p. 178.
Fig. » » HARV. Phyc. Brit. t. 63.
Exsicc. » » ARESCH. Alg. Scand. exsicc. N:o 156.
Syn. Chondrus crispus ARESCH. Phyc. Scand. p. 308.
» » KLEEN, Nordl. Alg. p. 18.
Fucus crispus GUNN. Fl. Norv. 2, p. 91.
» » WG. Fl. Lapp. p. 497.

Habitat. On exposed coasts this alga is litoral, growing scattered on rocks and stones, above low-water mark, or more usually in rock-pools between tide-marks. In the interior of Alten Bay it was found in the upper part of the sublitoral zone at a depth of about one metre. It has been found with tetrasporangia on the coast of the Norwegian Polar Sea in June and August.

Geogr. Distrib. Known only from the Atlantic region of the Polar Sea, in the southern part of which it is commonly distributed. The northernmost place where it has been met with is Gjesvær on the north coast of Norway, about Lat. N. 71 .

Localities: The Norwegian Polar Sea: Nordlanden, common and plentiful; Finmarken, pretty common, but not plentiful, as at Maasö, Gjesvær, Öxfjord, and Talvik.

Fam. CERAMIACEÆ (AG.) HAUCK.

Meeresalg. p. 15. AG. Syst. Alg. p. XXVIII; char. mut.

Gen. **Microcladia** GREY.

Alg. Brit. p. 99.

Microcladia glandulosa (SOLAND.) GREV.

l. c. Fucus glandulosus SOLAND. in TURN. Hist. Fuc. 1, p. 81.

Descr. Microcladia glandulosa J. G. AG. Epicr. p. 109.

Fig. » » HARV. Phyc. Brit. t. 29.

Addition to the description of this species. The figures of the structure of the plant given by HARVEY l. c. are very unsatisfactory. The corresponding figure in KÜTZING'S Tab. Phyc. 13, t. 21 is better. This seems to represent a transverse section of some younger portion of the plant. In transverse sections of older portions I have found the central cell as well as the large cells of the parenchyma much more thick-walled, the latter cells being moreover surrounded with more or less numerous small cells, which are rich in endochrome. The Scandinavian specimen of the species that I have seen was furnished with tetrasporangia. These were cruciate, lying without order as stated by J. G. AGARDH l. c., numerous in the outermost or next to outermost segment. The first tetrasporangia are formed at the external margin of the segment, but afterwards new ones are produced within these, until the whole surface of the segment is occupied.

Habitat. I have seen a single individual of this species, which is new to the Flora of Scandinavia. It had been brought home from Nordlanden by KLEEN, but had been overlooked by him. It was found amongst a great number of *Plocamium coccineum*. If it had really grown together with these, as is possibly the case, the present species would occur on exposed coasts in deep water on stony or shelly bottom. The specimen had been taken in summer.

Locality: The Norwegian Polar Sea: Nordlanden, without any note of its special locality.

Gen. **Ceramium** (LYNGB.) HARV.

Mon. p. 98; LYNGB. Hydr. Dan. p. 117; spec. excl.

Ceramium Deslongchampii CHAUV.

Alg. Norm. N:o 85.

Descr. Ceramium Deslongchampii J. G. AG. Epicr. p. 97.
Fig. » » HARV. Phyc. Brit. t. 219.
Exsicc. » » CHAUV. l. c.

Syn. Ceramium diaphanum KLEEN, Nordl. Alg. p. 20; ex parte.

Habitat. It grows scattered within the lower part of the litoral zone, attached to other algæ, both on exposed coasts and in the interior of bays. It bears tetrasporangia in abundance during August on the north coast of Norway.

Geogr. Distrib. It belongs exclusively to the Atlantic region of the Polar Sea, but is rather uncommon. The northernmost place where it has been met with is Öxfjord in Finmarken about Lat. N. 70°.

Localities: The Norwegian Polar Sea: Nordlanden, according to specimens in KLEEN's herbarium; Finmarken, at Öxfjord and Talvik, local and scarce.

Ceramium circinatum KÜTZ.

Hormoceras circinatum KÜTZ. Linnæa p. 733.

Descr. Ceramium circinatum J. G. AG. Epicr. p. 99.
Fig. » » KÜTZ. Tab. Phyc. 12, t. 70.
» decurrens » » » » » 71.

Syn. Ceramium diaphanum KLEEN, Nordl. Alg. p. 20; ex parte.

Remark on the determination of this species. I have only seen two fragmentary individuals of this plant from the Polar Sea. They are to be found in KLEEN's collections of algæ from Nordlanden under the name of *C. diaphanum*. It cannot be denied that they resemble this species in habit, but nevertheless they cannot be referred to it, on account of the distinct, decurrent, cortical layer. Their habit and the shape of the upper branches agree well with fig. *a* and *b* in the quoted plate 71 of KÜTZING's Tab. Phyc. 12. Fig. *d*, tab. 70 of the same work shows the form and cortication of the lower branches. These two plates referring, according to J. G. AGARDH, to *C. circinatum*, I have called the present alga from Nordlanden by that name.

Habitat. Unknown to me.

Geogr. Distrib. According to specimens in KLEEN's herbarium this species occurs in the southern part of the Norwegian Polar Sea at Nordlanden.

Ceramium rubrum (HUDS.) AG.

Disp. Alg. p. 16. Conferva rubra HUDS. Fl. Angl. p. 600.

f. *decurrens* J. G. AG.

Spec. Alg. 2, p. 127.

Descr. Ceramium rubrum α decurrens J. G. AG. Epicr. p. 100.
Exsicc. » decurrens ARESCH. Alg. Scand. exsicc. N:o 208.

f. *genuina.*

f. interstitiis et juvenilibus et adultis densius corticatis; ramis lateralibus paucis, conformibus, dichotomis.

f. *prolifera* J. G. AG.

Spec. Alg. 2, p. 127.

Descr. Ceramium rubrum β proliferum J. G. AG. Epicr. p. 100.
Fig. » secundatum LYNGB. Hydr. Dan. t. 37 A.
» botryocarpum HARV. Phyc. Brit. t. 215.

f. *pedicellata* DUBY.

sec. J. G. AG. Spec. Alg. 2, p. 128.

Descr. Ceramium rubrum η pedicellatum J. G. AG. Epicr. p. 101.
Fig. » » KÜTZ. Tab. Phyc. 13, t. 4, fig. a. et b.

f. *squarrosa* HARV.

Ner. Am. 2, p. 214.

Descr. Ceramium rubrum ε. squarrosum HARV. l. c.
Fig. » » f. squarrosa tab. nostra 15, fig. 7.
Syn. Ceramium rubrum J. G. AG. Spetsb. Alg. Progr. p. 2; Bidr. p. 11.
» » CROALL, Fl. Disc. p. 460.
» » GOBI, Algenfl. Weiss. Meer. p. 46.
» » KJELLM. Spetsb. Thall. 1, p. 25; Algenv. Murm. Meer. p. 23.
» » KLEEN, Nordl. Alg. p. 20.
» » Nyl. et Sæl. Herb. Fenn. p. 74.
» » virgatum POST. et RUPR. Ill. Alg. p. II.
Conferva diaphana WG. Fl. Lapp. p. 511.

Remark on the definition of the forms. Ceramium rubrum, in the Polar Sea as well as elsewhere, is multiform. The rather few specimens from the former region which I have had the opportunity of examining, appear however to be referable to the above-mentioned forms. The specimens determined by me as being f. *decurrens* are in all essential points similar to those from the northern portion of the Atlantic. This form is easily known by its articular cells being at first uncorticated, by the absence of prolification and by its poverty in lateral branches, which, if existing, branch in the same manner as the main axis. Besides, the frond is more strongly attenuated upwards than in any other form. The form designated by me as f. *genuina* resembles the former in habit, branching etc., but differs from it by its darker colour and the strongly corticated articular cells. Nearest this there stands a robust and large-sized form, well represented by fig. *a*, table 4 in KÜTZING's Tab. Phyc. 13 and certainly belonging to the same series of forms of *Ceramium rubrum* as the plant figured pl. 181 in HARV.

Phyc. Brit. This must be the form recorded by J. G. AG. under the combined names of *C. rubrum η pedicellatum*. It seems to be easy to recognize by having numerous, but short and not dichotomously compound lateral branches. This form leads over to f. *prolifera* with its more or less richly prolificating thallus. Some of the specimens from the Polar Sea, that I have seen, are most nearly related to the *Ceramium secundatum* figured by LYNGBYE, others to HARVEY'S *C. botrycarpum*. Of all the arctic forms that one which I have had delineated in table 15 fig. 7, is perhaps the most peculiar. I have thought to recognize in it a form of *C. rubrum* described by HARVEY under the name of f. *squarrosa*. It is a small-sized, rather slender form, most easily recognizable by the upper segments being strongly spreading from one another, those of the last order, which bear tetrasporangia, being recurvate.

Habitat. In the Norwegian Polar Sea this alga is mostly litoral, in the other parts of the Polar Sea that I have myself examined, it is sublitoral, belonging here to the Laminariaceæ-formation. It is generally attached to other algæ, growing scattered on exposed coasts, and more abundant in sheltered localities. At Nordlanden it has been found with sporocarps and tetrasporangia during all summer, at Spitzbergen with tetrasporangia at the end of July and the beginning of August. At Novaya Zemlya and Waygats in June and July, and at Finmarken in July, August, and September, I have met with only sterile individuals. At Russian Lapland, on the coast of Cisuralian Samoyedeland, and in the White Sea, it appears often to bear plenty of tetrasporangia in summer.

Geogr. Distrib. It belongs to the Atlantic as well as the arctic region of the Polar Sea, reaching its maximum of frequency within the southern part of the former and not being widely spread within the latter. Its northernmost known place of growth is on the west coast of Spitzbergen about Lat. N. 76° 30'.

Localities: *The Norwegian Polar Sea*: Nordlanden, common and plentiful; Finmarken, pretty local and scanty at Gjesvær and Talvik.

The Greenland Sea: the west coast of Spitzbergen, generally local and scanty, in one place abundant.

The Murman Sea: the coast of Russian Lapland and Cisuralian Samoyede-land; Kolgujew Isle; the west coast of Novaya Zemlya and Waygats, from M. Karmakul Bay to Jugor Shar, local and scarce.

Baffin Bay: the west coast of Greenland: Neuherrnhut, Godhavn.

At Nordlanden all the forms mentioned are to be found. North of this region only f. *decurrens* and f. *genuina* are met with, most commonly the former or transitions between this and the typical form.

Ceramium acanthonotum CARM.

in J. G. AG. Advers. p. 26.

f. *typica*.

Descr. Ceramium acanthonotum J. G. AG. Epicr. p. 103.
Fig. » » HARV. Phyc. Brit. t. 140.
Exsicc. » » ARESCH. Alg. Scand. exsicc. N:o 12.

f. *coronata* KLEEN.

Nordl. Alg. p. 19.

Descr. Ceramium acanthonotum var. coronata KLEEN l. c.
Fig. » » » » » t. 10, fig. 5.

Remark on the form coronata. In KLEEN's collections only the typical form of *Ceramium acanthonotum* is to be found, but some specimens have the spines in certain parts of the frond so arranged as in the variety distinguished by KLEEN. This variety resembling the typical form in all other respects: branching, colour, structure of the cortical layer, shape and disposition of the tetrasporangia etc., it may be considered as rather little independent, though it deserves perhaps to be specially mentioned.

Habitat. This alga is litoral, pelagic, and somewhat gregarious in the Polar Sea. It keeps chiefly to such places as are left dry at low-tide, but is also to be found in rock-pools between tide-marks. In the localities where it has been observed as yet, it bears plenty of tetraspores in July and August.

Geogr. Distrib. Known only from the southern part of the Atlantic region of the Polar Sea.

Locality: *The Norwegian Polar Sea*: Nordlanden, common and abundant. This is the northernmost known locality of the species. The form *coronata* has been found together with the typical form.

Gen. **Ptilota** (AG.) J. G. AG.

Spec. Alg. 2, p. 92; Ag. Syn. Alg. p. XIX: ex parte.

Ptilota elegans BONNEM.

Hydr. loc. p. 22, sec. J. G. Spec. Alg. 2, p. 94.

Descr. Ptilota elegans J. G. AG. Epicr. p. 74.
Fig. » sericea HARV. Phyc. Brit. t. 191.
Exsicc. » elegans ARESCH. Alg. Scand. exsicc. N:o 11.
Syn. Ptilota elegans KLEEN, Nordl. Alg. p. 20; excl. syn.

Habitat. The present species is litoral, pelagic, and somewhat gregarious, and usually grows on steep rocks between tides under beds of *Fucus* and *Ozothallia*, or on the walls of grotto-shaped cavities, which it covers, in company with *Delesseria alata*, with a dense mat. Even on the coast of Finmarken it reaches a considerable size. I have seen very luxuriant individuals more than 10 cm. high near North Cape. In the Norwegian Polar Sea it bears tetraspores during the summer, sporocarps in August.

Geogr. Distrib. This species belongs to the Atlantic region of the Polar Sea. Its most northern known locality is Gjesvær about Lat. N. 71°.

Localities: *The Norwegian Polar Sea*: Nordlanden, scarce. Finmarken: Maasö, Gjesvær, Öxfjord, local, but pretty abundant.

Ptilota plumosa (L.) Ag.

Syn. Alg. p. 39, excl. β. Fucus plumosus L. Mant. p. 134.
Descr. Ptilota plumosa J. G. Ag. Epicr. p. 75.
Fig. » » Harv. Phyc. Brit. t. 80.
Exsicc. » » Aresch. Alg. Scand. exsicc. N:o 160.

Syn. Fucus plumosus R. Br. in Scoresby, Account., 1, p. 132 et append V? Nonne Pt. pectinata?
» » Gunn. Fl. Norv. 2, p. 91.
» » α et β Wg. Fl. Lapp. p. 501.
» ptilotus Gunn. Fl. Norv. 2, p. 135.
Plumaria pectinata var. tenerrima Rupr. Alg. Och. p. 336.
Ptilota plumosa J. G. Ag. Spetsb. Alg. Progr. p. 3; Bidr. p. 11.
» » Aresch. Phyc. Scand. p. 319.
» » α typica Gobi, Algenfl. Weiss. Meer. p. 44.
» » Harv. Ner. Am. 2, p. 224.
» » Kjellm. Spetsb. Thall. 1, p. 26; Algenv. Murm. Meer. p. 22.
» » Kleen, Nordl. Alg. p. 20.
» » Lyngb. Hydr. Dan. p. 38.
» » var. tenuissima Schübeler in Heugl. Reise, p. 317.
» » Witth. in Heugl. Reise, p. 284.

Habitat. This species is generally sublitoral, occurring in 5—20 fathoms water on stony or pebbly bottom. Sometimes it is found in more shallow water, even within the litoral zone in rock-pools between tides. It is attached sometimes to other algæ, usually to stones, and grows on exposed coasts as well as in the interior of bays and at other sheltered places. I never found it gregarious. On the coasts of Spitzbergen and Novaya Zemlya it usually constitutes an element of the Laminariaceæ-formation; on the west coast of Norway I have generally found it in company with purely arctic algæ. Here it attains a high degree of luxuriancy. The largest specimens I ever saw, from the interior of Altenfjord, had a length of 20—25 cm. The plant has not been examined in winter in the Polar Sea. Individuals with sporocarps have been found at Nordlanden in July and August, at Finmarken at the end of August and in the earlier part of September, at Spitzbergen in July, on the west coast of Novaya Zemlya and Waygats in August and at the end of June. Specimens with tetrasporangia have been collected in most of these parts of the Arctic Sea in July and August.

Geogr. Distrib. The present species belongs both to the Atlantic and the arctic region of the Polar Sea, having its maximum of frequency within the former. Its most northerly known locality is the outer Norse Island off the north-western coast of Spitzbergen Lat. N. 79° 50'.

Localities: The Norwegian Polar Sea: Nordlanden, common and plentiful; Finmarken: Maasö, Gjesvær, Magerö Sound, Öxfjord, and Talvik, abundant, but local.

The Greenland Sea: According to my experience, local and scanty on the coasts of Spitzbergen.

The Murman Sea: the coast of Cisuralian Samoyede-land; the west coast of Novaya Zemlya and Waygats from Matotshin Shar to Jugor Shar, pretty common and abundant.

The White Sea, probably scarce.

The American Arctic Sea, see HARVEY l. c.

Baffin Bay: the west coast of Greenland, according to LYNGBYE and specimens in the collections of the Copenhague Museum without any special locality being noted.

Ptilota pectinata (GUNN.) nob.

Fucus pectinatus GUNN. Fl. Norv. 2, p. 122.

f. *typica.*

Descr. Ptilota serrata J. G. AG. Epicr. p. 76.
Fig. » plumosa var. serrata KÜTZ. Tab. Phyc. 12, t. 55.
Exsicc. » serrata KJELLM. in ARESCH. Alg. Scand. exsicc. N:o 406.

f. *integerrima* RUPR.

Alg. Och. p. 334.

Descr. Plumaria pectinata var. integerrima RUPR. l. c.
Fig. Ptilota pectinata f. integerrima tab. nostra 15, fig. 1.

f. *litoralis* nob.

F. laxe cæspitosa, minuta, 3—4 cm. alta, inferne diametro maximo 300—380 μ.; ramis confertis, alteris lanceolato-falciformibus, parce serratis vel integris, alteris multo brevioribus, linearibus, simplicibus vel infra apicem parce et irregulariter pinnulato-ramulosis. Tab. 15, fig. 2—5.

Syn. Fucus plumosus γ. tenerrimus WG. Fl. Lapp. p. 501.
Ptilota plumosa β. asplenioides LYNGB. Hydr. Dan. p. 38.
» » POST et RUPR. Ill. Alg. p. 11, sec. GOBI, Algenfl. Weiss. Meer. p. 44.
» serrata J. G. AG. Spetsb. Alg. Bidr. p. 11; Grönl. Alg. p. 111.
» » CROALL, Fl. Disc. p. 460.
» » DICKIE, Alg. Sutherl. 1, p. 143; Alg. Cumberl. p. 239.
» » EATON, List. p. 44.
» » β. arctica, GOBI, l. c.
» » HARV. Ner. Am. 2, p. 222.
» » KJELLM. Vinteralgv. p. 64; Spetsb. Thall. 1, p. 26; Algenv. Murm. Meer, p. 22; Kariska hafvets algv. p. 22.
» » KLEEN, Nordl. Alg. p. 20.
» » ZELLER, Zweite d. Polarf. p. 85.

Remark touching the name and the forms of the species. It seems to be beyond a doubt that the plant figured and described by GUNNERUS under the name of *Fucus pectinatus* l. c. and tab. 2 fig. 8, is the same that goes at present commonly under the name of *Pt. serrata* KÜTZ. For this reason I have adopted that denomination. GOBI denies its right to be regarded as a separate species, and I must allow that among the specimens, certainly several thousands in number, that have passed between my hands, there have been some few that were allied in character both to *Pt. plumosa* and *Pt. pectinata*, although it was scarcely ever difficult to decide to which of these species they ought with more reason to be referred. Having in general found *Pt. pectinata* independent, I do not hesitate to let it remain a separate species.

Of the form *integerrima* I have only seen a couple of specimens. Though I have no reason to suppose it to be any very constant or independent form, I have wished to call attention to it, because, as has already been pointed out by RUPRECHT, it shows a remarkable approximation to *Pt. asplenioides* and, on this account, might possibly be considered as a proof of phylogenetic connection between this species and *Pt. pectinata*. It differs from typical *Pt. pectinata* by its almost complete want of so-called *rami compositi* and by more spreading, generally perfectly entire *rami foliiformes*, which makes it very dissimilar to typical *Pt. pectinata*. Judging from some few cases observed, the sporocarps issue generally from the outer side of the leaf-shaped branches, sitting on distinct, unarticulated stalks, sometimes from the middle of the rhachis itself.

The pretty form *litoralis* I have met with on the north coast of Norway among *Pt. elegans*, which it resembles rather much on a cursory examination. As is shown by the figure given, it differs most considerably from typical *Pt. pectinata*; however, from its agreeing essentially with this species in structure and ramification, I suppose it to be a variety of it produced by different conditions of life. It differs from *Pt. pectinata* by being somewhat, though only slightly, tufted and by its smallness, slenderness and far denser branching. How considerable the difference of thickness is between f. *litoralis* and the typical form, is seen on comparing the figures 4 and 6 in tab. 15, which exhibit transverse sections of corresponding portions of the thallus of these two forms. These figures moreover show that there exists a certain difference in structure between these forms. In the typical *Pt. pectinata* the central cell of the transverse section in the fullgrown branches of the last order but one, is surrounded with a complete circle of large cells poor in endochrome, between which cells and the small cortical cells rich in endochrome there lie other smaller cells poor in endochrome. The structure of f. *litoralis* differs from this in so far that the central cell adjoins large cells poor in endochrome only in the direction of the longest axis of the transverse cut, whereas along the short axis it is immediately contiguous to small cells rich in endochrome. In consequence of this, in a fullgrown axis of the last order but one in a *Pt. pectinata* f. *litoralis* the row of axial cells along the middle of the axis is translucent, which is not the case in typical *Pt. pectinata*. A comparison of fig. 4 with fig. 5 shows the structure of older and younger portions of the frond in the former to differ rather much. It is to be seen also in the latter figure that the row of axial cells is not surrounded with a complete circle of large cells poor in endochrome, even in the older parts of the branches. Only sterile individuals are known.

Habitat. The typical form and f. *integerrima* are decidedly sublitoral or elitoral. I have usually found the former in 10—20 fathoms water within the arctic region; but it belongs to those Florideæ which descend to the greatest depths. On the coast of Spitzbergen I have dredged specimens in one place from a depth of 150 fathoms, in two other places at 80—100 fathoms. According to KLEEN it is a deep-water form at Nordlanden. I have met with it at Finmarken on exposed shores at a depth of 15—20 fathoms, in the interior of Altenfjord in 5—6 fathoms water. In the White Sea it occurs at pretty varying depths, from one and a half to 10 fathoms. In the eastern part of the Murman Sea it is generally found at 10—20 fathoms. It thrives on bottoms of various kinds, but appears to prefer solid rock or shingle. In the Norwegian Polar Sea it belongs to the so-called arctic formation, in the arctic region proper it forms a common element of the Corallineacæ-formation, although it is not wanting in other formations. Here it occurs sometimes gregarious in large masses. It prefers exposed coasts and in the glacial part of the Polar Sea seems not to enter into the interior of deep bays. However, on the north coast of Norway I have found it in the interior of Altenfjord. On the north coast of Spitzbergen it is found in full vigour throughout the year, developing plenty of propagative organs in winter. During this season I have found individuals with tetrasporangia in November, December, January, February, and March; individuals with sporocarps have been met with in November in plenty and in December. Besides, specimens with tetrasporangia have been observed on the coasts of Spitzbergen in June, July, August, and October. At Nordlanden individuals with tetrasporangia and sporocarps have been collected in July and August, on the coast of Finmarken in August; on the west coast of Novaya Zemlya specimens with sporocarps have been taken in July, with tetrasporangia in June and July.

The form *litoralis* belongs to the litoral zone, as is designated by the name. It grew together with *Ptilota elegans* on the sides of a shallow, grotto-shaped cavity in a rock.

Geogr. Distrib. The present species seems to have its maximum of frequency in the Greenland Sea. It grows here where so ever the locality is suitable, in great numbers and with large, luxuriant individuals. It is common also on the coasts of Novaya Zemlya, although not so much so as in the Greenland Sea, being replaced in the former region by *Pt. plumosa*. On the coast of Finmarken and at Nordlanden it is decidedly more scarce than *Pt. plumosa*, contrary to what seems to be the case in the White Sea. The most northerly place where it has been found up to the present time, is Treurenberg Bay on the north coast of Spitzbergen, Lat. N. 79° 56′. It is not known from the Siberian Sea.

Localities: *The Norwegian Polar Sea*: Nordlanden, common, according to KLEEN; Finmarken at Maasö, Gjesvær, the south coast of Magerö, Talvik, local, but pretty plentiful.

The Greenland Sea: the east coast of Greenland; on the coasts of Spitzbergen common and abundant.

The Murman Sea: the coast of Russian Lapland; on the west coast of Novaya Zemlya and Waygats from Matotshin Shar to Jugor Shar common, but less abundant.

The White Sea: common and abundant.

The Kara Sea: on the east coast of Novaya Zemlya, scarce.

The American Arctic Sea: taken by RICHARDSON, according to HARVEY.

Baffin Bay: Cumberland Sound, pretty common; the west coast of Greenland at Tessarmiut, Neuherrnhut, Godthaab, Sukkertoppen, Holstenborg, Claushavn, Jakobshavn, Godhavn, Rittenbenk.

The herbarium of the Copenhague Museum contains several specimens of this species without any locality being noted. The form *integerrima* is known to me only from the outer Norse Isle and on the north-western coast of Spitzbergen, f. *litoralis* from the Norwegian Polar Sea at Maasö.

Gen. **Callithamnion** (LYNGB.) THUR.

in LE JOL. List. Alg. Cherb. p. 17; LYNGB. Hydr. Dan. p. 123; lim. mut.

Callithamnion polyspermum BONNEM.

in Ag. Spec. Alg. 2, p. 169.

Descr. Callithamnion polyspermum J. G. AG. Epicr. p. 32.
Fig. Phlebothamnion polyspermum Kütz. Tab. Phyc. 11, t. 97.
Syn. Callithamnion polyspermum KLEEN, Nordl. Alg. p. 22.

Remark on the determination of this species. The individuals from the Polar Sea that I have seen, agree more with the quoted figure in KÜTZING, than with the figure on pl. 231 in HARV. Phyc. Brit. They differ from English specimens by being less decompound, so that those branch-systems which are called *plumulæ* in the descriptions are comparatively rare, being supplied by simple branches. From this cause those arctic specimens present an aspect differing from typical *C. polyspermum.* However, they ought probably to be referred to this species. This is also the opinion of KLEEN.

Habitat. Pelagic, growing scattered, fastened to litoral algæ, as *Polysiphonia fastigiata, Furcellaria fastigiata,* and *Rhodymenia palmata.* It has been found with scarce tetrasporangia at Nordlanden in August.

Geogr. Distrib. Found only within the Atlantic region of the Polar Sea.

Locality: The Norwegian Polar Sea: Nordlanden in several places. This is the northernmost place where it is known to grow.

Callithamnion Hookeri (DILLW.) AG.

Spec. Alg. 2, p. 190. Conferva Hookeri DILLW. Brit. Conf. t. 106 sec. AG. l. c.

Descr. Callithamnion Hookeri J. G. AG. Epicr. p. 33.
Fig. Phlebothamnion Hookeri KÜTZ. Tab. Phyc. 11, t. 94.
Exsicc. Callithamnion Hookeri ARESCH. Alg. Scand. exsicc. N:o 311.
Syn. Callithamnion Hookeri KLEEN, Nordl. Alg. p. 21.

Habitat. In that part of the Polar Sea where it has been as yet observed, this plant is litoral, growing attached to *Polysiphonia fastigiata.* It has been found here with tetrasporangia in June, with sporocarps in July.

Geogr. Distrib. Known only from the southern border of the Polar Sea.

Locality: The Norwegian Polar Sea: Nordlanden at Bodö.

Callithamnion arbuscula (DILLW.) LYNGB.

Hydr. Dan. 123. Conferva arbuscula DILLW. Brit. Conf. t. 85.

Descr. Callithamnion arbuscula J. G. AG. Epicr. p. 37.
Fig. » » HARV. Phyc. Brit. t. 274.
Exsicc. » » ARESCH. Alg. Scand. exsicc. N:o 14.

Syn. Callithamnion arbuscula KLEEN, Nordl. Alg. p. 21.

Habitat. The present species grows litoral on exposed coasts, partly on rocks between tides, partly in rock-pools. Occurring often in great masses, it contributes essentially to mark the character of the vegetation. KLEEN states that on the coast of Nordlanden it begins to appear at the beginning of July and develops tetrasporangia already in the same month, carpospores in August. It seems to follow from these statements that the species is annual here. On the coasts of Britain it is perennial according to HARVEY, attaining its highest development during the summer and the autumn.

Geogr. Distrib. Known only from the Atlantic region of the Polar Sea.

Locality: The Norwegian Polar Sea: Nordlanden, common and plentiful.

Callithamnion roseum (ROTH) HARV.

in Hook. Brit. Fl. 2, p. 341. Conferva rosea ROTH, Cat. Bot. 2, p. 182.

Descr. Callithamnion roseum J. G. AG. Epicr. p. 39.
Fig. » » HARV. Phyc. Brit. t. 230.

Syn. Callithamnion roseum KLEEN, Nordl. Alg. p. 22.

Habitat. It has been found by KLEEN in rock-pools between tide-marks. Taken with tetrasporangia in August.

Geogr. Distrib. Like the preceding species, this is known only from the Atlantic region of the Polar Sea.

Locality: The Norwegian Polar Sea: at Givær in Nordlanden, its most northerly place of growth.

Callithamnion corymbosum (SM.) LYNGB.

Hydr. Dan. p. 125. Conferva corymbosa SM. Engl. Bot. t. 2352.

Descr. Callithamnion corymbosum J. G. AG. Epicr. p. 40.
Fig. » » HARV. Phyc. Brit. t. 272.
Exsicc. » » ARESCH. Alg. Scand. exsicc. N:o 15.

Syn. Callithamnion corymbosum KLEEN, Nordl. Alg. p. 21.

Habitat. It grows scattered on exposed coasts, sublitoral, attached to old shells and deepwater algæ, as *Desmarestia aculeata, Ptilota pectinata* a. o. It attains a considerable size even in the Polar Sea. KLEEN'S collections contain specimens of very luxuriant growth and a height of even 6 cm. In the Polar Sea it has as yet been found only sterile.

Geogr. Distrib. Found only in the southern part of the Polar Sea on the north-west coast of Norway.

Locality: The Norwegian Polar Sea: Nordlanden, scarce.

Gen. **Antithamnion** (NÄG.) THUR.

in Le Jol. Liste Alg. Cherb. p. 111; NÆG. N. Algensyst. p. 200; char. mut.

Antithamnion floccosum (MÜLL.) KLEEN.

Nordl. Alg. p. 21. Conferva floccosa MÜLL. Fl. Dan. t. 828, fig. 1.

f. *atlantica* J. G. AG.

Descr. Callithamnion floccosum var. α atlanticum J. G. AG. Epicr. p. 22.
Fig. » » HARV. Phyc. Brit. t. 81.
Exsicc. » » HOHENACK. Alg. Mar. N:o 325.

Syn. Antithamnion floccosum KLEEN, l. c.

Habitat. The present plant has been found in the Polar Sea both litoral, growing in rock-pools, and sublitoral, in the lowest part of this zone, being attached here to *Lithothamnia*. On the coasts of Europe only a few sterile individuals have ever been found.

Geogr. Distrib. Known from the Norwegian Polar Sea and, according to J. G. AGARDH, from the sea off Greenland. The northernmost known locality of this species is Maasö on the north coast of Norway, about Lat. N. 71°.

Localities: The Norwegian Polar Sea: Nordlanden, local, very scarce; Finmarken exceedingly rare at Maasö.

Baffin Bay: Cp. J. AG. Epicr. p. 22 in the note under the species in question.

Antithamnion Pylaisæi (MONT.) nob.

Callithamnion Pylaisæi MONT. Pl. Cell. N:o 11, sec. J. G. AG. Epicr. p. 22.

Descr. Callithamnion Pylaisæi HARV. Ner. Am. 2, p. 239.
Fig. » » » » » t. 36 B.
» » » KÜTZ. Tab. Phyc. 11, t. 90.

f. *norvegica* nob.

Planta minuta, vix semipollicaris, articulis mediis axis primarii et ramorum quam in forma typica brevioribus, diametro vix 4-plo longioribus. Tab. 16, fig. 1.

Remark on the form norvegica. If one compares the above-quoted figures in HARVEY and KÜTZING with figure 1 in tab. 16 in the present work, representing an alga from the north coast of Norway, one cannot well doubt but that the plant called *C. Pylaisæi* exists on the coast of Scandinavia, though in a somewhat different form. I have seen the same form also in collections from Greenland. The form from the high North differs from the American by its smaller size and by its growing scattered, and, above all, by the middle cells of the main axis and the long branches being only half as long as in the latter. It can hardly be a distinct species. The length of the cells is subject to variation. Nevertheless I have thought fit to record it at present under a special name.

Habitat. Those few individuals, which I have myself collected, grew in 10—15 fathoms water, attached to *Lithothamnion soriferum.* On the coast of Greenland it appears to be litoral, as J. VAHL has noted on the label belonging to it »inter cæspites Gigartinæ subfuscæ». From the Polar Sea it is known only as sterile.

Geogr. Distrib. Found in the Norwegian Polar Sea and Baffin Bay. Its northernmost locality is Gjesvær on the north coast of Norway, Lat. N. about 71°.

Localities: The Norwegian Polar Sea: Finmarken at Gjesvær.

Baffin Bay: the west coast of Greenland at Julianeshaab and Godhavn, according to specimens in the herbarium of the Copenhague Museum.

Antithamnion boreale GOBI (nob.)

Antithamnion plumula var. boreale GOBI, Algenfl. Weiss. Meer, p. 47.

f. *typica* nob.

Descr. Antithamnion plumula var. boreale GOBI, l. c. p. 47 et sequent.
Fig. » boreale f. typica tab. nostra 16, fig. 2, 3.

f. *lapponica* RUPR. (nob.)

Descr. Callithamnion lapponicum RUPR. Alg. Och. p. 343. Cfr. GOBI, Algenfl. Weiss. Meer. p. 48—49 sub A. plumula var. boreali.

f. *corallina* RUPR. (nob.)

Descr. Callithamnion corallina RUPR. Alg. Och. p. 340—341.
Antithamnion corallina KJELLM. Algenv. Murm. Meer. p. 24.
Fig. Antithamnion boreale f. corallina Tab. nostra 16, fig. 4, 5.
Syn. Antithamnion corallina KJELLM. Algenv. Murm. Meer. p. 24.
» plumula KJELLM. Vinteralgv. p. 64; Spetsb. Thall. 1, p. 26; Algenv. Murm. Meer. p. 24; Kariska hafvets Algv. p. 23.
» KLEEN, Nordl. Alg. p. 21.
Callithamnion corallina RUPR. l. c.
» lapponicum RUPR. l. c.
» plumula J. G. AG. Spetsb. Alg. Progr. p. 2; Bidr. p. 11.

Remark on the determination of the species and its forms. I have mentioned already in my account of the Florideæ of Spitzbergen, that the plant collected here which I named *Antithamnion plumula* in accordance with J. G. AGARDH, differs in certain respects from the southern species of that name. I have found later in the eastern Murman Sea and in the Kara Sea a form resembling that from Spitzbergen, and, besides, in the first-mentioned sea another form that I have considered and still consider identical with RUPRECHT's *Callithamnion corallina.* GOBI has subsequently, in his account of the marine Flora of the White Sea, made a detailed and most excellent exposition of the arctic algæ that are most nearly related to *Antithamnion plumula.* He proves the form of *Antithamnion* found in the White Sea to differ in certain points from the typical *A. plumula*, and on this account he sets it down as a variety of this, by the name of var. *boreale*. He further elucidates *Callithamnion lapponicum*, which RUPRECHT has first described, though scarcely in such a manner as to make it possible to recognize it. This form GOBI regards as intermediate between *A. plumula* f. *typica* and f. *boreale*. At the same time, pointing out that *A. plumula* is closely allied to *A.* (*Callithamnion*) *americanum*, he utters the supposition that the latter alga is to be regarded only as a form of the former. GOBI ends his disquisition with the following words: »Es ist bekannt, dass *A. plumula* eigentlich dem Gebiete des Atlantischen Oceans und des Mittelmeeres angehört; im nördlichen Ocean kommt diese Form schon viel seltener vor und zwar vereinzelt in sehr dünnen Büscheln von unbedeutender Grösse ... Folglich erscheint sie für den nördlichen Ocean nicht als eine aborigene Stammform, sondern vielmehr als eine dahin eingewanderte und dabei sehr stark veränderte.»

I quite agree with GOBI in thinking *A. americanum* to be nearly related to *A. plumula*, especially to *A. plumula* var. *boreale* GOBI. But if the former alga should not be regarded as a distinct species, a great many other *Antithamnia* described as independent species would on perhaps as good grounds have to subsumed as various forms under the same species. There are to be found transitions between *A. americanum* and *A. Pylaisæi* according to FARLOW (New Engl. Alg. p. 123), and *A. plumula* var. *boreale* not seldom shows a strong tendency towards the latter species. *C. corallina* RUPR., which is connected by intermediate forms with *A. plumula* var. *boreale*, is plainly very closely allied to *A. cruciatum*, so that it might with as great reason be considered as a form of this species as of *A. plumula* or *A. americanum*. To this may be added that the distance between *A. Pylaisæi*, especially the form occurring on the coast of Norway, and *A. floccosum* is not great. In specimens of *A. Pylaisæi* from Finmarken I have seen branch-systems of the last order replaced here and there in the frond by simple subulate branches of the shape characteristic of *A. floccosum*, which makes me think that these two plant, considered hitherto as species, are connected by intermediate forms. Thus, if *A. americanum* be reduced, the reduction ought apparently to be extended to several other species. I am of opinion that the genus *Antithamnion* is a young genus whose species are in course of development, no marked differentiation being as yes established and the transitional forms not having disappeared. If the common practice with regard to such genera, *Salix*, *Rubus* a. o., is to be followed, according to which every form met with in greater quantity at different places and easy of recognition is considered as a species, both *Antithamnion plumula* var. *boreale* GOBI, commonly distributed in the Polar Sea, and *A. plumula*, *A. americanum*, *A. Pylaisæi*, *A. floccosum* and *A. cruciatum* ought to be regarded as distinct species. It is by these considerations that I have been led to set down the polar form as an independent species: *A. boreale* GOBI. On the other side, I cannot but hold, with GOBI, *Callithamnion lapponicum* RUPR. and also *C. corallina* RUPR. to be forms of *A. boreale*, as these certainly do not differ so much from *A. boreale*, as this does from *A. plumula* and other *Antithamnia*.

A. boreale differs from *A. plumula* by the different branching, by longer cells in the main axis and in the long branches of the frond, and by its sessile tetrasporangia. Among its forms f. *lapponica* approaches most nearly to *A. boreale* by the branches of the last and next to last order being more generally one-sided, by the systems of short branches being spreading or recurvate, and by the branches of the last order being courser and stiffer. The typical form of the species, according to my opinion, is that which has the branches of the last and next to last order longer and more slender, rarely one-sided, but sometimes opposite, sometimes alternate, or some being opposite or alternate, others one-sided. When now and then they are one-sided, they occur on the inside as well as the outside of their main axis. This form stands very near *A. americanum*, as has been correctly stated by GOBI. The principal differences between them are as follows: *A. americanum* is larger and more tufted, more violet in colour, with longer cells — even ten times as long as thick — and fewer, longer, and more flaccid, branch-systems of the last order, with longer and finer side-branches. The form *corallina* differs from the typical form by having branches and branch-systems densely crowded into dense, button-shaped fascicles at the tops of the main axis and the long branches, and by four branch-systems issuing from most of the articular cells of the main axis and the long branches. In these respects it approaches *A. cruciatum*, from which it differs by the longer cells and the finer, more flaccid and elongated branches of the last order, and by the tetrasporangia being situated not at the base of the secondary branch-systems supplying their branches of the first order, but on the branches of the first order representing the axes of the second order of these branch-systems.

From the supposition that *A. plumula* belongs properly to the Atlantic and the Mediterranean and that it occurs, as he thinks, less frequent and luxuriant in the Arctic Sea, GOBI concludes that this species has immigrated into the Arctic Sea and become strongly changed there. It is impossible, of course, to determine with certainty how this has been, but for my own part I should be more inclined to adopt a quite contrary opinion, that is to say, that *A. boreale* has originated within the Arctic Sea and that *A. plumula* and other species have issued from it and been developed in a southward direction. *A. boreale* is so widely distributed in the Arctic Sea, that it can hardly be assumed to be an immigrant. I have found it commonly diffused in all the parts of the Arctic Sea that I have investigated. It is, indeed, often but little luxuriant and seldom occurs in greater numbers; but I have pointed out above and already mentioned in Spetsb. Thall. 1, p. 27 that this is not always the case. Touching *A. plumula*, it may be remarked that is often difficult to arrive at any certain knowledge about the frequency of a species bymeans of the terms generally used, particularly with regard to the occurrence of the species in question I must admit that I know but very little. ARESCHOUG states that on the coast of Scandinavia it is »minime infrequens». On the ground of my own experience, I should translate this expression by saying that, like *A. boreale*, it is commonly diffused, but seldom or never appears in greater masses nor surpasses the last-mentioned species in luxuriancy of growth when this is most luxuriantly developed. On the coasts of Britain the present species according to HARVEY is »not uncommon», which need not mean anything else than that it is rather commonly spread. It is rare on the north-west coast of France at Cherbourg according to LE JOLIS, and on the north-east coast of America according to FARLOW. I conclude from these facts that *A. plumula* is not more commonly distributed nor more abundant in the northern Atlantic than *A. boreale* in the Arctic Sea, and that the latter can be at least almost as luxuriantly developed as the former.

Habitat. The species is sublitoral in the arctic region of the Polar Sea, descending to the lower limit of this zone. It is generally sublitoral even in the Norwegian Polar Sea, though it has been met with here also in the litoral zone. It is a common element in the formation of Corallinaceæ, although being of little importance for the characterization of this division of the vegetation. I have found it most abundant on half-dead bottom, attached to *Desmarestia aculeata* that lay loose on the bottom. It was also found once rather abundant among *Phyllophora interrupta*. Though preferring an exposed coast, it enters also into deep bays. Specimens are generally found scattered, but in Actinia Bay I found the species in pretty great, in Musselbay even in very great masses. On the north coast of Spitzbergen it occurs during all winter, retaining even at this season its ordinary appearance and continuing its development, which does not however become vigorous before the month of March. During the winter season it is always sterile. It has been found with tetrasporangia at Nordlanden in July and August, at Finmarken in August and September, at Spitzbergen in July and August, in the eastern part of the Murman Sea in June, July, and September. I have not seen specimens with sporocarps from the Polar Sea.

Geogr. Distrib. It belongs to the Atlantic as well as the arctic region of the Polar Sea. In the latter region it is widely distributed and it certainly has its maximum of frequency there. Part of that *Antithamnion* which is reported from the American Arctic Sea and Baffin Bay is possibly to be referred to the present species. If that should be the case, the species is circumpolar. The northernmost place where it has been found is Treurenberg Bay on the north coast of Spitzbergen, Lat. N. 79° 56′.

Localities: The Norwegian Polar Sea: Nordlanden according to specimens in KLEEN'S herbarium; Finmarken at Maasö and Gjesvær pretty common, but scarce, at Öxfjord and Talvik local and scarce.

The Greenland Sea: On the west and north coasts of Spitzbergen commonly diffused, but in general scanty; in Musselbay abundant.

The Murman Sea: the coast of Russian Lapland; the west coast of Novaya Zemlya and Waygats from Matotshin Shar to Jugor Shar, commonly distributed, but scanty.

The White Sea: rare.

The Kara Sea: the eastern coast of Novaya Zemlya at Uddebay in rather great number; in Actinia Bay pretty plentiful, but local.

The Siberian Arctic Sea: Koljutshin Isle, Pitlekay and Tjapka, pretty common, but scarce.

Of the forms mentioned above, f. *typica* is known from the Greenland Sea, the eastern Murman Sea, and the Siberian Arctic Sea; f. *lapponica* from the Greenland Sea and the western Murman Sea; f. *corallina* from N. Gusinnoi Cape on the west coast of Novaya Zemlya. The form observed in the Kara Sea, though most nearly allied to the last-mentioned, is not identical with it, but intermediate between it and f. *typica.*

Antithamnion americanum (HARV.) FARL.

New Engl. Alg. p. 123; Callithamnion americanum HARV. Ner. Am. 2, p. 238.

Descr. et Fig. Callithamnion americanum HARV. l. c. et t. 36 A.

Syn. Callithamnion americanum DICKIE, Alg. Walker. p. 86; Alg. Cumberl. p. 239.
" " CROALL, Fl. Disc. p. 460.

Habitat. According to existing statements it appears to occur both as litoral and as sublitoral at the places where it has been as yet observed in the Arctic Sea. In the former case it is attached to stones, in the latter to algæ: *Chætomorpha melagonium*.

Geogr. Distrib. It has been reported from the American Arctic Sea and Baffin Bay.

Localities: The American Arctic Sea: Port Kennedy.

Baffin Bay: Cumberland Sound, plentiful; the west coast of Greenland at or about Disco Island.

Gen. **Rhodochorton** NÄG.

Ceram. p. 355.

Subgen. 1. Thamnidium THUR.

in LE JOL. Liste Alg. Cherb. p. 110.

Rhodochorton intermedium KJELLM.

Spetsb. Thall. 1, p. 28.

Descr. Thamnidium intermedium KJELLM. l. c.

Fig. " " " " t. 1, fig. 10.
Rhodochorton intermedium tab. nostra 15, fig. 8.

Habitat. Litoral, growing gregarious on rocks exposed to the surge. Specimens gathered in July bear few tetrasporangia.

Locality: Found as yet only in the Greenland Sea on the west coast of Spitzbergen in the interior of Icebay.

Rhodochorton spinulosum (SUHR) nob.

Callithamnion spinulosum SUHR, Flora 1840, p. 292.

Descr. Callithamnion spinulosum J. G. AG. Epicr. p. 12.

Remark on this species. The present species resembles *Rh. Rothii* in habit, differing from it by the structure and arrangement of those branch-systems which bear tetrasporangia. With regard to those systems it approaches nearly to *Rh. floridulum* as determined by *Thuret* and figured by LE JOLIS, Liste Alg. Cherb. tab. 6. It differs from it in size, habit, ramification, etc.

Habitat. According to SUHR this plant grows epiphytic on other algæ. J. G. AGARDH doubts the correctness of this statement, and after having myself examined specimens in the herbarium of the Copenhague Museum I cannot but embrace the supposition of this algologist that the plant grows attached to stones. It is probably litoral.

Locality: Greenland.

Rhodochorton Rothii (TURT.) NÄG.

Ceram. p. 355. Conferva Rothii TURT. Syst. 6, p. 1806; sec. DILLW. Brit. Conf. t. 73.

f. *typica*.

Descr. Callithamnion Rothii J. G. AG. Epicr. p. 13.
Fig. Thamnidium Rothii THUR. in Le Jol. Liste Alg. Cherb. t. 5.
Exsicc. » » ARESCH. Alg. Scand. exsicc. N:o 259.

f. *globosa* nob.

Planta globosa, densissime intertexta, diametro vix 2 mm., plexu basali e filis repentibus ramosis, confertis constante, systemata ramorum, creberrima, fastigiata, dense radiatim disposita emittente; axi primario ramorum systematum paullo supra basim in fasciculo ramorum soluto, ramis raro simplicibus, vulgo præsertim supra medium ramulis plus minus crebris, elongatis, adpressis, approximatis, secundis, vel alternis obsessis; articulis inferioribus ramorum diametro fere æquilongis, circa 14 μ. crassis, summis ramulorum diametro saltem 3-plo longioribus, vix 5 μ. crassis; ramis tetrasporangiferis subapicalibus. Tab. 15, fig. 9—13.

Syn. Callithamnion floridulum LYNGB. Hydr. Dan. p. 130, tab. 41 D.
» » SOMMERF. Suppl. p. 193.
» Rothii CROALL, Fl. Disc. p. 460.
» » DICKIE, Alg. Sutherl. 1. p. 143; Alg. Cumberl. p. 239.
» » SOMMERF. Suppl. p. 193.
Thamnidium » KJELLM. Spetsb. Thall. 1, p. 27; Algenv. Murm. Meer. p. 25.
» » KLEEN, Nordl. Alg. p. 22.

Description of f. *globosa.* The plant forms almost globular, dense, solid tufts, which are about 2 mm. in diameter and whose colour inclines to violet (fig. 9). Its basal portion is composed of densely intertwisted, procumbent, branching filaments (fig. 10). From these there issue radially fastigiate, dense, very densely congested branch-systems (fig. 11), having a short, more or less curved, main axis about 14 μ. thick at the base and formed of slightly tun-shaped cells which are about as long as thick. This axis is divided into a more or less dense bunch of branches which at their base are of about the same thickness as the main axis or the secondary axes from which they arise, but taper equally and strongly towards the tip, so that they are here scarcely half as thick as downwards. The cells are also elongated upwards, so that, from being in the lower portions of these axes slightly tun-shaped and about as long as thick, they become quite cylindrical and three times longer than thick (fig. 12, 13). These branches are rarely simple, generally throwing out nearer the base one or two and above their middle two or more, erect, appressed side-axes composed of cylindrical cells, attaining the same height as their respective primary axis and tapering upwards like this, though only slightly. The branches bearing tetrasporangia seem to be subapical as in the typical form. The present form differs from this by its habit and richer branching and by the cells being different as to size, length, and thickness in the upper and lower part of the frond.

Habitat. The typical form is litoral or sublitoral. In the former case it forms a more or less dense mat of sometimes rather great extent on rocks between tide-marks, in the latter case it lives in 3—4, sometimes 5—15, fathoms water, sometimes covering stones as a mat, sometimes fastened to algæ in the shape of small tufts. It occurs both on exposed coasts and in the interior of deeper bays, sometimes gregarious in large masses, so as to determine the character of the vegetation for rather considerable stretches. I have never found it with fully developed tetraspores in the Polar Sea; it probably bears such organs at those seasons when I have not had an opportunity of examining them here, in winter, spring, or autumn. Judging from those few tetrasporangia abnormally developed and apparently produced after the proper season for tetrasporangia, which I have seen in specimens from Spitzbergen, the formation of tetrasporangia would seem take place here during the spring, i. e. in May or June.

I have found the form *globosa* in the upper part of the litoral zone at places exposed to a heavy surge.

Geogr. Distrib. The present species is known from the Atlantic province and the adjoining parts of the Polar Sea. According to my experience, its maximum of frequency is in the eastern part of the Spitzbergen province. The most northern locality where it has been collected is Fairhaven on the north-west coast of Spitzbergen, Lat. N. 79° 49'.

Localities: The Norwegian Polar Sea: Nordlanden, common, according to KLEEN; Finmarken: Maasö, Gjesvær, Öxfjord, and Talvik, local and rather scarce.

The Greenland Sea: the west and north-west coasts of Spitzbergen, local and scanty.

The Murman Sea: the west coast of Novaya Zemlya, pretty common, at some places plentiful.

Baffin Bay: Cumberland Sound, common; Cape Adaire; the west coast of Greenland at Neuherrnhut and Hunde Islands. LYNGBYE and CROALL report it also from Greenland without noting any special locality, and specimens collected there are to be found in the herbarium of the Copenhague Museum.

I know the form *globosa* only from the Norwegian Polar Sea, from Gjesvær.

Rhodochorton (?) sparsum (CARM.) nob.

Callithamnion sparsum CARM. in Hook. Brit. Fl. p. 348.

Descr. Callithamnion sparsum J. G. AG. Epicr. p. 14.
Fig. » » HARV. Phyc. Brit. t. 297.

Syn. Callithamnion sparsum DICKIE, Alg. Cumberl. p. 239.
Thamnidium sparsum KLEEN, Nordl. Alg. p. 23; Cfr KJELLM. Algenv. Murm. Meer. sub Th. Rothii p. 25.

Habitat. This most uncertain and little known species has been found sterile in the Polar Sea, fastened to stems of *Laminariaceæ* and to *Sphacelaria arctica.*

Localities: The Norwegian Polar Sea: Nordlanden.

Baffin Bay: Cumberland Sound.

Subg. 2. **Thamniscus** KJELLM.
Spetsb. Thall 1, p. 29.

Rhodochorton mesocarpum (CARM.) nob.
Callithamnion mesocarpum CARM. in Hook. Brit. Fl. 2, p. 348.

f. *rupicola* nob.

Descr. Thamnidium mesocarpum KLEEN, Nordl. Alg. p. 22.
Fig. Callithamnion » HARV. Phyc. Brit. t. 325.

f. *penicilliformis* KJELLM.
Spetsb. Thall. 1, p. 30.

Descr. Thamnidium mesocarpum f. penicilliformis KJELLM. l. c.
Fig. » » » » tab. nostra 16, fig. 6—7.
Syn. Thamnidium mesocarpum KLEEN, Nordl. Alg. p. 22.
» » f. penicilliformis KJELLM. l. c.; Algenv. Murm. Meer. p. 25; Kariska hafvets Algv. p. 23.

Addition to the description of f. *penicilliformis.* In general the appearance of this plant is that represented by fig. 6. I have always found it such within the Arctic Sea. On the coast of Finmarken among such individuals I have, besides, observed others whose primary branches bore a greater number of secund or alternate, appressed branches.

Habitat. The form *rupicola* is litoral or sublitoral, attached to stones, Bryozoa, muscles a. o. The form *penicilliformis* is sublitoral and epiphytic on several algæ, chiefly species of *Odonthalia*, *Delesseria*, and *Ptilota*, and *Chætomorpha melagonium*. It has been met with hitherto only on exposed coasts, always few specimens. Both the forms have been found with tetrasporangia in July and August, f. *penicilliformis*, besides, at Greenland in March.

Geogr. Distrib. The typical form is known only from the Atlantic province of the Polar Sea, f. *penicilliformis* also from the arctic region, within which it is widely diffused. Its most northern known locality is Fairhaven on the north-west coast of Spitzbergen, Lat. N. 79° 41'.

Localities: The Norwegian Polar Sea: Nordlanden (f. *rupicola*) local, scarce; Finmarken at Gjesvær (f. *penicilliformis*) local, scarce.

The Greenland Sea: the north-west coast of Spitzbergen at Fairhaven, local, scarce.

The Murman Sea: on the west coast of Novaya Zemlya at several places pretty plentiful.

The Kara Sea: Uddebay on the east coast of Novaya Zemlya.

Baffin Bay: the west coast of Greenland at Julianeshaab.

Rhodochorton spetsbergense KJELLM.
Spetsb. Thall. 1, p. 31.

Descr. Thamnidium spetsbergense KJELLM. l. c.
Fig. » » » t. 1, fig. 11, 12.

Habitat. Only a few specimens have been found within the sublitoral zone, attached to *Chætomorpha melagonium* in August. They bore plenty of tetraspores.

Locality: Known only from the Greenland Sea at Fairhaven on the north-west coast of Spitzbergen.

Fam. PORPHYRACEÆ (Kütz.) Thur.

in Le Jol. Liste Alg. Cherb. p. 16; Kütz. Phyc. gener. p. 382; char. mut.

Gen. **Diploderma** nob.

Thallus membranaceus duobus cellularum stratis constructus.

Diploderma amplissimum nob.

Planta initio aliis algis adnata, demum soluta in mari libera circumnatans, fronde usque 90 cm. longa, 30 cm. lata, ovata, ovato-cordata, oblongo-obovata, oblongo-lanceolata, crebre et profunde undulato-plicata, non lobata, juvenili intense violaceo-purpurea, ætate provectiore plus minus dilute violaceo-carnea, lubrica, chartæ arctissime adhærente; cellulis medii thalli plantæ adultæ sectione transversali quadratis vel verticaliter rectangularibus; organis reproductionis zonam marginalem subflavam occupantibus. Tab. 17, fig. 1—3; tab. 18, fig. 1—8.

Syn. Porphyra laciniata f. linearis et vulgaris Kleen, Nordl. Alg. p. 23.
» coccinea Kleen, Nordl. Alg. p. 24.
Ulva umbilicalis β purpurea Wg. Fl. Lapp. p. 506.

Description of the species. This alga is at first attached to other algæ by means of a feeble holdfast. At this time it has a strong, saturated, purplish-violet colour. After having attained a more considerable size, it is loosened and floats about on the surface of the water. In proportion as it grows larger and older, it bleaches more and more, passing finally to a livid, flesh-colour inclining to violet. The largest specimen I have found attached was 28 cm. long by 12 cm. broad at its broadest place. Drifting individuals reach a considerable size. I have measured one that was 90 cm. in length by 30 cm. in breadth. The shape of the frond is subject to great variation, but in general it is oblong, inclining to cordiform or ovate. It sometimes bends round the fastening-point by developing one side more strongly, so as to get, when this bent is at its strongest, an appearance resembling that of *Porphyra laciniata* f. *umbicalis*. It is densely folded, often so deeply that the folds extend to the middle line of the frond. The margin is either even or irregularly laciniate, sometimes, though rarely, beautifully crenulated. I have not seen any lobed specimens; tab. 17 fig. 1—3. The stipital portion of the frond is composed of claviform cells with the shafts directed downward and more or less obliquely outward, shooting beyond one another (tab. 18, fig. 1, 2). In fully developed individuals the cells at the middle of the frond, in cross section, are generally squarish, sometimes rectangular, considerably more high than long. I cannot determine at present whether this difference denotes different ages or different forms. The shape and disposition of the cells as seen from the surface, is shown in fig. 3. It should be remarked, however, that this figure as well as the others are drawn from dried and afterwards moistened specimens. Antheridia and sporocarps are sometimes, but apparently not always, developed on the same individual. The development begins at the margin, proceeding inwards. Of two cells in the same cross section either both may be developed into antheridia or sporocarps, or the one may become an antheridium, the other a sporocarp (fig. 7—8). The sporocarps contain only few spores.

Habitat. It grows, when attached, sublitoral, and usually scattered in 2—3 fathoms water. I have never found it but on exposed coasts. Specimens with sporocarps have been taken at the end of July and the beginning af August.

Geogr. Distrib. Known only from the Norwegian Polar Sea. Its most northerly known locality is Maasö in Finmarken about Lat. N. 71°.

Localities: The Norwegian Polar Sea: Nordlanden according to specimens in KLEEN's and WAHLENBERG's herbaria, Tromsö Amt near the town of Tromsö; Finmarken at Maasö, local, but abundant.

Diploderma miniatum (AG.) nob.

Ulva purpurea β miniata AG. Syn. Alg. p. 42.

Descr. Ulva miniata LYNGB. Hydr. Dan. p. 29.

Fig. » » » » » t. 6, D.

Porphyra miniata Fl. Dan. t. 2394.

» » KÜTZ. Tab. Phyc. 19, t. 81.

Diploderma miniatum Tab. nostra 18. fig. 9.

Syn. Porphyra miniata KJELLM. Spetsb. Thall. 1, p. 32.

» vulgaris CROALL, Fl. Disc. p. 461 (?).

» » DICKIE, Alg. Sutherl. 1, p. 144 (?).

Remark on this species. In the herbarium of the Copenhague Museum there are to be found under the name of *Porphyra (Ulva) miniata* a considerable number of specimens of the plant in question at different stages of development. The description given by LYNGBYE l. c. of the alga named by him *Ulva miniata* accords well with them. Thus I think we may safely assume that LYNGBYE's description, as well as C. A. AGARDH's description of *Ulva purpurea β miniata*, is founded on some of these specimens. The last author states expressly that the plant designed by him was from Greenland, communicated by WORMSKIÖLD; cp. Spec. Alg. 1, p. 407. However, this Greenland species is no *Porphyra*, but a species of *Diploderma*, most closely allied to the preceding one, though certainly specifically distinct from it. It has a different colour, more firmness, at least when older, and almost no folds. Besides, it is always dioecious, as far as my observations go.

Habitat. At Spitzbergen I have found the present species in the lower part of the sublitoral zone at a depth of 10—15 fathoms, attached to stones. I cannot state anything with certainty with respect to its occurrence at Greenland. On the labels appended to the specimens in the herbarium of the Copenhague Museum, we read: »in mari ad saxa, ad stipites L. saccharinæ (caule fistuloso) ad stipites L. saccharinæ»; from which it may be concluded that the plant is even here sublitoral, growing chiefly within the Laminariaceæ-formation. Most of the Greenland specimens are collected in March and April, some in October, which seems to indicate that the plant is to be found here all the year round. Judging from these specimens it also bears reproductive organs at different seasons.

Geogr. Distrib. Its maximum of frequency is no doubt in Baffin Bay. Besides, it has been observed in the eastern part of the Greenland Sea. The northernmost locality where it is at present known to grow, is Fairhaven, on the north-west coast of Spitzbergen, Lat. N. 79° 49'.

Localities: The Greenland Sea: the north-west coast of Spitzbergen, rare.

Baffin Bay: Kakertok, Tessarmiut Bay, Kangek (near the same bay), Julianeshaab, Egedesminde, Godthaab. If the *P. vulgaris* of CROALL and DICKIE is the present species, as I think it is, *D. minatum*, is known also from Disco Bay and Whale Sound.

Gen. **Porphyra** AG.

Syst. Alg. p. XXXII.

Porphyra laciniata (LIGHTF.) AG.

l. c. p. 190. Ulva laciniata LIGHTF. Fl. Scot. p. 974.

f. *typica*.

Descr. Porphyra laciniata THUR. in Le Jol. Liste Alg. Chorb. p. 100—101.
Fig. » » HARV. Phyc. Brit. t. 92.
Exsicc. » » ARESCH. Alg. Scand. exsicc. N:o 116.

f. *umbilicalis* (L.) KLEEN.

Nordl. Alg. p. 23. Ulva umbilicalis L. Spec. Pl. 2, p. 1163.

Descr. Ulva umbilicalis LYNGB. Hydr. Dan. p. 28.
Exsicc. Porphyra laciniata f. b. ARESCH. Alg. Scand. exsicc. N:o 260.
Syn. Porphyra laciniata ARESCH. Phyc. Scand. p. 404.
» » GOBI, Algenfl. Weiss. Meer. p. 50.
» » KLEEN, Nordl. Alg. p. 23; excl. var. lineari et vulgari.
» » umbilicata RUPR. Alg. Och. p. 393.
» » vulgaris Nyl. et Sæl. Herb. Mus. Fenn. p. 75.
Ulva umbilicalis GUNN. Fl. Norv. 2, p. 121.
» » WG. Fl. Lapp. p. 506; excl. var.

Remark on the species. THURET has pointed out that what is set down by algological authors under the name of *P. linearis*, *P. vulgaris* or *P. purpurea*, and *P. laciniata*, sometimes as separate species, sometimes as forms of the same species, is in fact nothing but one and the same plant at different stages of development; cp. THUR. l. c. This being so, the names *P. linearis*, *vulgaris*, and *purpurea*, ought to be struck out altogether. The plant described under the name of *Ulva umbilicalis* or *U. umbilicata* I think ought to be regarded as a special form of *P. laciniata*. It differs from the typical *P. laciniata* both biologically and morphologically, and is well known to Scandinavian algologists. LYNGBYE l. c. has described it so well that it is perfectly easy to identify it.

Habitat. This species grows between tides, f. *umbilicalis* near high-water mark, always attached to rocks and stones, f. *typica* farther down, often fastened to stones, sometimes to algæ. They occur at exposed as well as sheltered places, f. *umbilicalis* preferring, however, the former. Being somewhat gregarious, it occurs sometimes in so great masses as to influence the character of the vegetation. Both forms bear propagative organs on the arctic coast of Norway in July and August.

Geogr. Distrib. The species certainly belongs properly to the Atlantic province of the Polar Sea, having its maximum of frequency there, but it has been observed also in the adjoining parts of the Polar Sea. Its northernmost known locality is Gjesvær on the north coast of Norway, Lat. N. about 71°.

Localities: The Norwegian Polar Sea: Nordlanden, common and abundant; Finmarken, pretty common and abundant at Maasö, Gjesvær, the south coast of Magerö, Öxfjord, and Talvik.

The Murman Sea: the coast of Russian Lapland, probably pretty common and plentiful.

Baffin Bay: the coast of Greenland, according to specimens in the herbarium of the Copenhague Museum.

Both the forms occur on the coast of Norway; on the coast of Russian Lapland f. *umbilicalis* has been observed, on the coast of Greenland f. *typica*.

Porphyra abyssicola nob.

P. fronde elongato-obovata, late oblonga vel ovato-cordata, integra, parce at profunde undulata vel subplana, lubrica, chartæ arctissime adhærente, coccineo-violacea, dioica; organis fructificationis zonam marginalem occupantibus. Tab. 17, fig. 4; tab. 18, fig. 10—11.

Syn. Porphyra miniata J. G. AG. Grönl. Alg. p. 111; fide spec.
» » GOBI, Algenfl. Weiss. Meer. p. 51; fide syn.
» » KLEEN, Nordl. Alg. p. 23; fide spec.
» » Nyl. et Sæl. Herb. Fenn. p. 75.

Description of the species. I have seen only a few individuals of this species, among which only two were complete. It is attached by means of a small *callus radicalis*. Stipes wanting. Both the complete specimens are elongated-obovate, somewhat oblique. As far as I have been able to judge from the fragmentary specimens, the form of the alga is, however, often another than this. Some of those specimens seem to have been broadly oblong, others ovate with cordiform base. The largest specimen I have examined, was 15 cm. long and 5 cm. broad in its broadest part. The fragments seem also to indicate smallness of size. The plant is sometimes almost smooth, sometimes scantily, but deeply, plicate. It is more gelatinous than any other species of the genus, adhering closely to the paper in preserving, and contracting but little in drying. Younger individuals have a rather strong carmine colour inclining to violet. When becoming older, the plant pales, assuming a pallid, yellowish flesh-colour (tab. 17, fig. 4).

The lower part of the frond is composed of club-shaped cells. The tips of these cells imbedded in gelatine form the *callus radicalis*. The rest of the frond consists of cells that are rectangular in transverse section with the corners of the cell-rooms pretty much rounded. Sporocarps and antheridia are produced in different individuals and take up an irregular zone in the margin of the frond, broader than in P. laciniata. The sporocarps have few spores (tab. 18, fig. 10 and 11).

Habitat. As far as my experience goes, this species belongs to the deeper parts of the sublitoral region, on exposed coasts, growing scattered, attached to other algæ or *Hydromedusæ*. It has been found in deep water also by KLEEN. Specimens with antheridia have been taken at Nordlanden in July, with sporocarps in development at Finmarken in the beginning of August. Individuals from Greenland, collected at the end of September or beginning of October, possess antheridia and sporocarps.

Geogr. Distrib. I cannot decide where the centrum of distribution of this species is to be placed. It is not commonly spread in the Arctic Sea. The greatest number of individuals appears to have been met with on the coast of Russian Lapland. The most northern locality is Gjesvær on the north coast of Norway, Lat. N. 71° 5'.

Localities: The Norwegian Polar Sea: Nordlanden rare; Finmarken rare at Maasö and Gjesvær.

The Murman Sea: the coast of Russian Lapland, a great number of specimens collected.

The White Sea: probably scarce.

Baffin Bay: the west coast of Greenland at Sukkertoppen.

Gen. **Bangia** (LYNGB.) KÜTZ.

Phyc. gener. p. 248; LYNGB. Hydr. Dan. p. 82; lim. mut.

Bangia fuscopurpurea (DILLW.) LYNGB.

l. c. p. 83. Conferva fuscopurpurea DILLW. Brit. Conf. p. 92.

Descr. Bangia fuscopurpurea HAUCK, Meeresalg. p. 22.

Fig. » » KÜTZ. Tab. Phyc. 3, t. 29.

Exsicc. » » ARESCH. Alg. Scand. exsicc. N:o 118.

Syn. Bangia arctica FOSLIE, Arct. Havalg. p. 5. Cfr. HAUCK, l. c. et BERTH. Bangiaceen, p. 23.
» fuscopurpurea KLEEN, Nordl. Alg. p. 24.
» » CROALL, Fl. Disc. p. 461.
Conferva atropurpurea WG. Fl. Lapp. p. 515.

Habitat. It is litoral, growing between tide-marks either in rock-pools or on stones. I do not know at what time it bears propagative organs in the Arctic Sea.

Geogr. Distrib. The species is known with certainty from the Norwegian Polar Sea. It has been met with also in Baffin Bay; but not being attached, its occurrence here is uncertain. Its northernmost known locality is North Cape on the north coast of Norway.

Localities: The Norwegian Polar Sea: Nordlanden at Bodö; Finmarken at North Cape pretty plentiful.

Baffin Bay: found drifting in Davis Strait; no special place noted.

Gen. **Erythrotrichia** ARESCH.

Phyc. Scand. p. 435.

Erythrotrichia ceramicola (LYNGB.) ARESCH.

l. c. p. 436. Conferva ceramicola LYNGB. Hydr. Dan. p. 144.
Descr. Erythrotrichia ceramicola ARESCH. l. c.
Fig. Bangia ceramicola HARV. Phyc. Brit. t. 317.
Syn. Erythrotrichia ceramicola KLEEN, Nordl. Alg. p. 24.

Habitat. KLEEN has found this species growing on litoral algæ.
Locality: The Norwegian Polar Sea: Nordlanden, pretty common.

Series FUCOIDEÆ (AG.) J. G. AG.

Alg. Med. p. 24; AG. Syst. Alg. p. XXXV; lim. mut.

Fam. FUCACEÆ (AG.) J. G. AG.

Spec. Alg. 1, p. 180; AG. l. c. p. XXXVII; lim. mut.

Gen. **Himanthalia** LYNGB.

Hydr. Dan. p. 36.

Himanthalia lorea (L.) LYNGB.

l. c. Fucus loreus L. Syst. Nat. Ed. 12, 2, p. 716.
Descr. Himanthalia lorea J. G. AG. Spec. Alg. 1, p. 196.
Fig. » » HARV. Phyc. Brit. tab. 78.
Exsicc. » » ARESCH. Alg. Scand. exsicc. N:o 2 et 251.
Syn. Fucus loreus GUNN. Fl. Norv. 2, p. 125.
» » WG. Fl. Lapp. p. 499.
Himanthalia lorea J. G. Ag. Enum.
» » ARESCH, Phyc. Scand. p. 259.
» » KLEEN, Nordl. Alg. p. 31.
Ulva pruniformis GUNN. Fl. Norv. 2. p. 89; ex parte.

Habitat. This alga generally grows gregarious within the lower part of the litoral region. At Nordlanden it is sterile in May; at the beginning of June the receptacles commence to develop and within a few weeks attain more than one foot in length.

The proper time of fructification here is probably in the month of July. The plant is not to be found during the winter according to WAHLENBERG.

Geogr. Distrib. It belongs only to the southern part of the Atlantic region of the Polar Sea, not going so far northward as to Finmarken.

Localities: The Norwegian Polar Sea: Nordlanden common, »*Norvegia arctica*» (BERGGREN), according to J. G. AG. Enum.

Gen. **Halidrys** (LYNGB.) GREV.

Alg. Brit. p. XXXIV; LYNGB. Hydr. Dan. p. 37; lim. mut.

Halidrys siliquosa (L.) LYNGB.

l. c. Fucus siliquosus L. Spec. Plant. 2, p. 1160.
Descr. Halidrys siliquosa J. G. AG. Spec. Alg. 1, p. 236.
Fig. » » HARV. Phyc. Brit. t. 66.
Exsicc. » » ARESCH. Alg. Scand. exsicc. N:o 151.
Syn. Fucus siliquosus GUNN. Fl. Norv. 1, p. 83.
» » WG. Fl. Lapp. p. 498.
Halidrys siliquosa J. G. AG. Enum.
» » ARESCH. Phyc. Scand. p. 253.
» » KLEEN, Nordl. Alg. p. 31.

Habitat. It belongs to the upper part of the sublitoral zone, growing in from one to three fathoms water on stony or gravelly bottom. Besides this, I know nothing about its habitat in the Polar Sea.

Geogr. Distrib. It is known only from the Atlantic region of the Polar Sea. I cannot decide with certainty whether it occurs within the whole of this region or not. According to ARESCHOUG and WAHLENBERG it grows also in its northern and eastern parts at Finmarken. I have certainly found it here myself, though never attached, but only washed ashore.

Localities: The Norwegian Polar Sea: Nordlanden common, »*Norvegica arctica*» (BERGGREN); Finmarken (ARESCH. and WAHLENB.). At Maasö I found a number of robust, vigorous specimens washed ashore.

Gen. **Ozothallia** DCSNE et THUR.

Rech. Fuc. p. 13.

Ozothallia nodosa (L.) DCSNE et THUR.

l. c. Fucus nodosus L. Spec. Plant. 2, p. 1159.
Descr. Fucodium nodosum J. G. AG. Spec. Alg. 1. p. 206.
Fig. Fucus nodusus HARV. Phyc. Brit. t. 158.
Exsicc. Halicoccus nodosus ARESCH. Alg. Scand. exsicc. N:o 51.

Syn. Ascophyllum nodosum GOBI, Algenfl. Weiss. Meer. p. 52.
Fucodium nodosum J. G. AG. Spetsb. Alg. Till. p. 27, 28 et 31; Grönl. Alg. p. 110.
Fucus nodosus CROALL, Fl. Disc. p. 457.
» » DICKIE, Alg. Sutherl. 1, p. 140; Alg. Cumberl. p. 236.
» » GUNN. Fl. Norv. 1, p. 83.
» » POST. et RUPR. Ill. Alg. p. 11.
» » Nyl. et Sæl. Herb. Fenn. p. 73.
» » WG. Fl. Lapp. p. 499.
Halicoccus nodosus ARESCH. Phyc. Scand. p. 254.
» » KLEEN, Nordl. Alg. p. 31.
Halidrys nodosa LYNGB. Hydr. Dan. p. 37.
Ozothallia nodosa KJELLM. Spetsb. Thall. 1, p. 3.

Habitat. On the arctic coast of Norway this species is litoral, occupying nearly the middle part of the litoral zone. It occurs on exposed coasts as well as in the interior of deep bays, but seems to prefer such places as are sheltered from the violence of the waves. It is gregarious, occurring in some places in more considerable masses than other *Fucaceæ*. It generally grows mixed with such algæ. I have found specimens with plenty of receptacles in July and the earlier part of August on the coast of Finmarken. It was always sterile here in the latter part of August, but new receptacles began to be formed at the beginning of October. Accordingly, it appears to fruit here at least twice a year, probably even more often. LE JOLIS states that it bears receptacles in winter on the coast of France. At Bohuslän it commences to develop such organs in March and April, the period of fructification ending in June. Cp. ARESCH. l. c.

Geogr. Distrib. I cannot state at present any certain limits for its distribution in the Arctic Sea. It is known with certainty to occur in the Norwegian Polar Sea, the western Murman Sea, the White Sea, and Baffin Bay. The alga is also reported from the eastern Greenland Sea, but it does not appear to be quite sure that it really grows there. It was brought home from there by the Swedish expedition of 1868, and said to be collected at several places, as Icefjord, Smeerenberg Bay, Kobbebay, and the bays on the north coast. I have had the opportunity of examining the three first-mentioned localities, but I failed to detect any traces of this plant. But it was found washed ashore at South Cape, where however it had probably been carried from the south. I saw it floating on the surface of the sea at several places between Norway and Spitzbergen. The opinion that it does not grow on the coasts of Spitzberg is supported by the fact of its not having been observed on the west coast of Novaya Zemlya, whose Flora resembles that of Spitzbergen so nearly. Already in the White Sea this plant is rarer than on the north coast of Norway, and east of that Sea, on the coast of Cisuralian Samoyede-land, it has not been found at all. Certainly it reaches its maximum of frequency on the north coast of Norway. The most northern place where it is said to have been found, is the north coast of Spitzbergen.

Localities: The Norwegian Polar Sea: Nordlanden, common and abundant; Tromsö amt, common and abundant at the town of Tromsö, at Renö and Carlsö; Finmarken:

Mausö, Gjesvær, the south coast of Magerö, Öxfjord, pretty common and plentiful every-where.

The Greenland Sea: Beeren Eiland; the west and north coasts of Spitzbergen; compare above.

The Murman Sea: the coast of Russian Lapland.

The White Sea: pretty common.

Baffin Bay: Cumberland Sound; the west coast of Greenland at Smallesund, Fiskernæs, Sukkertoppen, Godhavn, Whale Islands, and drifting in the sea off Lichtenau and at Lat. N. 73° 50'.

Gen. **Fucus** (Tourn.) Dcsne et Thur.

Rech. Fuc. p. 13; Tourn. Inst. Herb. 3, p. 565; char. mut.

Fucus serratus L.

Spec. Plant. 2, p. 1158.

Descr. Fucus serratus Aresch. Fuc. et Pycnoph. p. 101.

f. *grandifrons* nob.

f. robusta, thalli segmentis plurimis ad costas reductis, summis membranaceo-coriaceis, 2 cm. latitudine excedentibus, argute et profunde serratis, receptaculigeris præsertim abbreviatis, apicibus rotundato-truncatis; cryptostomatibus sat numerosis.

f. *abbreviata* nob.

f. parvula, 15—20 cm. alta, thalli segmentis plurimis ad costas reductis, summis membranaceo-coriaceis circa 1 cm. latis, parce obsoletiusque serratis, abbreviatis, apicibus rotundato-truncatis; cryptostomatibus numerosis.

f. *arctica* J. G. Ag.

Fucus serratus var. arcticus J. G. Ag. Spetsb. Alg. Bidr. p. 9.

f. thalli segmentis complurium ordinum alatis, summis coriaceo-membranaceis, circa 1 cm. latis, parce obsoletiusque serratis, apicibus subrotundatis; cryptostomatibus fere nullis.

***typica* nob.**

f. elongata, thalli segmentis complurium ordinum alatis, summis subcoriaceis vel coriaceis, 1,5—2,5 cm. latis, plus minus crebre et argute serratis, elongatis, apicibus truncatis; cryptostomatibus numerosioribus.

Fig. Fucus serratus Harv. Phyc. Brit. t. 47.
Exsicc. » » Aresch. Alg. Scand. exsicc. N:o 55.

f. *angusta* nob.

f. elongata, thalli segmentis complurium ordinum alatis, summis subcoriaceis, circa 1 cm. latis, profunde, plus minus crebre et argute serratis, elongatis, apicibus truncatis; cryptostomatibus sat numerosis.

Syn. Fucus serratus J. G. Ag. Enum.; Spetsb. Alg. Bidr. p. 9, 11.
» » Aresch. Phyc. Scand. p. 258.
» » Gobi, Algenfl. Weiss. Meer. p. 57.
» » Gunn. Fl. Norv. 1, p. 28.
» » Kjellm. Algenv. Murm. Meer. p. 28.

Syn. Fucus serratus KLEEN, Nordl. Alg. p. 25.
» » LYNGB. Hydr. Dan. p. 5.
» » Nyl. et Sæl. Herb. Fenn. p. 73.
» » POST et RUPR. Ill. Alg. p. II.
» » WG. Fl. Lapp. p. 489.

Remark on the definition of the forms. Being of the opinion that the directions and limits of the variations of a very variable species are less sharply and distinctly brought into view by giving a general description comprising all the forms, than by establishing and characterizing certain typical forms, round which the others may be grouped, I have thought fit to call attention to the types of *Fucus serratus* described above. I think that all the forms contained in the collections of *Fucus serratus* from the Polar Sea may be pretty easily and naturally arranged round these types. I have set down as the typical *Fucus serratus* the form delineated by HARVEY in Phyc. Brit. tab. 47, which occurs commonly on the west coast of Sweden and also on the arctic coast of Norway. Near this there stands a form, called by me f. *angusta*, which has been found within the arctic region of the Polar Sea in the eastern part of the Murman Sea. It differs from the typical form by the considerably slenderer frond and the more densely serrate upper segments. In both forms the frond is leathery when dried, with winged segments of five or more orders. The segments, both the sterile and the fertile, are linear or wedge-shaped, with almost straight contour and truncate tips. The cryptostomata are pretty numerous and distinct. F. *angusta* is most closely allied to that form which may be considered the proper glacial forms, namely f. *arctica* J. G. AG. This has somewhat shorter segments, of a less firm consistency and with the tips more rounded, and few or no cryptostomata. Then there is a peculiar form, approaching the former, which KLEEN has found growing in rock-pools at Nordlanden, f. *abbreviata*. It has all the segments, except the ultimate, reduced to costæ; the segments are short, with a few shallow serratures on their curved margins and numerous cryptostomata. This is the smallest of all the forms, probably becoming only 15—20 cm. high. That form which I have named f. *grandifrons* possesses the largest dimensions of all. In this, as in f. *abbreviata*, only the segments of the last order are winged with curved margins and roundly truncate tips, and all the segments are short, particularly the upper ones which bear receptacles; but it differs by its more considerable size, especially in breadth, and by the upper segments being sharply and profoundly serrate.

Habitat. The species is generally litoral in the Norwegian Polar Sea, growing (f. *typica* and f. *angusta*) in the lower part of this zone, or (f. *abbreviata*) in rock-pools between tides. Sometimes (f. *grandifrons*) it descends into the upper part of the sublitoral zone. On the coast of Novaya Zemlya and probably also at Spitzbergen (f. *angusta* and f. *arctica*) it occurs in the sublitoral zone as an element of the formation of *Laminariaceæ*. It flourishes both in exposed and sheltered places of the coast; f. *grandifrons* prefers, however, more quiet localities. The typical form is gregarious, but of the others a few scattered individuals only are found to grow in the same place. The typical form and f. *grandifrons* are furnished with receptacles from July to the beginning of October on the north coast of Norway, f. *angusta* has been found with such organs in July and August at Nordlanden and Finmarken, in July on the west coast of Novaya Zemlya, and a form which is nearest related to f. *arctica* at the end of July at Novaya Zemlya. Of f. *abbreviata* I have seen only sterile specimens.

Geogr. Distrib. It is known both from the Atlantic and the arctic region of the Polar Sea. In the latter it has but a limited distribution, being probably a form that has immigrated there. It is not known from the Kara Sea, the Siberian Sea, and the American Arctic Sea. If it occurs in Baffin Bay, it must be very local and scarce. The only notice of its occurring there is by LYNGBYE, who alleges having seen specimens from that sea in the herbaria of FABRICIUS and GIESEKE. During the numerous expeditions which have in later times visited those waters, this plant has not been met with, as far as I know. Its maximum of frequency is on the north-west coast of Norway. The most northern locality where it has been observed is the coast of Spitzbergen.

Localities: The Norwegian Polar Sea: Nordlanden (f. *abbreviata*, f. *typica* and f. *angusta*) common and abundant; Tromsö amt (f. *grandifrons* and f. *typica*), the latter form abundant and common; Finmarken (f. *typica*) common and abundant, at Maasö, Gjesvær, the south coast of Magerö, and Öxfjord; f. *grandifrons* local and more scarce at Maasö and Gjesvær.

The Greenland Sea: the coast of Spitzbergen (f. *arctica*), local and probably scarce.

The Murman Sea: the coasts of Russian Lapland and Cisuralian Samoyede-land, local and scarce; the west coast of Novaya Zemlya and Waygats; probably everywhere forms most nearly allied to f. *angusta* and f. *arctica*. Cp. GOBI and KJELLMAN l. c.

The White Sea: common and plentiful, probably f. *angusta* and *arctica*.

Baffin Bay: the coasts of Greenland, according to LYNGBYE.

Fucus vesiculosus L.

Spec, Plant. 2. p. 1158.

Descr. Fucus vesiculosus ARESCH. Fuc. et Pycnoph. p. 102.

f. *vadorum* ARESCH.

l. c.

Descr. Fucus vesiculosus β vadorum ARESCH. l. c.
» » f. vadorum KLEEN, Nordl. Alg. p. 26.

f. *typica*.

Descr. Fucus vesiculosus α rupincola ARESCH. l. c.
» » sens. strict. KLEEN, l. c.
Fig. » » HARV. Phyc. Brit. t. 204.
Exsicc. » » ARESCH. Alg. Scand. exsicc. N:o 53.

f. *angustifrons* GOBI.

Algenfl. Weiss. Meer. p. 53.

Descr. Fucus vesiculosus f. pseudoceranoides KLEEN, l. c. p. 27.

f. *turgida* nob.

f. vesiculifera; segmentis summis tantum alatis, linearibus angustis, vix ultra 5 mm. latis; systematibus segmentorum fertilium quam steriles parum brevioribus; receptaculis globosis vel ellipsoideo-globosis, 1 cm. crassitudine superantibus, valde turgidis.

f. *sphærocarpa* J. G. AG.

Grönl. Lam. och Fuc. p. 29.

Descr. Fucus vesiculosus f. sphærocarpus KLEEN, l. c.

Hybr. F. serratus + vesiculosus.

Descr. et Fig. KLEEN, Nordl. Alg. p. 24; t. 9.

Syn. Fucus vesiculosus J. G. AG. Enum.; Grönl. Lam. och Fuc. p. 30; Grönl. Alg. p. 110—111.
» » ARESCH. Fuc. et Pycnoph. p. 102.
» » ASHM. Alg. Hayes. (?) p. 96.
» » CROALL, Fl. Disc. p. 457.
» » DICKIE, Alg. Sutherl. 1, p. 140; Alg. Cumberl. p. 236; ex parte (?).
» » GOBI, Algenfl. Weiss. Meer. p. 53.
» » GUNN. Fl. Norv. 1, p. 48.
» » KLEEN, Nordl. Alg. p. 26.
» » LYNGB. Hydr. Dan. p. 3; ex parte.
» » POST. et RUPR. Ill. Alg. p. II; ex parte.
» » WG. Fl. Lapp. p. 490; excl. var.

Remark on the forms of this species. The first two of the forms quoted are well known. I have preferred to call by the name of f. *angustifrons* GOBI that form common in the southern part of the Polar Sea whose frond is narrow with ovate, oblong-lanceolate receptacles and which KLEEN has recorded as f. *pseudoceranoides* ARESCH. KLEEN'S description of this form differs considerably from ARESCHOUG'S description of subspec. *pseudoceranoides*, and those specimens from Nordlanden which are to be found in KLEEN'S herbarium with the name of f. *pseudoceranoides* do not resemble either the figure cited by ARESCHOUG for this subspecies or the form occurring in Bohuslän with which ARESCHOUG'S description agrees. On this account I am of opinion that the forms to which the above-mentioned name has been applied by ARESCHOUG and KLEEN are not identical. On the other hand, KLEEN'S f. *pseudoceranoides* resembles that form which has been taken by BERGGREN in northern Norway and distributed by J. G. AGARDH under the name of *F. vesiculosus.* This is to be found also in the White Sea according to GOBI, who has proposed for it the name which I have employed above. The form I have called *turgida* and of which I have given the diagnosis is nearly related to the preceding as well as to f. *sphærocarpa.* It differs, however, from both by its large, much swollen receptacles, and seems to deserve attention, as on certain stretches of the coast of Norwegian Finmarken it is an essential constituent of the vegetation of *Fucaceæ.* The very pretty form from Greenland distributed by J. G. AGARDH under the name of f. *sphærocarpa* has been found by KLEEN at Nordlanden and by GOBI in collections brought home from the western Murman Sea and the White Sea. On the coast of Finmarken at the mouths of streamlets I have met with a form, the more robust specimens of which perfectly agree with the Greenland form. Further up in the streamlets where the water is but little brackish, it assumes a different aspect, becoming dwarfed, 2—3 inches high, with only the lower segments elongated, the upper ones being, on the contrary, very short and densely fasciculate; almost every segment of the last order bears a spherical receptacle, 2—3 mm. in diameter.

Habitat. This species is generally litoral, at least in the Norwegian Polar Sea and the immediately adjacent parts of the Arctic Sea. The form *vadorum*, however, is sublitoral, preferring sheltered localities. The others are to be found at exposed as well as sheltered places of the coast. I have already mentioned that I have found f. *sphærocarpa* at the mouths of streamlets on the coast of Finmarken. Some of the forms grow in society, in large masses, contributing strongly to the character of the vegetation for considerable stretches. This is especially the case with f. *typica* and f. *turgida* on the north coast of Norway. All the forms mentioned have been taken in the Norwegian Polar Sea with receptacles in summer, June—August. I have seen specimens from Greenland with receptacles, collected in the month of July. GOBI reports f. *angustifrons* with receptacles from the White Sea, probably taken in summer.

Geogr. Distrib. In consequence of the present species having been confounded with *F. evanescens*, it is impossible to decide its area of distribution by means of the accessible literature. I believe that it does not grow in any other parts of the Arctic Sea than the Norwegian Polar Sea, the western Murman Sea, the White Sea, and Baffin Bay, and that all *F. vesiculosus* which has been reported from other arctic regions is either *F. evanescens* or possibly some individuals of *F. vesiculosus* which have drifted there from more southern parts. Such is probably the case with those fragments of *F. vesiculosus* which were stated to J. G. AGARDH to have been collected at Spitzbergen. Cp. J. G. AG. Grönl. Lam. and Fuc. p. 30. The present species has no doubt its maximum of frequency in the Norwegian Polar Sea. Already in the White Sea *F. vesiculosus* is less plentiful than *F. evanescens;* see GOBI, l. c. p. 54. The extension of the species northwards is uncertain.

Localities: The Norwegian Polar Sea: Nordlanden, f. *vadorum* common, f. *typica* common, f. *angustifrons* scarce, f. *sphærocarpa* scarce; Tromsö amt: f. *typica* common and abundant at Tromsö, Renö, and Carlsö; Finmarken, f. *typica* common and abundant at Maasö, Gjesvær, the south coast of Magerö, Öxfjord, and Talvik; f. *angustifrons* local and scarce at Gjesvær; f. *turgida* local, but abundant, at Maasö and Gjesvær; f. *sphærocarpa* scanty at Gjesvær.

The Murman Sea: the coast of Russian Lapland (f. *sphærocarpa*).

The White Sea: f. *angustifrons* and f. *sphærocarpa* less abundant.

Baffin Bay: f. *sphærocarpa* with certainty on the west coast of Greenland, as at Julianeshaab, Sukkertoppen, Godhavn, Rittenbenk, probably also at Egedesminde and in Cumberland Sound. Other reported localities dubious.

The hybrid form *F. serratus + vesiculosus* has been found by KLEEN at Nordlanden in the Norwegian Polar Sea.

Fucus ceranoides L.

Spec. Plant. 2, p. 1158.

f. *typica* nob.

Descr. Fucus ceranoides HARV. Phyc. Brit. t. 271.
Fig. » » HARV. l. c. et KLEEN, Nordl. Alg. t. 10, fig. 4.

f. *Harveyana* DCSNE nob.

Voyage Venus t. 4.

Descr. Fucus Harveyanus J. G. AG. Spetsb. Alg. Bidr. p. 10 et Till. p. 43.
Fig. » ceranoides KLEEN, Nordl. Alg. t. 10, fig. 2.

f. *divergens* J. G. AG. nob.

Fucus divergens J. G. AG. Grönl. Lam. och Fuc. p. 28.

Descr. Fucus divergens J. G. AG. l. c.
Fig. » ceranoides KLEEN, Nordl. Alg. t. 10, fig. 1.

Syn. Fucus ceranoides LYNGB. Hydr. Dan. p. 5; fide syn. quoad spec. Groenl. (?).
» » SOMMERF. Suppl. p. 182; fide syn.

Remark on the forms of this species. Since I have seen the excellent and highly instructive series of forms of *Fucus ceranoides* which KLEEN has brought home from Nordlanden, and had the opportunity of comparing these forms with the two forms of Fucus found in the Arctic Sea proper, *F. Harveyanus* and *F. divergens*, I cannot but adopt KLEEN'S opinion that these two must not be regarded as independent species, but are, in fact, forms of *F. ceranoides*. However, these forms occurring independently in widely distant parts of the Arctic Sea, I have thought best to set them down under the names once attributed to them.

Habitat. In the Norwegian Polar Sea, this alga is litoral, growing in such places where salt water is mixed with fresh water. *F. divergens* appears in almost fresh water. About its mode of occurrence in the Arctic Sea I know only what is stated by J. G. AGARDH, who follows BERGGREN, namely that f. *divergens* grows on the west coast of Greenland »in scrobiculis». In KLEEN'S collections there are many individuals with receptacles. They have been collected in July and August. I have seen specimens with receptacles of f. *Harveyana* from Spitzbergen and of f. *divergens* from Greenland. These had also been gathered in summer, in July and August.

Geogr. Distrib. The plant is known from isolated parts of the Atlantic and the arctic regions of the Polar Sea. In neither of these seas it is commonly spread. The northernmost locality where it has been found is the coast of Spitzbergen.

Localities: The Norwegian Polar Sea: Nordlanden (ff. *typica*, *Harveyana*, and *divergens*) at several places.

The Greenland Sea: the coast of Spitzbergen; the exact place is not stated (f. *Harveyana*).

Baffin Bay: the west coast of Greenland at Rittenbenk (f. *divergens*). I doubt that LYNGBYE'S *Fucus ceranoides* from Greenland is the present species. In the herbarium of the Copenhague Museum there is no form to be found that can be referred to *F. ceranoides*.

Fucus spiralis L.

Spec. Plant. 2, p. 1159, sec. ARESCH. Fuc. et Pycnoph. p. 106.

Descr. Fucus Sherardi α spiralis ARESCH. l. c.
Fig. » spiralis Fl. Dan. t. 286; non bona.
Exsicc. » platycarpus ARESCH. Alg. Scand. exsicc. N:o 54.

f. *borealis* nob.

f. parvula, vix ultra 15 cm. alta, vulgo minor; segmentis plurimis ad costas reductis, supremis abbreviatis, plus minus crispis vel spiraliter tortis.

Syn. Fucus Sherardi KLEEN, Nordl. Alg. p. 29.
» spiralis GUNN. Fl. Norv. 2, p. 64.
» vesiculosus γ spiralis WG. Fl. Lapp. p. 490.

Habitat. In the Polar Sea this species, wherever it has been met with hitherto, always occurs in the litoral zone, usually at its upper margin, occupying a narrow border below *Pelvetia canaliculata.* It is certainly gregarious, like several other species of Fucus, but never occurs, at least not on the north coast of Norway, in so great numbers as some of these. I never found it but on exposed coasts. During the summer, even to the month of October, it bears receptacles in the Norwegian Polar Sea.

Geogr. Distrib. It is known only from the Atlantic region of the Polar Sea, attaining its maximum of frequency in its southern part. The most northern place where it has been found is Gjesvær in Finmarken about Lat. N. 71°.

Localities: The Norwegian Polar Sea: Nordlanden, common; Tromsö amt, pretty abundant at several places in the neighbourhood of Tromsö; Finmarken: Gjesvær and the south coast of Magerö, more local and scarce.

Fucus evanescens AG.

Spec. Alg. 1, p. 92.

f. *pergrandis* KJELLM.

Descr. Fucus evanescens f. pergrandis KJELLM. Spetsb. Thall. 2, p. 3.
Syn. Fucus evanescens, grandifrons J. G. AG. Grönl. Alg. p. 110; sec. spec.

f. *typica* KJELLM.

Descr. Fucus evanescens f. typica KJELLM. l. c. p. 3—4.
» » J. G. AG. Spetsb. Alg. Till. p. 40—41.
Syn. Fucus evanescens normalis J. G. AG. Grönl. Alg. p. 110; sec. spec.

f. *angusta* KJELLM.

Descr. Fucus evanescens f. angusta KJELLM. Algenv. Murm. Meer. p. 27.
Syn. Fucus evanescens f. elongata, angusta et f. minor, angusta J. G. AG. Grönl. Alg. p. 110.

f. *nana* KJELLM.

Descr. Fucus evanescens f. nana KJELLM. Spetsb. Thall. 2, p. 4.

f. *bursigera* J. G. AG. (KJELLM.)

Spetsb. Thall. 2, p. 4. Fucus bursigerus J. G. AG. Spetsb. Alg. Till. p. 41.

Descr. et Fig. Fucus bursigerus J. G. AG. l. c. et t. 3.

Adnot. 1. F. evanescentis f. angustæ proxime accedit Fucus miclonensis J. G. Ag. Spetsb. Alg. Till. p. 35, 39 et Grönl. Lam. et Fuc. p. 28, saltem quoad specimina Spetsbergensia et Groenlandica.

Adnot. 2. In grege formarum inter f. bursigeram et f. typicam intermediarum J. G. Agardhii Fucus evanescens, minor receptaculis inflatis ad F. bursigerum tendens, me judice est adnumerandus.

Syn. Fucus ceranoides Pall. Reise 3, p. 34.
» » Post. et Rupr. Ill. Alg. p. II; Cfr. Gobi, Algenfl. Weiss. Meer. p. 55.
» » Schrenk, Ural Reise p. 546.
» evanescens J. G. Ag. Spetsb. Alg. Till. p. 27, 35, 40; Grönl. Alg. p. 110; Cfr. supra.
» » Gobi l. c.
» » Kjellm. Vinteralgv. p. 64; Spetsb. Thall. 2, p. 3; Algenv. Murm. Meer. p. 26; Kariska hafvets Algv. p. 23.
» » Quercus Pall. Reise 3, p. 34. Cfr. sub Delesseria sinuosa.
» vesiculosus J. G. Ag. Spetsb. Alg. Progr. p. 2; Bidr. p. 11.
» » Ashm. Alg. Hayes, p. 96 (?) Cfr p. 199 sub. F. vesiculoso.
» » Croall, Fl. Disc. p. 457; ex parte.
» » Dickie, Alg. Sutherl. 1, p. 140; ex parte(?); Alg. Cumberl. p. 236; ex parte.
» » Eaton, List. p. 44.
» » Lindbl. Bot. Not. p. 157.
» » Martin, Met. Observ. p. 313.
» » Post. et Rupr. Ill. Alg. p. II; saltem ex parte.
» » Schübeler, in Heuglin Reise p. 317.
» » Scoresby, Account 1, p. 132.
» » Sommerf. Spitsb. Fl. 233.
» » Zeller, Zweite d. Polarf. p. 85.
» » Cfr. Martens Voyage Spitsb. p. 77, t. F, fig. b.

Remark on the determination of the forms. Fucus evanescens has of late become ever better known, and the algologists who have had an opportunity of studying it have adopted J. G. Agardh's opinion that it is to be considered an independent species. It was formerly confounded with *F. vesiculosus*, although it is probably less closely allied to this species than to *F. edentatus*. It differs from the former species by its branching, the shape of its segments, its colour and consistency, and above all by the different structure of its scaphidia. Of *F. edentatus* I have, on the contrary, seen forms very nearly approaching *F. evanescens*, and, on the other side, forms of *F. evanescens* much resembling *F. edentatus* in the shape and size of the receptacles. Nevertheless I believe the two species can be distinguished by certain differences in the ramification of the frond, in consistency and in the nature of the costa. I have set down here the same forms that I have before endeavoured to distinguish and to define. They are certainly connected by numerous intermediate forms, but they deserve however to be mentioned specially, because they show the limits and directions of the variations of the species and differ somewhat with regard to biology and geographical distribution. I have arranged under them the forms mentioned by J. G. Agardh in his works on the marine Flora of the Arctic Sea. I cannot possibly distinguish *Fucus miclonensis* J. G. Ag. from Spitzbergen and Greenland, of which I have seen specimens determined by J. G. Agardh, from low-sized *F. evanescens* f. *angusta* and from intermediate forms between this and f. *nana*. Farlow says of f. *miclonensis* De la Pyl. »F. miclonensis of De la

PYLAIE is probably a small form of the present» (*F. evanescens*); FARL. New Engl. Alg. p. 102. However, this and *Fucus distichus* KLEEN var. *miclonensis* KLEEN (Nordl. Alg. p. 30) may, I think, be regarded as a distinct species; more on this point below.

Habitat. The comparatively rare forms, f. *nana* and f. *bursigera*, are litoral, the others are always sublitoral, as far as my experience goes. The form *pergrandis* descends deepest of all. I have generally found the form *nana* in such places where salt and fresh water is mixed together. The present species is only little gregarious and occurs in exposed as well as in sheltered localities. I have collected specimens of f. *nana* with receptacles at Spitzbergen and Novaya Zemlya in July, of f. *bursigera* at Spitzbergen in July, of f. *angusta* in the eastern part of the Murman Sea in July, in the Kara Sea at the end of August, in the Siberian Sea (scarce) in the earlier part of July, of f. *pergrandis* at Spitzbergen in July, August, and September, at Novaya Zemlya in July. The typical form bears receptacles all the year round. On the coast of Spitzbergen I have seen specimens with receptacles in November (abundantly), December (abundantly), January, February and March. At the same place I observed germinating spores the 30 December, the 2 and 10 (abundantly) January, the 3, 17, 20 (abundantly) February, and the 29 March. On the west coast of Novaya Zemlya and Waygats I have found the same form in fruit during June and July. Specimens collected in July and August on the west coast of Greenland bear receptacles.

Geogr. Distrib. This species is circumpolar, but does not occur in the Norwegian Polar Sea, being replaced here by *F. edentatus*. Though abounding also in other parts of the Arctic Sea, it has its maximum of frequency in the Greenland Sea. Its northernmost locality is Musselbay on the north coast of Spitzbergen, Lat. N. 79° 53'.

Localities: *The Greenland Sea*: all the forms mentioned, except f. *angusta*, common and abundant on the coasts of Spitzbergen; the east coast of Greenland.

The Murman Sea: (ff. *grandifrons*, *typica*, *angusta* and *nana*) common and abundant on the west coast of Novaya Zemlya and Waygats from Matotshin Shar to Jugor Shar.

The White Sea: more common and abundant than *F. vesiculosus*.

The Kara Sea: (f. *typica* and especially f. *angusta*) Uddebay abundant; Kara Bay.

The Siberian Sea: (f. *angusta* and forms most nearly related to this) Kolyushin Bay abundant. Observed also on the coast east of this point.

The American Arctic Sea: I have thought I might refer to the present species that *Fucus* which is reported from here under the name of *F. vesiculosus*.

Baffin Bay: (ff. *pergrandis*, *typica*, *angusta*, *nana*, and transitions to f. *bursigera*) with certainty at Smallesund, Claushavn, Godhavn and Rittenbenk on the west coast of Greenland. Probably also at Hunde Islands, in Whale Sound, Cumberland Sound a. o.

Fucus edentatus DE LA PYL.

Fl. Terre neuve p. 84.

Descr. Fucus furcatus KLEEN, Nordl. Alg. p. 29.

f. *typica* nob.

Descr. Fucus furcatus ARESCH. Fuc. et Pycnoph. p. 107.
Exsicc. » » KJELLM. in ARESCH. Alg. Scand. exsicc. N:o 401.

f. *contracta* nob.

f. parvula, thallo circa 10 cm. alto, coriaceo, denso dichotomo; segmentis inferioribus ad costas validas, firmas reductis, superioribus alatis, costa distincta vel subdistincta, 3—4 mm. latis; receptaculis cylindrico-fusiformibus, simplicibus vel rarius furcatis 1,5—2 cm. longis, diametro 3 mm.; scaphidiis creberrimis, minutis.

Syn. Fucus furcatus ARESCH. Fuc. et Pycnoph. p. 107, quoad spec. Norvegica et Groenlandica.

Remark on this species. J. G. AGARDH in Spec. Alg. identifies *Fucus furcatus* AG. with *Fucus edentatus* DE LA PYL. To this view J. E. ARESCHOUG accedes on account of a specimen communicated by HARVEY; Fuc. et Pycnoph. p. 109. It is however contested by RUPRECHT who shows *F. furcatus* AG. and the alga described by AGARDH under this name to be two rather distant forms which differ even so much that they ought to be referred to different species; Alg. Och. p. 346. From this cause J. G. AGARDH in his survey of the species of *Fucus* abandons his former view and sets down *Fucus edentatus* DE LA PYL. as specifically distinct from *F. furcatus* AG. Spetsb. Alg. Till. The identity of the *Fucus* in question, occurring on the north coast of Norway, with *Fucus edentatus* DE LA PYL. appears to me to be beyond a doubt. As I have not, among the great number of specimens from Norway and the north-eastern coast of North America examined by me, found any that agrees fully with AGARDH's figure of *F. furcatus*, I have thought best to follow J. G. AGARDH's later exposition, regarding *F. edentatus* as specifically distinct from *F. furcatus* AG.

As respects the form that I have described above under the name of f. *contracta*, it is incontestably united by intermediate forms with the typical form, that is to say, that delivered by me for distribution in ARESCHOUG's Alg. Scand. exsicc. Being however pretty unlike this and at the same time resembling in many respects other species of *Fucus*: *F. Fueci*, *F. miclonensis* J. G. AG. and *F. distichus*, I have thought fit to point it out specially, lest it should be described as a separate species by some one who has not had an opportunity of seeing the transitions between it and *F. edentatus* f. *typica*, or lead to some sort of unnatural combination of *Fucus edentatus* with some of the species mentioned above.

Besides these forms, f. *typica* and f. *contracta*, there is to be found on the north coast of Norway even a third form which is perhaps worthy of attention, being analogous to f. *grandifrons* of *F. serratus*, f. *vadorum* of *F. vesicolosus* and f. *pergrandis* of *F. evanescens*. However, all the specimens on which this opinion is founded being sterile, I cannot here enter into details, but only wish to draw the attention of future investigators to the subject.

Habitat. The species is exclusively litoral, and lives within the lower part of this zone, occupying a region situate between *Fucus serratus* on the one side and *F. vesiculosus* and *F. spiralis* on the other, as has been pointed out already by KLEEN. It seems to prefer exposed localities and is found gregarious in large numbers on extensive areas of the bottom on the coast of Norwegian Finmarken. Farther southwards in the Norwegian Polar Sea this is not the case according to KLEEN. On the coast of Finmarken it is profusely furnished with receptacles during the month of July. In August and at the beginning of September I found only sterile specimens.

Geogr. Distrib. The present species is known only from the Norwegian Polar Sea and Baffin Bay. Its maximum of frequency is certainly on the north coast of Norway. The northernmost locality where it is known with certainty to live is Gjesvær on the north coast of Norway about Lat. N. 71°.

Localities: The Norwegian Polar Sea: Nordlanden pretty common and abundant; Tromsö amt at Tromsö and Renö common and abundant; Finmarken common and very abundant at Maasö, Gjesvær, and Öxfjord.

Baffin Bay: the west coast of Greenland at Julianeshaab and Godhavn, according to specimens in the herbarium of the Copenhague Museum [1]).

Fucus miclonensis DE LA PYL.

Fl. Terre neuve p. 90.

Descr. Fucus miclonensis J. G. AG. Spetsb. Alg. Till. p. 39.
» distichus var. miclonensis KLEEN, Nordl. Alg. p. 30.
Fig. » miclonensis tab. nostra 19, fig. 1, 2.

Remark on the species. In my collections from Finmarken there is a number of specimens of a form of Fucus, with which the above-quoted description of *F. miclonensis* by AGARDH agrees so completely as to leave no doubt that they are to be referred to this species. I have had such a specimen delineated on tab. 19, fig. 1. With these specimens that plant agrees very well which KLEEN has brought home from Nordlanden in great number and named *F. dictichus* var. *miclonensis* in his account of the marine vegetation of that region. One of his specimens is figured in tab. 19 fig. 2. In my opinion it is to be regarded as a distinct species, approaching certainly nearly *F. filiformis*, but differing from it by spreading or even patent segments, considerably broader — even 3 mm. broad — upper segments, more solid structure, shorter, coarser, and less swollen receptacles, which are often united in pairs at the base and in that case widely distended from one another, and by large, very patent scaphidia.

That *Fucus* from Spitzbergen and Greenland which J. G. AGARDH has distributed under the name of *F. miclonensis* does not, as far as I can judge, belong to this species as described by himself, but is to be considered as a form of *F. evanescens*, according most nearly with its f. *angusta*. This form of Fucus is distinguished from *F. miclonensis* as understood by me, by considerably firmer consistency, less distinct costa in the upper segments, more numerous and more distinct cryptostomata, and above all by smaller and less swollen receptacles which differ less in shape from the sterile segments and reach the same height as these. The scaphidia are small and numerous. The plant becomes very black in drying, whilst *F. miclonensis*, if handled carefully in

[1]) It is by some mistake that FARLOW reports this species as taken at Spitzbergen; New Engl. Alg. p. 102.

the preserving, retains its dark-brown colour at least in the upper segments. *F. miclonensis* comes more near *F. linearis*, from which it is scarcely to be distinguished except by the different shape of the upper segments. It also approaches *F. edentatus* f. *contracta*, as KLEEN has rightly noticed, but may be recognized from it by means of the characteristics given by KLEEN; Nordl. Alg. p. 29 note.

Habitat. It grows scattered in rock-pools in the litoral zone. I have met with it only at exposed localities. The proper season of fructification on the north coast of Norway is at the end of July and the beginning of August. However it is to be found with ripe receptacles even in September.

Geogr. Distrib. Known only from the Norwegian Polar Sea, where it is local and not abundant. The northernmost locality is Maasö, about Lat. N. 71°.

Localities: *The Norwegian Polar Sea:* Nordlanden scarce; Finmarken, local and scarce at Maasö.

Fucus linearis Fl. Dan.

t. 351.

Descr. Fucus linearis J. G. AG. Spetsb. Alg. Till. p. 39.
Fig. » » Fl. Dan. l. c.

Syn. Fucus distichus GOBI, Algenfl. Weiss. Meer. p. 52; ex parte.
» » KLEEN, Nordl. Alg. p. 30; ex parte?
» linearis J. G. AG. Enum.; Grönl. Alg. p. 110; Grönl. Lam. och Fuc. p. 29.

Remark on the species. The plant delineated in Fl. Dan. l. c. under the name of *F. linearis* is no doubt so well-marked a form as to deserve to be registered as a separate species. In general it is to be known without difficulty from the cognate species *F. filiformis* and *F. miclonensis* by the considerable difference in breadth between the segments of the last order and those which form the middle part of the frond, and by the shape and disposition of the receptacles. Part of the specimens collected by me at Finmarken agree well with the figure in Fl. Dan., others approach *F. filiformis* more nearly, so that I thought at first they were a broader and more robust form of this species. But J. G. AGARDH, who has kindly examined some of them, has pronounced the opinion that they should rather be referred to *F. linearis*, and after having instituted a more careful comparison between them and a greater number of *F. filiformis* from different localities I cannot but accede to his view. Other specimens resemble *F. miclonensis* in the form of their receptacles, but are easily distinguished from this species by the peculiar form of the segments. As far as I can see, there is accordingly, no reason, to unite this species with any one of the cognate forms.

Habitat. I have only once found this species growing. It was met with in rather little number in rock-pools within the litoral zone. At the end of August at Finmarken it bore receptacles in course of dissolution. Accordingly its fruiting season is here probably in July and at the beginning of August.

Geogr. Distrib. The plant is recorded from the Norwegian Polar Sea, the western Murman Sea, and Baffin Bay. KLEEN reports it from the coast of Nordlanden, but I have not found it in his rich collections of Fuci made here. I cannot decide where it reaches its maximum of frequency. According to my own experience it is rare at Finmarken. The most northerly place where it has been taken is Gjesvær on the north coast of Norway, about Lat. N. 71°.

Localities: The Norwegian Polar Sea: Nordlanden (?); Finmarken at Gjesvær local and scarce; *Norvegia arctica* BERGGREN according to J. G. AGARDH.

The Murman Sea: the coast of Russian Lapland.

Baffin Bay: the west coast of Greenland at Godhavn (?) and Sukkertoppen. Cp. with regard to the locality Godhavn J. G. AG. Grönl. Alg. p. 110 with Grönl. Lam. och Fuc. p. 29.

Fucus filiformis Gmel.

Hist. Fuc. p. 72.

f. *Gmelini* J. G. AG.

Spetsb. Alg. Till. p. 38.

Descr. Fucus filiformis a. Gmelini J. G. AG. l. c.
Fig. » » Gmel. l. c. t. 1.
» » f. Gmelini tab. nostra 19, fig. 3.

f. *Pylaisæi* J. G. AG.

l. c.

Descr. Fucus filiformis b. Pylaisæi J. G. AG. l. c.
Fig. » linearis KÜTZ. Tab. Phyc. 10, t. 15.
Exsicc. » distichus ARESCH. Alg. Scand. exsicc. N:o 201.

Syn. Fucus ceranoides WG. Fl. Lapp. p. 490; ex parte.
» distichus ARESCH. Phyc. Scand. p. 257.
» » GOBI, Algenfl. Weiss. Meer. p. 52; ex parte.
» » GUNN. Fl. Norv. 2, p. 125; ex parte(?).
» » KLEEN, Nordl. Alg. p. 30; ex parte.
» » Nyl. et Sæl. Herb. Fenn. p. 73; ex parte(?).
» » POST. et RUPR. Ill. Alg. p. II; ex parte(?).
» filiformis J. G. AG. Enum.; Grönl. Alg. p. 110; Grönl. Lam. och Fuc. p. 28.

Remark on the species. Every algologist who has studied or intends to study the northern Fuci will no doubt gratefully acknowledge the great service done by J. G. AGARDH towards elucidating their mutual relations in his account of them in Spetsb. Alg. Till. As far as I can judge, AGARDH, as was to be expected from such an experienced, sharp-eyed, and learned algologist, has almost everywhere hit upon the truth, unravelled the confused knot in which those plants were formerly entangled, highly promoted a clear view of the series of forms, and laid a sure foundation for the future study of these series. Among the most difficult forms are those which belong to the species *F. filiformis*, *F. linearis*, and *F. miclonensis* characterized by J. G. AGARDH. They stand pretty distant from certain species, but nearly approach others in certain respects, and in some of their forms resemble each other so closely as to make it sometimes doubtful to which of them a given form is rightly to be referred. But from this fact that certain forms of Fucus resemble one another, it does not in my opinion follow necessarily that they are phylogenetically allied. For although dwarfed forms of for inst. *F. vesiculosus*, *F. evanescens*, and *F. ceranoides* happen sometimes to be so like as to make it hardly possible to establish any limits between them, yet these forms may be traced by intermediate ones into such forms as *F. vesiculosus* f. *vadorum*, *F. evanescens* f. *pergrandis*, and *F. ceranoides* f. *typica*, which will certainly not be regarded by any one as forms of one and the same species. I believe that the above-mentioned species of Fucus hitherto only little attended to, either are links of different series of forms or of the same series — perhaps one culminating in *F. edentatus* f. *typica* — or else are really different species. This point has not as yet been cleared up, and before this is done, it would be inconsiderate to unite the species in question into one. In such a case much more would have to be added to this collective species, and consistently all the northern Fuci must be thrown together into a chaotic whole — a mode af proceding which is certainly very convenient with regard to the systematization of a group rich in forms, but can hardly be considered satisfactory.

Although I acknowledge sincerely that the plant, which I have called *F. filiformis* and which according to the kind communication of J. G. AGARDH is identical with the alga thus named by him, is difficult to define sharply from other species, I do not think, however, that I am justified either by the experience acquired by examining a great number of living and preserved individuals or by reasons adduced by others to refer it as a subform to any other species. Accordingly, I retain it at present as a separate species.

Habitat. This alga in company with *F. distichus* forms sometimes the principal vegetation in rock-pools within the upper part of the litoral zone. I have met with it both in exposed and sheltered localities. Its proper season of fructification on the north coast of Norway seems to be in June and July. It was found sterile in August and October on the coast of Finmarken.

Geogr. Distrib. Known from the southern part of the Polar Sea north of the Atlantic. Its maximum of frequency is on the coast of Norway. The northernmost place where it has been found is Maasö, about Lat. N. 71°.

Localities: *The Norwegian Polar Sea:* Nordlanden common; Tromsö amt, abundant at several places about the town of Tromsö; Finmarken: Maasö, Gjesvær, and the south coast of Magerö, pretty common and plentiful. Besides, it has been brought home from Krogönäs by BERGGREN according to J. G. AG.

The Murman Sea: the coast of Russian Lapland.

The White Sea: at the Solowetzki Isles.

Baffin Bay: the west coast of Greenland at Fridrikshaab and Rittenbenk.

Of the two forms f. *Gmelini* is in my experience the more common one.

Fucus distichus L.

Syst. Nat. Ed. 12, 2, p. 716.

f. *robustior* J. G. AG.

Spctsb. Alg. Till. p. 37.

Descr. Fucus distichus a. robustior J. G. AG. l. c.
Fig. » » KÜTZ. Tab. Phyc. 10, t. 15, fig. d.

f. *tenuior* J. G. AG.

l. c.

Descr. Fucus distichus b. tenuior J. G. AG. l. c.
Fig. » » TURN. Hist. Fuc. t. 4.

Syn. Fucus ceranoides WG. Fl. Lapp. p. 490; ex parte, fide syn.
» distichus J. G. AG. l. c. non KLEEN, Nordl. Alg. p. 30 nec. GOBI, Algenfl. Weiss. Meer. p. 52.

Remark on the species. By the examination I have made of the Fuci of northern Norway I feel justified in adopting J. G. AGARDH's opinion that the alga named by him *F. distichus* ought to be considered a valid species. KLEEN has denied its being different from *F. filiformis* and *F. linearis*; but he has apparently never known that form which J. G. AGARDH understands by the name of *F. distichus*. At least it is not to be found in his collections, forms of *F. filiformis* passing there under that name. Nor does GOBI seem to have met with it in those collections from the White Sea and the western Murman Sea which he has examined. On the coast of Norway it is in my experience much more rare than *F. filiformis*, from which it is pretty easily distinguished on closer examination.

Habitat. The same as the preceding species. On the north coast of Norway it still bears receptacles, though rather scarce, at the end of July. In August it was only found sterile. June and the beginning of July seem to be its proper fruiting season here.

Geogr. Distrib. Known only from the Norwegian Polar Sea. The northernmost point where it has been found is Gjesvær about Lat. N. 71°.

Localities: The Norwegian Polar Sea: Nordlanden at Bodö (f. *robustior*) scarce; Finmarken (f. *tenuior*): Maasö and Gjesvær, at both places rather local and not abundant.

Gen. **Pelvetia** DCSNE et THUR.

Rech. Fuc. p. 12.

Pelvetia canaliculata (L.) DCSNE et THUR.

l. c. Fucus canaliculatus L. Syst. Nat. Ed. 12, 2, p. 716.

Descr. Fucodium canaliculatum J. G. AG. Spec. Alg. 2, p. 204.
Fig. Fucus canaliculatus HARV. Phyc. Brit. t. 229.
Exsicc. » » ARESCH. Alg. Scand. exsicc. N:o 202.

Syn. Fucodium canaliculatum KLEEN, Nordl. Alg. p. 31.
Fucus canaliculatus ARESCH. Phyc. Scand. p. 258.
» » POST. et RUPR. Ill. Alg. p. II.
» » WG. Fl. Lapp. p. 495.
Pelvetia canaliculata GOBI, Algenfl. Weiss. Meer. p. 51.

Habitat. It forms a narrow girdle in the upper part of the litoral zone, scattered specimens advancing even above high-water mark. Though preferring exposed coasts, it is met with also in sheltered localities, nay, even in the interior of deep bays. It is gregarious, but does not occur any where in great number. On the arctic coast of Norway it bears receptacles during the whole summer.

Geogr. Distrib. Known from the Norwegian Polar Sea, the White Sea, and Baffin Bay. The only authority for its existence in the last-mentioned region is a specimen in the herbarium of the Copenhague Museum, stated to have been taken at Greenland by WORMSKIOLD. It probably reaches its maximum of frequency on the northern coast of Norway. The most northerly point from where it is reported is Gjesvær on the north coast of Norway, about Lat. N. 71°.

Localities: The Norwegian Polar Sea: Nordlanden common and abundant; Tromsö amt, common and abundant at the town of Tromsö and at Renö; Finmarken, common and abundant at Maasö, Gjesvær, the south coast of Magerö, and Öxfjord, scarce at Talvik.

The White Sea: Cp. GOBI l. c.

Baffin Bay: the west coast of Greenland, according to a specimen in the herbarium of the Copenhague Museum, the particular locality not being noted.

Fam. TILOPTERIDEÆ THUR.

in LE JOL. Liste Alg. Cherb. p. 16.

Gen. **Scaphospora** KJELLM.

Algenv. Murm. Meer. p. 29.

Scaphospora arctica KJELLM.

l. c. p. 31.

Descr. et Fig. Scaphospora arctica KJELLM. l. c. et t. 1, fig. 1—15.

Habitat. It has been found only once and in little number within the sublitoral zone on gravelly bottom at a depth of 5—10 fathoms in a sound with strong current. At the beginning of August it was richly provided with propagative organs.

Geogr. Distrib. Known only from the eastern Murman Sea.

Locality: the western mouth of Jugor Shar.

Gen. **Haplospora** KJELLM.

Skand. Ect. och Tilopt. p. 3.

Haplospora globosa KJELLM.

l. c. p. 5.

Descr. et *Fig.* Haplospora globosa KJELLM. l. c. et t. 1, fig. 1.

Syn. Haplospora globosa KJELLM. Spetsb. Thall. 2, p. 9; Algenv. Murm. Meer. p. 29.

Habitat. It grows sublitoral on gravelly bottom in 5—10 fathoms on exposed or sheltered coasts. The individuals occur scattered. In the Arctic Sea it bears spores in July and at the beginning of August.

Geogr. Distrib. Found in the Greenland Sea and the eastern Murman Sea. It is here more luxuriant than in the Atlantic. The northernmost locality where it has been taken is Smeerenberg Bay on the north-west coast of Spitzbergen Lat. N. 79° 45'.

Localities: *The Greenland Sea*: the west coast of Spitzbergen, local and scarce.

The Murman Sea: the west coast of Novaya Zemlya and Waygats, local and scarce.

Fam. LAMINARIACEÆ (AG.) ROSTAF.

in GOBI, Algenfl. Weiss. Meer. p. 74; AG. Syst. Alg. p. XXXVI; lim. mut.

Gen. **Alaria** GREV.

Alg. Brit. p. XXXIX.

Alaria esculenta (L.) GREV.

l. c. p. 25. Fucus esculentus L. Mant. 1, p. 135.

f. *australis* nob.

Descr. Alaria esculenta J. G. AG. Grönl. Lam. och Fuc. p. 22.

Fig. Fucus esculentus Turn. Hist. Fuc. t. 117.

Exsicc. Alaria esculenta ARESCH. Alg. Phyc. Scand. exsic. N:o 19.

f. *musæfolia* DE LA PYL. (nob.)

Laminaria musæfolia DE LA PYL. Fl. Terre neuve, p. 31.

Descr. Alaria musæfolia J. G. AG. l. c, p. 23; excl. syn.

Fig. Laminaria esculenta var. platyphylla DE LA PYL. Prod. Terre neuve, t. 9, fig. D.

Syn. Alaria esculenta ARESCH. Phyc. Scand. p. 342.

» » KLEEN, Nordl. Alg. p. 32; ex parte.

Fucus esculentus WG. Fl. Lapp. p. 494.

Remark on the species. I have at present comparatively very rich collections of Alaria from northern Norway at my disposal. In these there are several specimens which agree very well in all points with J. G. AGARDH's excellent description of *A. esculenta* and with the figure of *Fucus esculentus* given by TURNER. Again, other specimens, which have come from more northern localities, are distinctly marked with the characteristics set down by J. G. AGARDH for *A. musæfolia*. Some of them might very well have served as original for DE LA PYLAIE'S figure of *Laminaria esculenta* var. *platyphylla*, which is regarded by J. G. AGARDH as the typical *A. musæfolia*. Between these two plants, *A. esculenta* and *A. musæfolia* as understood in this sense, no fixed limits can be drawn. The one, *A. musæfolia*, is a northern form, the other, *A. esculenta*, a southern form of the same species. I have been able to lay out a complete series of transitions between them.

In order to show what the dimensions of this beautiful species are in our seas, I shall give the measures of the largest specimens at my disposal. The stipe is 15 cm. long, one cm. in diameter downwards; the rhachis has a length of 5 cm., the naked costa above the rhachis a length of 2 cm.; the lamina is one metre and a half long and 12 cm. broad at its broadest part, which is situate 35 cm. above the base of the lamina. The leaves are 12 cm. in length and 1 cm. in breadth. That *Fucus pinnatus* GUNN. belongs to this species, seems to me very dubious. I believe it is to be identified with *Alaria Pylaii* J. G. AG. I shall return to this question in my account of the last-mentioned species.

Habitat. The present species lives gregarious at low-water mark or somewhat below that line in the uppermost part of the sublitoral zone, principally on solid rocks in exposed localities. It is found still flourishing at the end of August, but its highest development seems to be attained during the earlier part of the summer. On the arctic coast of Norway I have found individuals with mature zoosporangia in July.

Geogr. Distrib. Undoubtedly it reaches its maximum of frequency in the southern part of the Norwegian Polar Sea; at Finmarken it is scarce, being replaced here by *A. membranacea*. It is also reported from the Greenland Sea by J. G. AGARDH, but I have failed to detect it here myself.

Localities: The Norwegian Polar Sea: Nordlanden common; Finmarken: Maasö local and scarce.

The Greenland Sea: the coast of Spitzbergen according to AGARDH.

Alaria Pylaii (DE LA PYL.) J. G. AG.

Grönl. Lam. och Fuc. p. 24. Laminaria Pylaii DE LA PYL. Fl. Terre neuve p. 29.

Descr. Alaria Pylaii J. G. AG. l. c.

Syn. Alaria Despreauxii J. G. AG. Grönl. Alg. p. 110.
» esculenta KLEEN, Nordl. Alg. p. 32; ex parte, fide syn.
» Pylaii CROALL, Fl. Disc. p. 457.
Fucus pinnatus GUNN. Fl. Norv. 1, p. 96 (?). Acta Nidros. t. 8, fig. 1 (?).

Remark on the species. Besides *Alaria esculenta* as I have understood above, there is to be found on the north and west coasts of Norway at least northward from Aalesund another *Alaria* which is certainly specifically distinct from the former. I possess a considerable number of specimens of it, both young ones bearing their first sporophylla and older ones with scars of fallen leaves. They are easily distinguished from specimens of *A. esculenta* by the stipe thickening upwards towards the rhachis, by the upper part of the stipe being somewhat flattened as well as the rhachis and broadly elliptical in transverse section, by broader and longer sporophylls which are distinctly stalked and the basal parts of which are somewhat thickened downwards and united by a thin margin, by the distinctly wavy ovate-lanceolate lamina whose base especially in older individuals is far more rounded, sometimes almost heart-shaped and always less decurrent than in *A. esculenta*, and by the costa being lower and less sharply marked against the lamina than in *A. esculenta*. In drying the plant becomes more dark-coloured than the last-mentioned alga. I consider the species in question identical with *A. Pylaii* J. G. AG. The specimens agree in all essential points with J. G. AGARDH's description, and on comparing them with Greenland specimens of *A. Pylaii* no constant essential differences can be detected. However, the Norwegian specimens are often narrower than those from Greenland and provided with narrower sporophylls. But on the other side there exists on the coast of Norway a litoral form of the plant, which resembles the specimens from Greenland with regard to the breadth of the lamina as compared with the length and surpasses them in the breadth of the sporophylls. It should be remarked also that even among the specimens from Greenland distributed by J. G. AGARDH under the name of *A. Pylaii* there are to be found several that have a more elongated lamina and narrower sporophylls. Between these and the sublitoral form from the north-west coast of Norway I have not been able to detect any differences. In all the young individuals from the Norwegian coast that I have seen, the stipe is very short, 5 cm. in length at the most. In some of them that part of the frond which is below the sporophylls is even 20 cm. long, but this is plainly no stipe proper, but the stipe together with the rhachis which elongates as the plant grows older, developing new sporophylls above the old ones which fall off after having served their purpose. In one of these older specimens whose axial portion below the collection of sporophylls is 15 cm. long, there is to be seen on either side of the axis a ridge which becomes more and more indistinct downwards, but can be traced with certainty to a distance of 5 cm. from the rhizines. These ridges obviously mark the part that has once borne sporophylls. All that part of the cauloid portion which is provided with those two ridges is accordingly to be regarded strictly as belonging to the rhachis, not to the stipe, so that the stipe itself is really short even in those old individuals in which the cauloid portion is long. The rhachis is long, on the contrary, longer than in *A. esculenta* and even longer than in f. *musæfolia*, in which I have never found any muricate margin, but only a short row of cicatrices of fallen sporophylls, depressed in a furrow.

I think GUNNER's *Fucus pinnatus* should be referred to the present species rather than to *A. esculenta* f. *musæfolia*. If the proportions between the length and the breadth of the lamina are at least approximately correct in the figure quoted, I cannot see how such an *Alaria* could possibly be referred *A. musæfolia*. Also with regard to the shape of the lamina the plant figured agrees more nearly with *A. Pylaii* than with any *A. esculenta* that I have seen. To these facts is to be added the form of the rhachis which seems to me to point very decidedly towards the identifying of *Fucus pinnatus* with *A. Pylaii*, not with *A. esculenta* f. *musæfolia*. Supposing the figure to be delineated from a dried specimen, it represents in this point pretty accurately *A. Pylaii*, in which according to the quite correct account of J. G. AGARDH »bases pinnarum quasi margine tenuiore conjunguntur», but it does not at all resemble *A. esculenta* f. *musæfolia*, in which I as well as J. G. AGARDH have always found the sporophylls (pinnæ) »quasi e canaliculo impresso egredientes» (cp. J. G. AG. Grönl. Lam. och Fuc. p. 25).

Habitat. This species sometimes occurs in rock-pools in the litoral zone, being then more or less dwarfed with very broad sporophylls. But on the coast of Norway it is generally found at or somewhat below low-water mark, between the litoral and the sublitoral region. It is gregarious as the preceding species, but does not grow in any very great numbers together. Specimens taken on the coast of Norway in July and August bear ripe zoosporangia. The individuals brought home from Greenland by Swedish expeditions, most of which seen by me are provided with zoosporangia, appear to have been collected in July, August, and September.

Geogr. Distrib. Known from the Norwegian Polar Sea and Baffin Bay. I cannot decide where it attains its maximum of frequency. The most northern point where it is known with certainty to occur is Maasö on the north coast of Norway about Lat. N. 71°.

Localities: The Norwegian Polar Sea: Nordlanden according to KLEEN; Tromsö amt near the town of Tromsö, local and scarce; Finmarken at Maasö, local, but rather plentiful.

Baffin Bay: the west coast of Greenland at Julianeshaab, Sukkertoppen, Jakobshavn, Claushavn, probably rather abundant. Cp. J. G. AG., l. c. p. 20—21.

Alaria membranacea J. G. AG.

Grönl. Lam. och Fuc. p. 26.

Descr. Alaria membranacea J. G. AG. l. c.

Syn. Alaria esculenta DICKIE, Alg. Sutherl. 1, p. 140 (?).
» » KLEEN, Nordl. Alg. p. 32; ex parte, fide syn.
» » Nyl. et Sæl. Herb. Fenn. p. 73 (?).
» » POST. et RUPR. Ill. Alg. p. II. Cfr. GOBI, l. c. p. 78.
» musæfolia KJELLM. Algenv. Murm. Meer. p. 35.
» Pylaii J. G. AG. Spetsb. Alg. Till. p. 28 et 30; Grönl. Alg. p. 110. Cfr. Grönl. Lam. och Fuc. p. 24 et 26.
Orgyia pinnata GOBI, Algenfl. Weiss. Meer. p. 77.

Remark on the species. In the collections of *Alariæ* which I have brought home from the coast of Finmarken in Norway, there are several young specimens that agree very closely with the Greenland alga distributed by J. G. AGARDH under the name of *A. membranacea*. They belong undoubtedly to another series of forms than *A. esculenta*. But they come very near *A. Pylaii* on the one side, *A. grandifolia* on the other, and might be considered as an intermediate form between these two species or as a more southern form of *A. grandifolia*. However, the plant occurring in an almost identical form at two places so widely distant as the coast of Finmarken in Norway and the west coast of Greenland, this may be regarded as a sufficient reason to consider it a fixed form and to set it down under a special name.

Mr FOSLIE at my demand has made a large collection of *Alariæ* from the southern part of the Norwegian Polar Sea, which he has had the kindness to send me. Besides plenty of *A. esculenta* f. *musæfolia*, it contained a great number of individuals of another *Alaria* that must be determined as *A. membranacea*, as far as I can see. However, it much approaches *A. Pylaii*, but the stipe is longer and coarser than in this species and the lamina is longer in proportion to its breadth and much thinner. A considerable part, several inches, of the cauloid portion is terete or almost terete. But judging from distinct cicatrices, this part has borne sporophylls, although these have been few in number and very thinly spread. The proper rhachis is compressed, provided upwards with sporophylls and downwards showing dense cicatrices of such organs, that have fallen off.

Already at Nordlanden the present species is large, though it does not by far attain the dimensions of *A. grandifolia*. At Finmarken I could get only younger individuals. These however give occasion to suppose that the species grows very large there. In one of these individuals the lamina is 20 cm. broad. I believe it was probably a thicket of this species that I saw once on the coast of Finmarken, but was unable to get hold of on account of stormy weather and unsuitable instruments. I estimate the length of the individuals of which the thicket was composed at least at five or six feet.

It seems to me probable that DICKIE's *A. esculenta* from Whale Sound is the present species. Possibly *A. esculenta* ASHM. from Smith Sound also belongs here.

Habitat. This species belongs to the sublitoral zone. On the coast of Norway it grows in the upper part of this zone in one or two fathoms water, but in other parts of the Polar Sea it descends to greater depths. It prefers exposed coasts and lives sometimes gregarious, sometimes scattered. Specimens taken at Nordlanden in April are furnished with zoosporangia.

Geogr. Distrib. The species appears to belong to those parts of the Polar Sea which lie north of the Atlantic. Here it has been observed at several different places. At present it is uncertain where its maximum of frequency is attained. The most northern locality where it has been found with certainty is the North Cape of Spitzbergen, Lat. N. 80° 31'.

Localities: The Norwegian Polar Sea: Nordlanden according to specimens communicated by FOSLIE and KLEEN; Finmarken local and rather scarce at Maasö and Gjesvær.

The Greenland Sea: Beeren Eiland; the coasts of Spitzbergen.

The Murman Sea: the west coast of Novaya Zemlya, local, scarce.

The White Sea: common and abundant according to GOBI, l. c.

Baffin Bay: the west coast of Greenland, Claushavn and Jakobshavn, Whale Sound (?). In the herbarium of the Copenhague Museum there are found specimens collected by different persons, localities unknown.

Alaria grandifolia J. G. Ag.

Grönl. Lam. och Fuc. p. 26.

Descr. Alaria grandifolia J. G. Ag. l. c.

» » Kjellm. Spetsb. Thall. 2, p. 10.

Syn. Alaria esculenta Eaton, List, p. 44 (?).

» » Kleen, Nordl. Alg. p. 32 (?); fide syn.

» grandifolia Kjellm. l. c. et Algenv. Murm. Meer. p. 35.

Laminaria esculenta Lindbl. Bot. Not. p. 157 (?); Cfr. J. G. Ag. Spetsb. Alg. Till. p. 30.

Habitat. This species is sublitoral, living generally at a depth of 2—15 fathoms. It is met with in the interior of deep bays as well as on exposed coasts, in the latter case sometimes near the shore, sometimes, when the bottom is favourable, several miles off. It is on rocky bottom that it attains its greatest size. Living sometimes alone in rather great numbers, sometimes in company with other *Laminariaceæ*, it constitutes an essential element of the formation of Laminariaceæ on the coasts of Spitzbergen and the west coast of Novaya Zemlya. I often met with young, sometimes very young, individuals during the winter on the north coast of Spitzbergen. Only once or twice, in December and January, I succeeded in finding some older specimens. These then bore zoosporangia fully developed. Besides, I have collected specimens provided with zoosporangia at Spitzbergen in July, August, and September, on the west coast of Novaya Zemlya in July.

Geogr. Distrib. This species has its maximum of frequency on the coasts of Spitzbergen. It is found with certainty also in the eastern part of the Murman Sea. But the statement that it occurs on the north-west coast of Norway seems to me doubtful. It is possibly a large *A. membranacea* that has been identified by Kleen with J. G. Agardh's *A. grandifolia*. The northernmost known locality of the species is Treurenberg Bay on the north coast of Spitzbergen, Lat. N. 79° 56'.

Localities: The Norwegian Polar Sea: Nordlanden (?).

The Greenland Sea: the north and west coasts of Spitzbergen, common and plentiful.

The Murman Sea: the west coast of Novaya Zemlya, local and probably rather scarce.

Alaria dolichorhachis nob.

A. stipite brevi, vix ultra pollicari sæpe breviore, terete; rhachide elongata, in planta senili fere pedali, apicem versus incrassata, supra sporophylla denuo contracta, in costam abrupte abeunte, compressa, in sectione transversali elliptica, usque ad basim cicatricibus sporophyllorum abjectorum elevatis, superne in marginem muriculatum confluentibus insigni; costa prominula, in sectione transversali elliptica; lamina angusta, lanceolato-lineari, in costam breviter decurrente, rigidiuscula, undulato-plicata; sporophyllis primariis distantibus, ceteris approximatis, in planta adulta numerosis, fasciculatis, elongatis, pluripollicaribus, angustis, lineari-spathulatis, vulgo crispis, undulatis vel spiraliter contortis; soro sporophyllorum apicem longe marginemque anguste sterilem relinquente. Tab. 20, 21 et 25, fig. 11—18.

Syn. Alaria esculenta Harv. Fl. West. Esk. p. 49 (?).

Description. The *rhizines* issue pretty regularly and in older individuals form two sometimes three, alternating whorls. Their length depends on the solidity of the substratum, the rhizines being shorter as the substratum is more solid.

The *stipe*, i. e. that part of the cauloid portion which does not produce sporophylls, is always short, usually 1—4 cm., seldom even 6 cm. long, terete, or upwards very slightly compressed, in old individuals even 8 mm. in diameter (tab. 25. f. 12).

The *rhachis*, or that part of the cauloid portion which produces sporophylls, becomes longer as the plant grows older. In certain specimens I have examined, it has a length of even 15 cm. It is compressed in its whole length, most so in the part that bears sporophylls at the time; there a transverse cut shows the narrowest ellipse with the longest longitudinal axis. Above the sporophylls it tapers swiftly, passing abruptly into the costa. Cp. tab. 25 fig. 13—15. Even in older specimens the rhachis is easily distinguished from the stipe, because through all its length it bears cicatrices of fallen sporophylls, the cicatrices being downwards thin, upwards very dense and forming small elevations on a ridge running on each side of the rhachis.

The *lamina* is linear-lanceolate. In regard of the form of its base the lamina resembles that of *A. esculenta* f. *musæfolia.* I have not seen specimens with longer lamina than about one metre. In older individuals the lamina is usually 4—8 cm. broad at its broadest part; however, I have seen individuals in which the breadth of the lamina attained 11—12 cm. In younger specimens the lamina is richly plicate and wavy, in older it is more plane and at the same time of firmer consistency. The lamina is thinner than in *A. esculenta* f. *typica*, resembling that of *A. membranacea* (tab. 20 and 21).

The *costa* is broad, and very prominent, sometimes rising equally on both sides, sometimes more on the one side, passing immediately into the lamina (tab. 25, fig. 10).

The *sporophylls* are elongatedly linear-linguiform, in younger specimens shorter, in older even 20 cm. long, usually 1,5 cm. broad below the top, sometimes even 2,5 cm. They taper strongly, but evenly, towards the base, passing almost imperceptibly into a pedicel 2—4 mm. long. While sterile they are membranaceous in the greater part of their length, and even after the zoosporangia have been developed, they are far more slender and less firm than in *A. esculenta.* They are always richly wavy and almost always twisted spirally several times downwards. Those first developed are thinly placed and distinctly separate, the following are densely crowded. In older individuals they are very numerous, densely clustered, sometimes 60 or more in one individual. The sorus reaches from the base of the sporophylls to about two thirds of their length and is surrounded at the sides by a narrow, finely wavy margin. The zoosporangiferous part of a sporophyll is membranaceous-coriaceous, the rest of it is membranaceous (tab. 21).

The *zoosporangia* are cylindrically spindle-shaped with blunt ends, about 40 μ. long and 10—12 μ. in diameter. As to these measurements it ought however to be remarked that I have had only a little number of specimens with fully developed zoosporangia at my disposal.

The *paraphyses* are elongatedly wedge-shaped in optical longitudinal section, about 10 μ. in diameter at the top. The outside of the membrane is thick and strongly gelatinized.

Structure of the frond. In anatomical structure the present species offers several differences from *A. esculenta.* A cross cut of the stipe and the lower part of the rhachis shows the central fibrous layer to be lanceolate or sickle-shaped in *A. esculenta*, but more plainly linear in *A. dolichorhachis*. The surrounding cell-layer, a tissue most nearly related to the collenchyme and constituting together with the central layer the mechanical system, possesses longer and wider elements in *A. dolichorhachis* than in *A. esculenta*. The tissue covering this layer shows a greater or less number of concentric zones in the former species, but not in the latter, as far as my observations go. Besides, it is of a looser consistency in *A. dolichorhachis* than in *A. esculenta* and has a strong tendency to burst radially in drying. The outside of the cauloid portion of *A. dolichorhachis* is composed of a tissue whose cells are transversely rectangular or square, arranged in pretty regular radiating rows and furnished with thin brown walls. This tissue, which I have never found in *A. esculenta*, is assuredly most closely allied with cork tissue. In older individuals it attains a considerable thickness, 150 μ. or even more. Here and there in it cavities are formed extending lengthwise as well as radially and finally opening outwards. These sometimes much resemble the lacunæ muciferæ that occur in certain species of the genus *Laminaria*. On this account the surface of the cauloid portion upwards to the rhachis is usually fissured and uneven in older specimens.

I have collected a considerable number of the Alaria now described of different ages at different rather widely distant points on the north coast of the Tshutsh-land. It is undoubtedly specifically distinct from the species of Alaria occurring in the Atlantic Sea and in the Arctic Sea north of the Atlantic. It might be taken at first sight for an *A. esculenta* f. *typica*, because it resembles this species most with regard to the form of the lamina. It is however decidedly distinguished from it by several strongly marked characteristics, as the form of the rachis and the costa a. o. In the shape of the rhachis it agrees most nearly with the group *A. Pylaii*, *A. membranacea*, and *A. grandifolia*, but it is known from all of these by the shape of the lamina as well as of the costa. That it differs in the shape of the costa both from these species and from *A. esculenta*, is shown by a comparison between fig. 16, 19, and 20 in tab. 25.

Older individuals of the present species have a very characteristic aspect on account of the cauloid portion being large and coarse in proportion to the lamina, the rhachis thickening upwards, the sporophylls being very numerous, clustered, often spirally twisted, long and thin. It can hardly be confounded with any species known to me. It exists also in the Behring Sea and is probably that alga which according to J. G. AGARDH was distributed by RUPRECHT under the name of *Phasganon alatum*. Cp. J. G. AG. Grönl. Lam. och Fuc. p. 23. It is possibly the same plant that was brought home by SEEMAN from the north coast of western Eskimaux-land and has been called *A. esculenta* by HARVEY in his list of the algæ collected by SEEMAN.

Habitat. It grows within the sublitoral zone in 2—3 fathoms water. It prefers rocky bottom, but is found also on pebbly bottom, though less richly developed. It is gregarious, forming in company with other *Laminariaceæ*, *Laminaria cuneifolia* and *L. solidungula*, a well-marked formation of *Laminariaceæ*. I have seen specimens taken in the Arctic Sea in April, May, and June, all with the lamina preserved and plenty of sporophylls. At the end of April I got some specimens with the sorus developed.

However, the proper season for producing zoospores appears to be later, probably in the month of June.

Geogr. Distrib. At present this species is known with certainty in the Arctic Sea only from the eastern part of the Siberian Sea. Here it was everywhere found abundant.

Localities: The Siberian Sea: Koljushin Isle, and at two places eastward of the mouth of Koljushin Bay, abundant.

The American Arctic Sea: the north coast of Western Eskimaux-land (?).

Alaria oblonga nob.

A. stipite perbrevi, vix ultra pollicari, terete; rhachide demum pluripollicari, apicem versus incrassata, compressa, in sectione transversali elliptica, residuis sporophyllorum abjectorum in plantis senilibus longe deorsum plus minus distincte muriculata, abrupte in costam abeunte; lamina elongata, lineari-oblonga, basi ovata, vix decurrente vel ovato-cordata, usque 25 cm. lata, tenue membranacea, undulata; costa prominula, in sectione transversali elliptica; sporophyllis numerosis, subdistantibus, petiolatis, lanceolato-spathulatis, augustis 1,5—2 cm. latis, sterilibus tenue membranaceis, parte fertili subpergameis, margine undulato-crispatis, longe infra apicem sorum formantibus. Tab. 22 et 25 fig. 21—24.

Description. The *rhizines* issue in alternate whorls and are subdichotomously branched. The branches in the present species as in other Laminariaceæ are shorter and more robust as the substratum is more solid.

The *stipe* is very short, sometimes almost imperceptible, on account of the cauloid portion bearing or, as is shown by the cicatrices, having borne sporophylls immediately above the rhizines. Its length in older individuals usually varies between 0,5 and 2 cm. In very young individuals the stipal part may be distinguished from the rhachis by the latter being thicker. In very old specimens the limit between these two parts of the cauloid portion is indistinct on account of the cicatrices of the fallen first sporophylls being effaced. In outline the stipe is terete downwards, somewhat, but only slightly, compressed upwards, where it passes into the rhachis. In the oldest individuals I met with, the stipe was 5 mm. in diameter.

The *rhachis* increases in length as the plant grows older, attaining a length of at least 14 cm. It is almost terete downwards, but upwards it is flattened and thicker. Above the uppermost sporophylls it tapers swiftly, soon passing into the costa. The part with sporophylls is flatly elliptical in profile, at least 8 mm. in its longest diameter. The difference in shape and thickness between the stipe and the rhachis in the same individual is shown by fig. 21 and 22 in tab. 25.

The *lamina*, as shown by the figures of tab. 22, in younger as well as older specimens has almost perfectly the same shape, elongatedly linear-oblong, or in other words linear-lanceolate, with rounded base which is not at all or almost imperceptibly decurrent. It sometimes tapers towards its extremity somewhat more swiftly and strongly than in the specimens delineated, and sometimes in very old specimens the base is almost cordate. The lamina attains a considerable size, but it is only seldom that a specimen is found with the whole lamina preserved. In my collections I have an older specimen with almost complete lamina; this is — measured dry — 90 cm. long and 20 cm. broad. In some fragmentary specimens the breadth amounts to nearly 25 cm. Even in older specimens with ripe zoosporangia the lamina is thinly membranaceous, densely wavy, rather brittle when dried. In drying it assumes a dark-brown colour with a soot-brown tinge.

The *costa* is prominent, sometimes almost planely convex, usually biconvex, elliptical in transverse section, in older individuals about 7 mm. broad, 3 mm. thick (tab. 25, fig. 23, 24).

The *sporophylls* are linear spade-shaped, with short pedicels, when sterile thinly membranaceous, about 10—20 cm. long and scarcely 2 cm. broad near the rounded top. The sorus occupies half or two thirds of the length, being surrounded upwards by a narrow margin. The soriferous part is comparatively thin, membranaceously pergameneous. The sporophylls are developed in rather little number (20—30) at the same time in the same individual, and are somewhat thinly arranged, those which are first developed every time being thinner than the later.

The *zoosporangia* are cylindrically club-shaped, 35—60 μ. long, 10—12 μ. thick. The paraphyses are club-shaped with a distinct pedicel somewhat enlarged downwards. The outside of the membrane is much thickened. The length of the pedicel is usually 45 μ., that of the head of the club 60 μ., the thickness of the latter about 10 μ.

In anatomical structure the present species is most closely allied to the preceding one. It agrees with this also with regard to the shape of the cauloid part, the sporophylls, and the costa, but differs decidedly and, as far as I can see, constantly as to the shape of the lamina. Even in young individuals that shape of the lamina which is charateristic of the present species, is found well marked and it remains the same through all its ages. The only difference in this respect I have been able to detect is the base's becoming more cordate as the plant grows older.

Thus this *Alaria* appears to show that at least in certain groups of this genus more importance may be attached to the shape of the lamina in distinguishing the various species than I myself and several algologists have been inclined to suppose.

Habitat. At the only place where this species has been hitherto met with, it grew on the exposed coast within the sublitoral zone in 4—5 fathoms water, on rocky and stony bottom, constituting here together with other *Laminariaceæ* a formation of considerable extent and richness. The plant when collected in the middle of September bore scarce zoosporangia.

Geogr. Distrib. Known only from the Siberian Sea.

Locality: Irkaypi on the north coast of Tshutshland.

Alaria elliptica nob.

A. stipite brevissimo, tereti; rhachide compressa, apicem versus incrassata; lamina elliptica, basi distincte decurrente, usque 40 cm. lata, fere papyracea, undulata; costa elevata, in sectione transversali elliptica; sporophyllis lanceolato-spathulatis, infra apicem rotundatum 2—3,5 cm. latis, papyraceis, parte sorifera submembranaceis, margine crispatis; soro dimidiam partem occupante; Tab. 23 et 25 fig. 25, 26.

Remark on the species. I have seen the *Alaria* the diagnosis of which I have given above, in a pretty considerable number of individuals younger as well as older, though none were very old. They had been collected at two points on the north coast of Siberia. The present species resembles the two preceding in the cauloid part and the costa, but differs from both of them in the shape of the lamina. This shape being constantly the same in older and younger specimens, I cannot but regard the plant as specifically distinct from both the preceding. The breadth of the lamina as compared with its length is considerably greater than in any of the species mentioned before, and the lamina is apparently always very thin. Even in specimens that are not more than 30 cm. long, the breadth is 10 cm. at the broadest part, immediately below the middle of the lamina. An incomplete specimen in my collections has the lamina 40 cm. broad, thin as paper, and distinctly angular at the base. I do not know the approximate maximum of the length of the rhachis, not having found any older individual with the cauloid part preserved whole. The costa is somewhat less prominent than in the preceding species, but in other respects it is similar, though more often plane-convex. The sporophylls are pretty numerous; I have found even 20 developed at the same time in one individual. Although densely crowded, they are not so clustered as in *A. dolichorhachis*, and they are broader than in the last-named species and *A. oblonga*. Most of the specimens collected were sterile. In some the sporophylls were provided with developed sori occupying little more than the lower half of the sporophylls. The sterile part is very thin and wavy; the part bearing zoosporangia is also comparatively thin, almost membranaceous. Zoosporangia and paraphyses have the same shape and size as in the two preceding species. The length of the zoosporangia is 50—75 μ., their thickness 10—15 μ.

Habitat. The present species is sublitoral and grows gregarious in 2—3 fathoms water on stony bottom on exposed coasts. In the month of July it was furnished with scarce zoosporangia on the north coast of Tshutsh-land.

Geogr. Distrib. Known only from the Siberian Sea.

Localities: the north coast of Tshutsh-land at Koljutshin Isle and at Pitlekay, the wintering station of the Vega expedition.

Gen. **Agarum** (Bory.) Post. et Rupr.

Ill. Alg. p. 11; Bory. Dict. Class 9, p. 193; spec. excl.

Agarum Turneri Post. et Rupr.

Ill. Alg. p. 12.

Descr. Agarum Turneri J. G. Ag. Grönl. Lam. och Fuc. p. 18.
Fig. » » Post. et Rupr. l. c. t. 22.
Exsicc. » » Farl. Ands. and Eat. Alg. Amer. N:o 12.

Syn. Agarum Turneri J. G. Ag. l. c.; Grönl. Alg. p. 110.
» » Croall, Fl. Disc. p. 457.
» » Dickie, Alg. Sutherl. 1, p. 141; Alg. Sutherl. 2, p. 191; Alg. Walker. p. 86; Alg. Cumberl. p. 237.
Laminaria Agarum Lyngb. Hydr. Dan. p. 24.

Habitat. It occurs in the greatest number and most richly developed on rocky bottom within the sublitoral zone at a depth of 3—15 fathoms, and is met with in exposed as well as sheltered localities. Specimens taken in the latter part of the summer on the west coast of Greenland bear zoosporangia.

Geogr. Distrib. It belongs to the American Arctic Sea and Baffin Bay. Its most northern known locality is Whale Island on the west coast of Greenland about Lat. N. 77° 30'. I cannot determine where its maximum of frequency is situate.

Localities: *The American Arctic Sea:* Port Kennedy, Union Bay, Assistance Bay, and Hudson Bay.

Baffin Bay: Cumberland Sound, the west coast of Greenland at several places, as Lichtenau, Julianeshaab, Sukkertoppen, Holstenborg, Egedesminde, Hunde Island, Godhavn, Claushavn, Jakobshavn, Rittenbenk, Melville Bay and Whale Island.

Gen. **Phyllaria** (Le Jol.)

Exam. p. 59; spec. excl.

Phyllaria dermatodea (De la Pyl.) Le Jol.

l. c. Laminaria dermatodea De la Pyl. Prod. Terre neuve, p. 180.

Ph. stipitis tela centrali mere parenchymatica a cellulis membrana valde incrassata eminentibus, longissimis, plus minus ramosis circumdata; lamina obscure olivaceo-fusca, demum coriacea vel coriaceo-membranacea; cryptostomatibus in planta juvenili paucis; Tab. 25 fig. 1—4.

f. *typica* nob.

f. cryptostomatibus in planta adulta numerosioribus, foveas profundiores, margine elevato circumdatas constituentibus.

Descr. Saccorhiza dermatodea Farl. New Engl. Alg. p. 95.
Fig. Laminaria dermatodea De la Pyl. l. c, t. 9, fig. G.
Exsicc. » lorea Aresch. Alg. Scand. exsicc. N:o 213.

f. *arctica* nob.

f. cryptostomatibus in planta adulta æque ac in planta juvenili perpaucis vel fere nullis, parum immersis, margine nullo circumdatis.

Syn. Laminaria Bærii Nyl. et Sæl. Herb. Fenn. p. 73.
» » Post. et Rupr. Ill. Alg. p. II, sec. Gobi, Algenfl. Weiss. Meer, p. 75.
» dermatodea J. G. Ag. Spetsb. Alg. Till. p. 28.
Phyllaria dermatodea Gobi, l. c.
Saccorhiza dermatodea Aresch. Obs. Phyc. 3, p. 11.
» » J. G. Ag. Enum.; Spetsb. Alg. Till. p. 31; ex parte; Grönl. Alg. p. 110.
» » Kjellm. Spetsb. Thall. 2, p. 19; ex parte; Algenv. Murm. Meer. p. 36; ex parte.

Remark on the species. J. G. Agardh in *Spec. Alg.* has kept up and characterized as distinct species two formerly known *Phyllariæ*, *Ph. lorea* and *Ph. dermatodea.* The same algologist in *Symbolæ ad Laminarieas* points out the existence of decided differences between these algæ, leaving it undecided whether these are differences of species or of age. However, in *Bidrag till kännedomen om Spetsbergens Alger, Tillägg*, he seems to have abandoned his earlier view and to regard the algæ in question as different forms of the same species. His description may be applied to both of them. In my works on the marine vegetation at Spitzbergen and in the eastern Murman Sea I have adopted the latest view of J. G. AGARDH and accordingly referred all the *Phyllaria* occurring in those regions to one and the same species, *Saccorhiza dermatodea*. I must now relinquish this opinion. After having seen on the coast of Finmarken a great number of *Phyllariæ* in different stages of development and having, on account of the observations I made there, again carefully examined my collections from other parts of the Polar Sea, I have arrived at the conclusion that there exist in the Polar Sea two species of *Phyllaria*, the one identical with the plant recorded by J. G. AGARDH (Spec. Alg.), J. E. ARESCHOUG, and FARLOW as *Laminaria (Saccorhiza) dermatodea*, the other identical with *Laminaria lorea* J. G. AG. I have been compelled to this opinion most of all by the fact that even among very young plants — what might almost be called germinating plants — there are to be observed two sharply distinct species. In some the stipe is longer, more or less distinctly marked from the lamina, the lamina is oblong, ovato-oblong or broadly lanceolate, darkbrown in colour, only little translucent, with very few short-haired cryptostomata. These are young plants of *Phyllaria dermatodea*. The young plants of the other species — I possess such plants in considerable number from Spitzbergen as well as from the west coast of Novaya Zemlya — have a very short stipe passing without definite limit into a narrow, sometimes almost filiform, linear, or more usually lanceolate lamina. Their lamina is thin, very light brown, almost yellowish-brown, perfectly pellucid with numerous long-haired cryptostomata. These plants belong to the alga described by J. G. AGARDH under the name of *Laminaria lorea*, which even when older has the same shape and colour of the lamina as the young plants and whose lamina is pellucid with numerous cryptostomata. Older specimens of the two species are easily distinguished from each other by several good characteristics. In *Ph. lorea* the stipe collapses in drying, and becomes flat, thin, almost membranaceous, and brittle; even in very large specimens it has the same colour as the lamina, being pellucid like this. In older specimens of *Ph. dermatodea* the stipe is far more solidly built, darkbrown, opaque, flat upwards, but almost terete downwards. With this outward difference of the stipe there is connected a difference in its structure, which, as far as my observations go, is essential and lasting during the whole life of the plant. In *Ph. dermatodea* the stipe is composed of 1:o a layer of cortical cells which are square or tangentially rectangular in transverse section, very rich in endochrome, with the outer wall cuticulated; 2:o inside this a thick layer of large thin-walled cells which increase inwards in length and also in width, and farthest in are several times longer than wide; 3:o a central layer formed of almost isodiametrical cells of different sizes and with thinner walls than the cells of the middle layer; 4:o very long tubular cells, with very thick walls, sometimes simple, sometimes branched, which in transverse section are seen to be arranged circularly on the limit between the middle and the central layer. These tubular cells occur even in very young individuals though in little number, but become later more numerous, sometimes so numerous as to constitute the greater mass of the central part of the stipe. By their very thick walls they contrast strongly with the adjoining cells; (tab. 25, fig. 1—4).

In general this is the structure of the stipe also in *Ph. lorea:* but in transverse sections of the stipe both in younger and older individuals the above-mentioned tubular cells are not visible. On closer examination they are certainly detected and are found to occupy the same position as in the preceding species, but their walls are always thin, not differing in thickness from the walls of the adjoining parenchymous cells (tab. 25, fig. 5). With regard to the cryptostomata the species *Ph. dermatodea* varies considerably. They are sometimes pretty numerous in older individuals, csarce in younger ones, sometimes very rare or almost absent both in younger and older specimens. The former is usually the case in specimens from the north coast of Norway, the latter in specimens from other parts of the Arctic Sea. The shape and structure of the cryptostomata is accurately described by J. E. Areschoug Obs. Phyc. 3, p. 12.

Habitat. In the Arctic Sea proper the present species occurs in company with other *Laminariaceæ* and is usually met with here at a depth of 2—10 fathoms on rocky or stony bottom. On the coast of Norway it does not belong to the proper formation of *Laminariaceæ*, but descends deeper than this, even to a depth of 20 fathoms. But it is most common here in shallow, rather exposed bays on gravelly bottom in 4—5 fathoms. On the north coast of Spitzbergen young individuals were common in the winter, nor were older ones wanting during the same season. Of the other species, *Ph. lorea,* on the contrary, young specimens were most common during the summer. On the coast of Norway younger and older specimens are of rather the same frequency during the summer-months, in July and August. At Spitzbergen I have found specimens with zoosporangia in July and August, at Novaya Zemlya in July, on the north coast of Norway in the latter part of August and at the beginning of September. However, the proper season for the formation of the zoospores on the last-named coast appears to come somewhat later, towards September or the beginning of October.

Geogr. Distrib. It is known from those parts of the Polar Sea which extend northwards of the Atlantic. Its maximum of frequency seems to be on the north coast of Norway, although even here it does not occur in such numbers as are comparable in any way with other *Laminariaceæ.*

Localities: The Norwegian Polar Sea: Finmarken at Maasö, Gjesvær, the south coast of Magerö, Öxfjord and Talvik, usually local and rather scarce; at Maasö pretty abundant.

The Greenland Sea: Beeren Eiland; on the north and west coasts of Spitzbergen local and scarce.

The White Sea: According to Gobi it is probably this species which K. v. Baer has taken at Tri-Ostrowa.

Baffin Bay: the west coast of Greenland at Claushavn.

With regard to the distribution of the two forms I can only say that the *Phyllaria dermatodea* which I found on the north coast of Norway belongs to f. *typica*, but that from the Greenland Sea and the Murman Sea to f. *arctica.*

Phyllaria lorea (BORY.) nob.

Laminaria lorea BORY in J. G. AG. Spec. Alg. 1, p. 130.

Ph. stipite breviore vel longiore, complanato, toto e cellulis membrana tenui contexto, in laminam lanceolato-ellipticam, basi cuneatam, usque 30 cm. latam vel lanceolatam, angustam circa 5--7 cm. latam, tenue membranaceam, e fusco lutescentem sensim abeunte; cryptostomatibus et in planta adulta et juvenili numerosis, parum immersis, nullo margine elevato circumdatis; pilis numerosioribus, longe persistentibus. Tab. 24 et 25 fig. 5—6.

Syn. Saccorhiza dermatodea J. G. AG. Spetsb. Alg. Till., p. 31; ex parte.
» » KJELLM. Spetsb. Thall. 2, p. 14; ex parte; Algenv. Murm. Meer. p. 36; ex parte.

Description. Figure 1 in tab. 24 represents a young specimen of the commonest habit in its natural size. The frond is attached by a depressed conical callus radicalis without any trace of rhizines. The stipe is flat, yellowish-brown, pellucid, 6 mm. long, passing into the lamina without any definite limit. The lamina is linear-lanceolate, 8 mm. broad at the middle, even, bearing at its top a fragment of an older lamina in a state of dissolution. Its lower part has the same colour as the stipe, the rest of it is of a lighter yellowish-brown. Cryptostomata are numerous, about 15 in a surface of 20 square mm. In the uppermost part such organs are wanting. Still younger individuals than the figured one have the same conformation as this, but somewhat fewer cryptostomata. However, I have seen specimens that were longer, but much more narrow, almost linear, being 1,5—2 mm. in breadth. These have few cryptostomata or none at all.

Figure 2 shows an older individual in natural size. Here the basal disk has some coarse rhizines. The individual is larger, as shown by the figure, but in other respects resembles that delineated in fig. 1. Other individuals of the same development and the same size have the stipe much longer, even 25 cm. in length, but narrow; again others have the stipe only about twice as long as in the figured specimen, but broader, upwards 0,5 cm. or more in breadth. The largest specimen I have seen, that was with certainty to be referred to the present species, is delineated in a third of its natural size in fig. 3 tab. 24. Its stipe is 40 cm. long, flat almost in its whole length, membranaceous in its dried state, pellucid, upwards where it passes into the lamina almost 2 cm. broad. The lamina (fragmentary) has been more than 80 cm. long, 30 cm. broad at its broadest part. It is pellucid, light brown, almost thin as paper in its dry condition, richly furnished with long-haired cryptostomata. All specimens that I have seen were sterile. In one, however, the sorus was in course of development. Its lamina is almost lanceolate, about 10 cm. broad at the middle. The stipe is 20 cm. long, upwards nearly one cm. broad.

As to the anatomical structure of the stipe, I have already described it under the preceding species. The long tubular cells never, even in the very largest specimens that I have seen, have thicker walls than the adjoining cells and on this account are never clearly visible in transverse section. All individuals that have attained at least the size shown by figure 1, have very numerous cryptostomata. These form shallow pits that are never surrounded with a prominent margin. The hairs are numerous and remain long. They taper much toward the base, and, as is usual in the hairs of the *Fucoideæ*, their lower cells are short and richly provided with endochrome. The sorus occupies the same position as in *Ph. dermatodea*.

Habitat. The species is sublitoral, growing in small numbers on rocky and stony bottom in 8—10 fathoms water. It has been found only on exposed coasts. I have collected young specimens in rather considerable number at Spitzbergen in August and at Novaya Zemlya during the latter part of July. At the same season, however, also older individuals are to be found. I have taken one specimen with zoosporangia in course of development on the coast of Spitzbergen at the end of July.

Geogr. Distrib. Known only from the eastern part of the Greenland Sea and the eastern part of the Murman Sea. At no point here it attains any greater degree of abundancy. The northernmost place where it has been found is Fairhaven on the north-west coast of Spitzbergen Lat. N. 79° 49′.

Localities: *The Greenland Sea:* the west and north coasts of Spitzbergen at Belsound, Smeerenberg Bay and Fairhaven, local and rather scarce.

The Murman Sea: the west coast of Novaya Zemlya at S. Gusinnoi Cape, local, but rather abundant; Rogatshew Bay, local and scarce.

Gen. **Laminaria** (LAMOUR.) J. G. AG.

Lam. p. 7; LAMOUR. Ess. p. 40; char. mut.

Laminaria solidungula J. G. AG.

Spetsb. Alg. Bidr. p. 3.

Descr. Laminaria solidungula J. G. AG. Lam. p. 8.
» » KJELLM. Spetsb. Thall. 2, p. 15.
Fig. » » J. G. AG. Spetsb. Alg. Bidr. t. 1.
» » KJELLM. l. c. t. 1, fig. 1.

Syn. Laminaria solidungula J. G. AG. l. c.; Grönl. Alg. p. 110; Grönl. Lam. och Fuc. p. 18.
» » KJELLM. l. c.; Vinteralgv. p. 64; Algenv. Murm. Meer. p. 36; Kariska hafvets algv. p. 24.

Habitat. This is a sublitoral alga that seems to prefer a bottom composed of small stones, but occurs also on rocks. It grows only seldom in company with other *Laminariaceæ*, and is then usually to be found in small number. Within the *Lithoderma*-formation it is more common and plentiful. However, it is never met with in large masses. It flourishes both on exposed and sheltered coasts, in localities with strong current as well as in such where the current is feeble. On the coast of Siberia the development of a new lamina begins at the end of March or the commencement of April, and already at the end of April it has attained a considerable size.

On the north coast of Spitzbergen the zoospores are developed most abundantly in the month of January. However, specimens with zoosporangia are met with as early as in November and the formation of zoospores is vigorously continued as late as in

February. During the latter part of the winter the sorus is formed of paraphyses and void or abnormously developed zoosporangia. In summer I have found only sterile specimens both on the coasts of Spitzbergen and on the west coast of Novaya Zemlya. I have collected individuals with young zoosporangia at the end of August in the Kara Sea and in the middle of September in the Siberian Sea.

Geogr. Distrib. The present species is probably circumpolar in the Arctic Sea proper. However, it is not known as yet from the American Arctic Sea. I have found it in the greatest abundancy on the north-west coast of Spitzbergen. Here it attains a high degree of luxuriancy. In the southern part of the Siberian Sea it is small of size. The most northern point where it has been found is Musselbay on the north coast of Spitzbergen Lat. 79° 53'.

Localities: The Greenland Sea: the north and west coasts of Spitzbergen, local, but rather abundant.

The Murman Sea: the west coast of Novaya Zemlya from Matotshin Shar to Karmakulbay, local and rather scarce.

The Kara Sea: Uddebay and Actinia Bay, at both places scarce.

The Siberian Sea: Irkaypi and Koljutshin Isle, pretty plentiful.

Baffin Bay: the west coast of Greenland: Jakobshavn and Rittenbenk, scarce.

Laminaria cuneifolia J. G. Ag.

Lam. p. 10.

Descr. Laminaria cuneifolia J. G. Ag. l. c. et Grönl. Lam. och Fuc. p. 14.

Syn. Laminaria caperata Dickie, Alg. Nares, p. 6 (?).
» cuneifolia J. G. Ag. Grönl. Alg. p. 110.
» » Gobi, Algenfl. Weiss. Meer. p. 75 (?).
» saccharina (?) Asum. Alg. Hayes, p. 96 (?).
» » Croall, Fl. Disc. p. 457 (?). Cfr. J. G. Ag. Grönl. Lam. och Fuc. p. 14.
» » Dickie, Alg. Sutherl. 1, p. 140.(?) Alg. Cumberl. p. 237.(?) Cfr. J. G. Ag. l. c.

Remark on the species. In the eastern part of the Siberian Sea I found a *Laminaria* abundant which I think belongs to *L. cuneifolia* J. G. Ag. In the shape, colour, consistency, and anatomical structure of the lamina and in the shape and position of the sorus it accords with specimens from Greenland determined by J. G. Agardh. It differs from them, it seems, by smaller size and somewhat longer stipe. The length of the stipe varies between 4 and 15 cm., in most cases not exceeding 10 cm. In almost all the Siberian specimens that I succeeded in collecting, the lamina was in course of being exchanged, so that I am not clear as to what size it attains in the Siberian Sea. Judging from remaining fragments of the old lamina and those few specimens with developed lamina that I have seen, it does not become so large here as on the coasts of Greenland. When the new lamina has made some progress in growth, the plant with regard to this part bears a delusive resemblance to *L. solidungula*. From other species of *Laminaria* the present species appears to be well differentiated. It is distinguished from the common arctic *L. Agardhii* by possessing lacunæ muciferæ in the lamina, which are wanting in *L. Agardhii*. It can hardly be confounded with *L. saccharina*, being quite different in appearance.

It is impossible at present to disentangle its synonymy. But I believe that that *L. saccharina* and *L. caperata* which is reported from Baffin Bay may be supposed to belong to the present species and not to *L. saccharina* as it must be now restricted. On the contrary, the *L. cuneifolia* reported from the White Sea appears to be referable rather to *L. Agardhii* than to the species named *L. cuneifolia* by J. G. AGARDH[1]), and the *L. saccharina* recorded from the American Arctic Sea may apparently be regarded as *L. longicruris*.

Habitat. The present species grew on the north coast of Siberia in exposed localities within the sublitoral zone in 2—5 fathoms water on stony and gravelly bottom. It was gregarious here. All specimens taken from the end of April to the beginning of July were in course of changing their lamina. The development of the new lamina appears to commence at the beginning of April. On the remaining part of the old lamina in some individuals collected at the end of April there was found a sorus with sporangia containing zoospores. Specimens from Greenland probably collected in August are richly provided with zoosporangia.

Geogr. Distrib. In the Arctic Sea this alga is known from Baffin Bay and the eastern part of the Siberian Sea. GOBI states it to occur also in the White Sea. Its most northern known locality is Jakobshavn on the west coast of Greenland Lat. N. 69° 15'.

Localities: The White Sea (?). Cp. GOBI, Algenfl. Weiss. Meer and above.

The Siberian Sea: Irkaypi, Koljutshin Isle, Pitlekay, and the coast eastward of this point, common and abundant.

Baffin Bay: Cumberland Sound (?), the west coast of Greenland at Jakobshavn. If the alga named *L. saccharina* by CROALL, DICKIE, and ASHMEAD should belong to the present species, it is probably common and plentiful along the whole western coast of Greenland up into Smith Sound.

Laminaria saccharina (L.) LAMOUR.

Ess. p. 42. Fucus saccharinus L. Spec. Plant, 2, p. 1161.

f. *linearis* J. G. AG.

Lam. p. 12.

Descr. Laminaria saccharina *a* linearis J. G. AG. l. c.
» » » f. prima J. G. AG. Spec. Alg. 1, p. 132.

f. *oblonga* J. G. AG.

Lam. p. 12.

Descr. Laminaria saccharina *b* oblonga J. G. AG. l. c.

---

[1]) GOBI has kindly sent me for examination a fragmentary specimen of that plant from the White Sea which he has called *L. cuneifolia*. This appears to me referable to *L. Agardhii*, although it is extremely difficult to decide the question definitely. However, I have tried in vain to find any lacunæ in the lamina.

f. *grandifolia* nob.

f. plantæ adultæ stipite elongato, 15—70 cm. longo, digiti minoris crassitudine; lamina obscure olivacea; subopaca, lineari-lanceolata, basi late cuneata, 125—280 cm. longa, 25—70 cm. lata, media parte zoosporangifera circa 1 mm. crassa, dense bullata, at non rugosa, disco lineari sublævi, margine angustiore undulato; lacunis muciferis in stipite nullis in lamina magnis, distinctis, at parcis; soro vittam elongatam, circa 10 cm. latam in parte media et superiore laminæ formante.

f. *latissima* nob.

f. stipite prælongo usque tripedali, digiti crassitudinem attingente; lamina plantæ junioris sublineari, basi ovata, plantæ adultæ late elliptica, basi ovato-cordata, 75 cm. lata, coriaceo-membranacea, olivacea, subpellucida, parte media 1—2 mm. crassa, scrobiculata vel rugoso-bullata, margine amplo tenui undulato; lacunis muciferis in stipite nullis, in lamina magnis, distinctis at parcis, soro vittam elongatam in parte media et superiore laminæ formante.

Syn. Fucus saccharinus GUNN. Fl. Norv. 1, p. 52.
» » WG. Fl. Lapp. p. 493; excl. syn. sec. SOMMERF. Suppl. p. 183.
Laminaria caperata KLEEN, Nordl. Alg. p. 32.
» saccharina J. G. AG. Enum.
» » ARESCH. Phyc. Scand. p. 343.
» » KLEEN, Nordl. Alg. p. 32.
» var. septentrionalis ROSTAF. in GOBI, Algenfl. Weiss. Meer. p. 78.
Ulva longissima GUNN. l. c. 2, p. 128, t. 7.
» maxima (?) » » » » 127 » »

Remark on the definition of the forms. The plant which I have referred to f. *linearis* J. G. AG. has a longer stipe in proportion to the breadth of the lamina than is stated in the diagnosis of the form. In some specimens the stipe is even 45 cm. long and even five times longer than the greatest breadth of the lamina. Even in specimens from the coast of Finmarken the laminais often more lanceolate than it seems to have been in those individuals on which the description of J. G. AGARDH was founded. The form is easily recognizable by the narrow, thick, coriaceous, rugose, little or not at all wavy, lamina, and by the proportionately long stipe. In the lamina there are distinct, large, but thinly scattered lacunæ muciferæ. Certain individuals from southern Norway that I have seen, agree better with J. G. AGARDH's description. However, there are to be found also here individuals nearly approaching the arctic form or even identical with it.

I have seldom met with *L. saccharina* f. *oblonga* in the Polar Sea. It is smaller here and generally has longer stipes than farther south. The rugæ are numerous and distinct, encompassing rounded patches. Even this form possesses large, distinct, and pretty numerous lacunæ in the lamina. *L. saccharina* f. *grandifolia* reminds one much of the arctic *L. Agardhii*, approaching it nearly in size and in the dimensions of the several parts. It differs from it by the colour being darker and the lamina less pellucid, less wavy, thicker especially in the middle, with distinct depressions there. Besides, it is distinguished from *L. Agardhii* by the large distinct lacunæ muciferæ of the lamina. It is known from the other forms of *L. saccharina* by its considerably larger dimensions, especially the great length of the stipe, by the absence of rugæ, which are replaced by deep large pits, by its less solidity and lighter colour. The following measurements show its proportions.

Total length of the alga.	Length of the stipe.	Length of the lamina.	Greatest breadth.
310	70	240	50 cm.
307	57	250	66
285	35	250	56
241	31	210	50
236	26	210	51
220	25	195	52
216	16	200	48
183	55	128	28

With regard to structure f. *grandifolia* differs also from the before-mentioned forms by having much wider and more thin-walled elements. This is especially the case with the collenchyme and the adjoining parenchyme of the stipe and with the parenchyme of the lamina. The central layer of the stipe is composed of thinner cells with more swollen membranes.

The pits of the lamina are sometimes very numerous, sometimes rare. In one specimen I have found a few rugæ in the middle part of the lamina.

In the lamina three parts are distinguishable: the middle part which is smooth or almost smooth, the intermediate part with many depressions, and the marginal part which is thin, wavy, with few depressions.

L. saccharina f. *latissima*. This is not the preceding form at a different age, as one might be inclined to suppose. For I have examined young specimens of both the forms that plainly showed the same shape of the lamina as in older individuals. But though I have not seen any transitions between them, they resemble each other in so many respects that they are probably to be considered as less strongly differentiated forms of the same species. Whether this is really *L. saccharina* or some other species different from it, is a question I must leave undecided at present. Just as f. *grandifolia* corresponds to the southern *L. saccharina* f. *membranacea* and replaces it in the North, f. *latissima* may be regarded as a northern form corresponding to a *L. saccharina* existing at Bohuslän which is distinguished by its short stipe, and thin almost membranaceous lamina, that is linear with rounded base and wants rugæ. From f. *grandifolia* f. *latissima* is distinguished almost exclusively by the shape of the lamina. This is in younger individuals almost linear with rounded base, or elongated linear-ovate. When older, it increases considerably in breadth and becomes broadly elliptical with ovato-cordate or cordate base. The surface is sometimes smooth, sometimes covered more or less densely with pits. I have seen one specimen with low rugæ. In structure it accords nearly with the preceding form, showing the same differences as this from f. *linearis* and f. *oblonga*. The lacunæ muciferæ in the lamina are sometimes scarce, sometimes numerous, always large, confined in a greater or less extent of their periphery by cells that are smaller and of another shape than the other cells of the parenchyme. The form in question resembles *L. Agardhii* even more than f. *grandifolia* does. It is distinguished from it by the same characteristics as f. *grandifolia*. It is this form *la-*

tissima that KLEEN has identified with *L. Agardhii* (*L. caperata*), and it is possibly the same that GUNNERUS has named *Ulva maxima*.

Habitat. In the Polar Sea *L. saccharina* is sometimes litoral, sometimes sublitoral. The f. *linearis* lives at low-water mark together with *L. digitata*. The f. *oblonga* occurs both farther down and higher up than f. *linearis*. The two other forms I have met with only in the lower part of the sublitoral zone, in 15—20 fathoms. The present species prefers rocky bottom, but is found richly developed also on pebbly ground. It occurs in the open sea and at exposed points of the coasts as well as in the interior of deep bays and in other sheltered localities. It is gregarious. On the coast of the Polar Sea I have only had the opportunity of examining it during July and August, and at the beginning of September. During this time f. *linearis* and f. *oblonga* were sterile, the two other forms were furnished with zoosporangia.

Geogr. Distrib. I have myself found the present species in the Polar Sea only on the coast of Norway, and on that account am of opinion that, with the exception of the *L. saccharina* var. *septentrionalis* reported by ROSTAFINSKI from the White Sea, all the *L. saccharina* which has been stated to occur in the Arctic Sea is to be referred to other species. It has its maximum of frequency on the north coast of Norway. The northernmost point where it is certainly known to occur is Gjesvær about Lat. N. 71°.

Localities: The Norwegian Polar Sea: Nordlanden common and abundant; Tromsö amt at Tromsö, Renö, and Carlsö, common and abundant; Finmarken at Maasö, Gjesvær, Magerö Sound, Öxfjord, and Talvik.

The White Sea: Cp. GOBI l. c.

Of the forms mentioned f. *oblonga* appears to be commonest south of Tromsö; f. *linearis* was plentiful at Gjesvær, f. *grandifolia* at Maasö and Gjesvær; f. *latissima* was abundant at Talvik in Altenfjord, more rare at Maasö and in the sound south of Magerö.

Laminaria longicruris DE LA PYL.

Prod. Terre neuve p. 177.

Descr. Laminaria longicruris J. G. AG. Spec. Alg. 1, p. 135.
Fig. » » DE LA Pyl. l. c. t. 9, fig. A et B.
» » HARV. Phyc. Brit, t. 339.
Exsicc. » » FARL. Eat. and Ands. Alg. Amer. N:o 117.

Syn. Laminaria longicruris J. G. AG. Grönl. Lam. och Fuc. p. 15; Grönl. Alg. p. 110.
» » ASHM. Alg. Hayes, p. 96.[1]).
» » BROWN, Fl. Disc. p. 457.
» » DICKIE, Alg. Cumberl. p. 237; Alg. Sutherl. 1, p. 141; Alg. Nares, p. 6.
» saccharina DICKIE, Alg. Walker, p. 86; Alg. Sutherl. 2, p. 191 (?).

Remark on the species. I cannot at all share the opinion of GOBI that »die Art Lam. longicruris ganz zu streichen ist und die unter diesem Namen verstandenen Formen

[1]) That *L. phyllitis* which is stated by ASHMEAD to occur in Smith Sound, is probably young individuals of this species. Cp. ASHM. Alg. Hayes, p. 96.

zu der Laminaria caperata zugezählt werden müssen». GOBI says that he has arrived at this conclusion by having found in the herbarium of the Botanical Museum of the Petersburg University two specimens of *Alaria* collected in the North by POSTEL (both named *A. esculenta*) which agree in all respects with each other, except that one specimen has the costa inflated and jointed. GOBI considers these two specimens as belonging to the same species and regards this as a proof that the cavity of the costa, or, what is the same thing, of the stipe cannot be accounted of any value whatever as a specific characteristic. From this consideration he forms the conclusion above quoted, that *L. longicruris* should be struck out, »weil man im entgegengesetzten Falle die erwähnte *Orgyia* mit einer tonnen-artig gegliederten Rippe dann ebenfalls als eine besondere Art ansehen müsste». It is, as far as I can judge, a specimen of *A. fistulosa* acknowledged as a valid species by all algologists, that has happened to receive the name of *A. esculenta* in the above-mentioned herbarium. I have myself in the Behring Sea observed a great many *Alariæ fistulosæ* of different ages, and have thus acquired the decided conviction that, if this alga is not acknowledged as a separate species distinct from *A. esculenta* and other *Alariæ*, then there can be no question about any species at all either among *Laminariaceæ* or algæ in general. I am perfectly sure however that GOBI, if he were made better acquainted with *A. fistulosa* which is one of the greatest and most splendid marine plants, would arrive at a quite different conclusion. Apparently his judgment, otherwise so clear and sure, has been misled by a small, badly preserved, dried specimen. (Cp. GOBI, Algenfl. Weiss. Meer. p. 76 and 78).

L. longicruris is distinguished from *L. caperata* (*L. Agardhii*), besides by the solidity of the stipe and many other characteristics, also by the stipe of the latter alga wanting lacunæ muciferæ, which are, on the contrary, to be found in a dense circle in the stipe of the former species. It is curious enough that this fact has escaped the observation of LE JOLIS. At least he refers *L. longicruris* to that group of *Laminariaceæ* which is characterized by »canales muciferi in stipite nulli, sub epidermide autem frondis numerosi, parvi» (LE JOLIS, Exam. p. 589—590).

Habitat. I have not had access to any certain statements about the mode of growth of the present species in the Arctic Sea. Probably it lives gregarious, as other *Laminariaceæ*, within the sublitoral zone on rocky and stony bottom.

Geogr. Distrib. It is known with certainty to occur in Baffin Bay, going here far to the north. According to ASHMEAD it is to be found in Smith Sound between Lat. N. 78° and 82°. The expedition of Nares also met with it here north of Lat. N. 78°. Probably it lives also in the American Arctic Sea. On the west coast of Greenland it appears to be plentiful.

Localities: The American Arctic Sea: I suppose the present species to be the alga reported under the name of *L. saccharina* at Port Kennedy and in Assistance Bay.

Baffin Bay: Cumberland Sound; the west coast of Greenland common (according to J. G. AG. Grönl. Alg. p. 110); Godhavn, Melville Bay, Whale Sound, Cape Saumarez, Smith Sound north of Lat. N. 78°.

Laminaria Agardhii KJELLM.

Spetsb. Thall. 2, p. 18.

Descr. Laminaria caperata J. G. AG. Lam. p. 13.
» Agardhii KJELLM. l. c.
Fig. » » » » t. 1, fig. 2—3.

Syn. Fucus saccharinus Pall. Reise, 3, p. 34.
» » SCORESBY, Account 1, p. 132.
Laminaria Agardhii KJELLM. Spetsb. Thall. 2, p. 18; Algenv. Murm. Meer. p. 37; Kariska hafvets algv. p. 24.
» caperata J. G. AG. Spetsb. Alg. Bidr. p. 5 et 11; Till. p. 28.
» » GOBI, Algenfl. Weiss. Meer. p. 76.
» » KJELLM. Vinteralgv. p. 64.
» longicruris J. G. AG. Spetsb. Alg. Progr. p. 2; Bidr. p. 11.
» ophiura LINDBL. Bot. Not. p. 157.
» phyllitis Nyl. et Sæl. Herb. Fenn. p. 73 (?).
» » POST. et RUPR. Ill. Alg. p. II (?).
» » ZELLER, Zweite d. Polarf. p. 84 (?).
» saccharina (?) J. G. AG. Spetsb. Alg. Progr. p. 2; Bidr. p. 11.
» » EATON, List. p. 44.
» » POST. et RUPR. Ill. Alg. p. II (?).
» » SCHRENK, Ural Reise, p. 546.
» » WITTR. in Hengl. Reise, p. 284.
Ulva latissima MARTIN, Met. Observ. p. 313. [2])

Remark on the species. It should be added here to the description of this species given by J. G. AGARDH and by myself, that it differs in the structure of the lamina from *L. saccharina*, from which certain forms of it are only with great difficulty to be distinguished by outward marks. I have examined a great many specimens from different localities, at different seasons, and in different stages of development, but I have never been able to detect any muciferous canals in the lamina. Thus I believe I may say that in the alga called by J. G. AGARDH *L. caperata*, which name I have thought fit for reasons already stated to replace by *L. Agardhii*, muciferous canals either are wanting in the lamina, or, if they really exist there, are very difficult to detect, and at least different in position, shape and size from the same organs in Scandinavian *L. saccharina*. I ought to point out however that it has been stated by LE JOLIS who has made the structure of the *Laminariaceæ* the subject of accurate investigations, that the lacunæ muciferæ (»canales muciferi») in the lamina of *L. saccharina* are very small — »très petits» — and that »leur extrême petitesse par rapport aux énormes cellules irrégulières, qui constituent le tissu de la fronde, les rend très difficiles à apercevoir». (LE JOL. Exam. p. 548), which observations do not at all accord with the results of my own researches on Scandinavian *L. saccharina*. In all the *L. saccharina* from our coasts that I have examined, I have found these cavities large, very distinct, and often confined in a greater or less part of their circumference by cells differing in shape and

---

[2]) The species is also recorded at Spitzbergen by MARTENS, by whom it is described and figured. See MARTENS, Voyage Spitzb. p. 79, tab. J, fig. C.

size from the other cells of the parenchyme of the lamina; (tab. 25, fig. 7). This seems to show that the *L. saccharina* which occurs on our coasts is not the same as the alga of the same name which is found on the coast of France. With respect to my list of synonyms, it may be remarked that it is founded in certain cases on mere suppositions. It is impossible without having access to the original specimens to make out with certainty what the different authors have meant by their different *Laminariæ*. I hope however to have in general hit upon the truth. If that is the case, *L. saccharina* in its present limitation does not live within the Arctic Sea proper.

Habitat. This species is sublitoral, growing in 2—10 fathoms water. As most other *Laminariaceæ*, it prefers rocky bottom and is gregarious. It occurs on exposed as well as sheltered coasts. In Musselbay on the north coast of Spitzbergen I found during the winter both younger and older specimens, from such as were microscopical in size to fully developed ones. Germinating plants were particularly numerous during December. On the coast of Spitzbergen the present species bears ripe zoosporangia as well during the winter, in November, December, January, February, and March, as during the summer. However the formation of zoospores is at its richest in July and August. Even on the west coast of Novaya Zemlya I have taken it with zoosporangia in July and August. On the coasts of Spitzbergen the change of the lamina appears to take place in May and June.

Geogr. Distrib. This species is known with certainty only from the Spitzbergen province of the Arctic region. In the Siberian province it is replaced by *L. cuneifolia*, in the American by *L. longicruris*, and in the Atlantic by *L. saccharina*. It has been found most abundant and luxuriant on the coasts of Spitzbergen. Its northernmost known locality is the North Cape of Spitzbergen Lat. N. 80° 31'.

Localities: The Greenland Sea: the east coast of Greenland at Sabina Isle (?); the coasts of Spitzbergen, common and abundant; Beeren Eiland.

The Murman Sea: the west coast of Novaya Zemlya and Waygats, common and abundant.

The White Sea: probably plentiful.

The Kara Sea: Uddebay on the east coast of Novaya Zemlya, Cape Palander and Actinia Bay pretty plentiful, Cape Tscheljuskin scarce.

Laminaria atrofulva J. G. Ag.

Grönl. Lam. och Fuc. p. 16.

Descr. Laminaria atrofulva J. G. Ag. l. c.

Syn. Laminaria atrofulva J. G. Ag. Grönl. Alg. p. 110.

Addition to the description of the species. I consider this alga, which has been described in detail by J. G. Agardh l. c., to be one of the most characteristic *Laminariæ* of the Arctic Sea. I have only to add to the author's description that this species, as well as the next following ones, has the middle layer of the lamina compact, sharply defined from the intermediate layer. The latter layer is composed of angular or rounded-angular, thin-walled cells of middle size, larger than in *L. digitata*, somewhat smaller than in *L. Clustoni*. In no arctic *Laminaria* I have found the lamina to be so rich in muciferous canals as in this species, and in none those canals are so small and so close to the surface. J. G. AGARDH does not notice the presence of those organs in the stipe. However, in the specimens I have examined, they occur also in this part, being here as in the lamina small in size and situate immediately beneath the cortical layer.

Habitat. In this respect nothing is known. According to J. G. AGARDH the specimens collected by BERGGREN at Sukkertoppen on the west coast of Greenland were furnished with zoosporangia. The expedition in which BERGGREN took part stayed at Sukkertoppen from 21 September to 21 October. Thus the species bears developed zoosporangia in the earlier part of the autumn on the west coast of Greenland.

Geogr. Distrib. Known only from Baffin Bay.

Locality: the west coast of Greenland at Sukkertoppen.

Laminaria fissilis J. G. AG.

Lam. p. 18.

Descr. Laminaria fissilis J. G. AG. l. c.
» » » Spetsb. Alg. Till. p. 28.
Syn. Laminaria fissilis KJELLM. Algenv. Murm. Meer. p. 40.

Remark on the species. As far as I can judge from my examination of some few individuals, all of which were young, this alga is to be held a distinct species. It is distinguished from *Laminaria digitata* by the intermediate layer of the lamina being formed of large, rounded-angular, thin-walled cells and sharply defined from the dense middle layer. By this characteristic it approaches *L. nigripes*, but it is known from this species by the stipe wanting muciferous canals. In the lamina those organs are smaller and more indistinct than in *L. nigripes.*

Habitat. Those few specimens I have myself found grew in the interior of deep bays on gravelly bottom in company with other *Laminariaceæ*. In the collections of algæ brought together at Spitzbergen by the Swedish summer-expedition of 1868 specimens with zoosporangia were found according to J. G. AGARDH.

Geogr. Distrib. Known from the eastern part of the Greenland Sea and the Murman Sea. The most northern locality where it has been met with is the coast of Spitzbergen.

Localities: The Greenland Sea: the coast of Spitzbergen.

The Murman Sea: rare at Karmakulbay and at N. Gusinnoi Cape on the west coast of Novaya Zemlya.

Laminaria nigripes J. G. AG.

Spetsb. Alg. Till. p. 29.

Descr. Laminaria nigripes J. G. AG. l. c.
» » » Grönl. Lam. och Fuc. p. 17.
Fig. » » Tab. nostra 25, fig. 8—10.

f. *reniformis* nob.

f. lamina reniformi, in lacinias numerosas usque ad basim digitato-fissa; lacunis muciferis stipitis numerosissimis in orbem regularem infra stratum corticale dispositis.

α. *longipes* nob.

f. stipite longiore usque 1—2 pedali, infra apicem complanato; laciniis laminæ numerosis, sat latis. Forma a J. G. AGARDHIO descripta.

β. *brevipes* nob.

f. stipite brevi, 2—6 pollicari; laciniis numerosissimis, angustis.

f. *oblonga* nob.

f. lamina late oblonga, basi sæpe obliqua, integra vel in lacinias latas, pauciores plus minus profunde fissa.

α. *compressa* nob.

f. stipite longiore usque tripedali, superne compresso, sæpe infra apicem valde complanato; lacunis muciferis stipitis irregulariter infra stratum corticale dispositis, paucioribus, magnis.

β. *subteres* nob.

f. stipite breviore, subterete; lacunis stipitis minoribus, numerosis infra stratum corticale in orbem fere regularem dispositis.

Syn. Laminaria digitata f. latifolia KJELLM. Spetsb. Thall. 2, p. 26.
» » forma KJELLM. l. c. p. 26—27.
» » f. typica (vera) KJELLM. Algenv. Murm. Meer. p. 38 et Kariska hafvets algv. p. 25; ex parte [1]).
» » KJELLM. Vinteralgv. p. 64; ex parte.

Remark on the species. In the collections of arctic algæ that I have had the opportunity of examining, there is to be found a *Laminaria* of the digitata group taken in different parts of the Arctic Sea, which I have formerly confounded with *L. flexicaulis* LE JOL. and recorded under the name of *L. digitata.* J. G. AGARDH has separated it from *L. digitata* and described it as *L. nigripes.* By a more anatomical investigation of a greater number of *Laminariæ* from different regions I have arrived lately at the conclusion that the structure of the lamina and the presence or absence of muciferous lacunæ in the various parts of the frond ought to be accounted more important in distinguishing the species of this group of algæ, than I supposed formerly, relying on the authority of J. G. AGARDH. Consequently I must acknowledge *L. nigripes* as a well differentiated species. It resembles closely *L. digitata* (*L. flexicaulis* LE JOL.) in habit and possesses a series of forms similar with this, but in structure it accords more nearly with *L. Clustoni* LE JOL. As in this species, there are muciferous lacunæ in the stipe, the middle layer of the lamina is very dense, sharply defined from the intermediate layer and the latter layer composed of large cells with thin walls. By these marks it is plainly distinguished from *L. digitata* (*L. flexicaulis* LE JOL.). On the other side it differs decidedly from *L. Clustoni* by no cork layer

[1]) J. G. AGARDH supposes that the Laminaria from the west coast of Greenland mentioned by BROWN in Fl. Disc. under the name of L. digitata is to be referred to the present species or to L. atrofulva. Cp. J. G. AG. Grönl. Lam. och Fuc. p. 18.

being developed in the stipe, in consequence of which the surface of the stipe is as even and smooth as in *L. digitata*, and by the muciferous lacunæ of the lamina being thinly scattered in the outermost part of the intermediate layer, small, and not surrounded with small cells of a peculiar shape, as is the case with the large lacunæ in *L. Clustoni* situate near the middle layer. Thus *L. nigripes* in the shape, size, and position of the lacunæ resembles *L. digitata* more closely than *L. Clustoni*.

It is seen by the list of forms given above that the present species is rather variable with regard to its habit and outward appearance. The external differences of different individuals are so great that it might be questioned whether there are not included several species in *L. nigripes* as here understood. Nevertheless, not having observed any constant difference in anatomical structure, I have thought best to regard the existing differences as constituting different forms, not different species.

It is fit that these differences should be set forth here.

The *rhizines* (root-like fibers) issue in basifugal, alternate, rather regular whorls. They are sometimes long and fine, sometimes short and coarse.

The *stipe* is always smooth, pliable, black or blackish-brown when dried, never perceptibly thicker at the base than at the apex. It is sometimes of almost equal thickness, being then nearly terete, sometimes thicker at the top, being then more or less compressed upwards, even 2¼ cm. broad in its longest diameter. It is sometimes short, about an inch long, sometimes longer, attaining in larger specimens a length of 2—3 feet.

The *lamina* is of two different types. It is either reniform in outline, split to the base into much spreading segments, which are sometimes very numerous, repeatedly separated and 1—2 cm. broad, sometimes fewer, 3—4 cm. broad; or else it is broadly oblong and in this case sometimes entire, resembling the lamina of *L. digitata* f. *integrifolia*, sometimes split into a small number of broad, appressed segments, which are separated to the base of the lamina. Both these forms of lamina may be combined with short, almost terete, or long, more or less flattened stipe. In structure the lamina varies scarcely at all, the stipe is somewhat more variable, especially with regard to the muciferous lacunæ. In certain individuals these are very numerous, forming in transverse section a dense, regular circle immediately beneath the cortical layer, in others they are fewer and at the same time larger, less regularly arranged and situate somewhat deeper in the intermediate layer of the stipe.

The *sorus* is developed at the base of the lamina, where it forms a coherent girdle reaching quite or almost quite down to the lower margin of the lamina. In individuals that were in course of changing their lamina, I have seen the sorus expanded in the shape of bands in the zone between the old and the new lamina.

Habitat. The present species occurs sublitoral, together with other *Laminariaceæ*, on rocky and stony bottom, at a depth of 5—15 fathoms. It lives both on exposed and on sheltered coasts. I have collected specimens in course of changing their lamina at Spitzbergen in July and September, specimens with zoosporangia in July (f. *reniformis*), at the end of August, and in September and December (f. *oblonga*). In winter I found a *Laminaria* of the digitata group on the north coast of Spitzbergen furnished with zoosporangia 9 and 16 January (richly), 2, 8, 14 February and 5 March. Having called this alga by the name of *L. digitata* in my notes, I am not sure as to what species the plant which was then furnished with zoospores did really belong, but I suppose it was *L. nigripes* of which I possess in my collections several specimens taken in Musselbay, all of them appertaining to f. *oblonga*. It is thus possible that the development of zoospores takes place at different seasons in the different forms on the coast of Spitzbergen, during spring and the earlier part of summer in f. *reniformis*, during autumn and winter in f. *oblonga*. From the north coast of Tshutsh-land I have one specimen of f. *reniformis* taken in May, which is richly provided with zoosporangia, and from the west coast of Novaya Zemlya one specimen of f. *oblonga* taken in July with mature zoosporangia.

Geogr. Distrib. This species is probably circumpolar within the Arctic Sea proper. I have found it most abundant and luxuriant on the north coast of Spitzbergen. The most northern point where it has been taken is Treurenberg Bay on the north coast of Spitzbergen Lat. N. 79° 56′.

Localities: The Greenland Sea: the west and north coasts of Spitzbergen in Smeerenberg Bay, Fairhaven, Musselbay, and Treurenberg Bay (f. *reniformis* and f. *oblonga*).

The Murman Sea: the west coast of Novaya Zemlya at Karmakulbay (f. *reniformis*).

The Kara Sea: Uddebay (f. *reniformis*).

The Siberian Sea: the north coast of Tshutshland — some miles east of the wintering-station of the Vega (f. *reniformis*).

Baffin Bay: the west coast of Greenland according to J. G. AG. Grönl. Lam. och Fuc. p. 18 sub *L. atrofulva*.

Laminaria Clustoni EDM.

Fl. Shetl. p. 54, sec. LE JOL. Exam p. 577.

f. *typica* FOSLIE.

Laminaria digitata f. typica FOSLIE, Digitata-Lam. p. 15.

Descr. Laminaria Clustoni LE JOL. l. c.

Fig. Laminaria digitata HARV. Phyc. Brit. t. 223.

f. *longifolia* FOSLIE.

l. c. p. 19.

Descr. Laminaria digitata f. longifolia FOSLIE. l. c.

Syn. Fucus digitatus WG. Fl. Lapp. p. 492; ex parte.

» hyperboreus GUNN. Fl. Norv. 1, p. 34, t. 3; saltem ex parte sec. FOSLIE, l. c. p. 11.

Laminaria digitata FOSLIE, l. c. p. 14.

Habitat. I have never myself met with this species in the Polar Sea. According to FOSLIE who has devoted himself to the elucidating of the Scandinavian *Laminariæ* of the digitata group, it grows exceptionally within the litoral zone, usually gregarious within the sublitoral zone at places exposed to the sea or in such more sheltered localities of the coast where there is a strong current. It is not found in the interior of bays and in quiet recesses. It is most vigorously developed at such places where there is a strong surge. It lives at a depth of 2—8 fathoms, on even, not suddenly sloping, rocky bottom, or on sandy bottom with large boulders. The season when it bears reproductive organs in the Polar Sea is not known.

Geogr. Distrib. In the Polar Sea it is only known from the north coast of Norway.

Localities: The Norwegian Polar Sea: common and abundant at Nordlanden and Finmarken on exposed coasts. Of the two forms f. *typica* is more common; f. *longifolia* has been found only washed ashore at Berlevaag in East Finmarken. Besides cp. FOSLIE l. c.

Laminaria digitata (L.) LAMOUR.

Ess. p. 42. Fucus digitatus L. Mant. p. 134.

f. *genuina* LE JOL.

Laminaria flexicaulis α genuina LE JOL. Exam. p. 580.

Descr. Laminaria flexicaulis f. genuina FOSLIE, Digitata-Lam. p. 20; excl. syn.

f. *valida* FOSLIE.

Descr. Laminaria flexicaulis f. valida FOSLIE, l. c.

f. *latilaciniata* FOSLIE.

Descr. Laminaria flexicaulis f. latilaciniata FOSLIE, l. c.

f. *ensifolia* LE JOL.

Laminaria flexicaulis δ. ensifolia LE JOL. l. c.

Descr. Laminaria flexicaulis f. ensifolia FOSLIE, l. c. p. 22.

f. *cucullata* LE JOL.

Laminaria flexicaulis γ cucullata LE JOL. l. c.

Descr. Laminaria cucullata f. typica FOSLIE, l. c. p. 24.

f. *ovata* LE JOL.

Descr. Laminaria flexicaulis β ovata LE JOL. l. c.

Exsicc. Laminaria digitata b. ARESCH. Alg. Scand. exsicc. N:o 167.

f. *complanata* KJELLM.

Kariska hafvets algv. p. 26.

Descr. et *Fig.* Laminaria digitata f. complanata KJELLM. l. c. et t. 1 et Algenv. Murm. Meer. p. 38.

Syn. Fucus digitatus WG. Fl. Lapp. p. 492; ex parte.
Laminaria cucullata FOSLIE, Digitata-Lam. p. 24.
» digitata ARESCH. Phyc. Scand. p. 344.
» » J. G. AG. Spetsb. Alg. Progr. p. 2; Bidr. p. 11; Till. p. 28.
» » DICKIE, Alg. Sutherl. 1, p. 141. (?)
» » GOBI, Algenfl. Weiss. Meer. p. 76.
» » KJELLM. Vinteralgv. p. 64; Spetsb. Thall. 2, p. 25; Algenv. Murm. Meer. p. 38; Kariska hafvets algv. p. 25; ubique ex parte.

Syn. Laminaria digitata KLEEN, Nordl. Alg. p. 33; ex parte; excl. f. stenophylla.
» » LINDBL. Bot. Not. p. 57; ex parte. (?)
» » POST. et RUPR. Ill. Alg. p. II. (?)
» » SCHRENK, Ural. Reise, p. 546; ex parte. (?)
» » SCHÜBELER, in Heugl. Reise. p. 317.
» » SOMMERF. Spitsb. Fl. p. 232. (?)
» » ZELLER, Zweite d. Polarf. p. 85. (?)
» flexicaulis FOSLIE, l. c. p. 19; excl. L. digitata var. stenophylla HARV.
» » Nyl. et Sæl. Herb. Fenn. p. 73.

Remark on the species. After baving lately had the opportunity of examining a considerable number of *Laminaria digitata* auct. from different parts of the Scandinavian coast, I cannot but admit unconditionally that there are to be found here at least two well marked and easily distinguished species, on the one hand that which Scandinavian algologists have been accustomed to call *L. digitata* and which is the commonest, on the other hand that described by LE JOLIS under the name of L. *Clustoni.* The former species is partly identical with the *L. flexicaulis* of the just-mentioned author, which includes however also HARVEY'S *L. digitata* var. *stenophylla.* As to the last-named alga, I am of opinion, for reasons to be stated afterwards, that it ought as yet to be considered a separate species. *L. Clustoni* has only lately, by the comprehensive researches of J. E. ARESCHOUG, been known with certainty as a Scandinavian species. Since he has called attention to it by private communications, it has been observed at several parts of the Scandinavian coast, in some localities even abundant. It is, however, far more rare on the west coast of Sweden than that species which has hitherto passed under the name of *L. digitata.* This name I think fit to retain for this species, while I call the other species by the name of *L. Clustoni,* which it received when it was first decidedly discerned as a distinct species. To replace the name of *L. Clustoni* by *L. digitata,* as FOSLIE has proposed to do, and rebaptize the plant called *L. digitata* by recent Scandinavian algologists, J. G. AGARDH, J. E. ARESCHOUG a. o., by the name of *L. flexicaulis,* can hardly be justified by the fact of earlier authors having described and quoted *L. Clustoni* LE JOL. under the name of *L. digitata.* For it certainly is pretty probable that these authors have called or at least would have called even *L. flexicaulis* by the name of *L. digitata,* and it is impossible to determine at the present time whether Linnæus understood by his name of *L. digitata* both the algæ in question or only the one of them and in such a case which. The change of names proposed by FOSLIE would scarcely lead to anything but to throw the already entangled nomenclature into still greater confusion.

The introduction of the name *flexicaulis* by LE JOLIS is hardly justifiable. It cannot be allowed, by the laws established for names-giving, to reject altogether the Linnæan name of *L. digitata,* and this should have been done in this case so much the less because EDMONSTON retained that name for the plant which LE JOLIS regards as identical with his own *L. flexicaulis,* while he employed the name of *L. Clustoni* adopted by LE JOLIS for the species separated from the old *L. digitata.*

In my opinion EDMONSTON'S proceeding is perfectly lawful and consistent, even if it should appear that the alga called by him *L. digitata* must also be detached as a separate species from the collective species *L. digitata* auct.

That I have referred both *L. flexicaulis* and *L. cucullata* FOSLIE to *L. digitata* is caused by my having in the former part of this work, that was printed before I had access to FOSLIE'S paper, assumed on the authority of LE JOLIS and J. E. ARESCHOUG the *Laminariæ* meant by these names to be forms of *L. digitata*. Accordingly my treatment of the algæ in question does not by any means imply that I think FOSLIE'S dividing of *L. digitata* (*L. flexicaulis*) into two distinct species is not justified. I have thought best to adopt the forms discerned by FOSLIE. But it appears to me doubtful whether f. *valida* is really a proper variety and not only a condition of f. *genuina* of a different age. However, it is possible that this is not the case.

That form of *L. digitata* which I have described under the name of f. *complanata* seems to deserve to be specially mentioned quite as well as the others. Older larger individuals, by their very broad stipe that is much flattened upwards, assume an aspect very different from other forms [1]).

As mentioned before, the present species is plainly distinct from *L. Clustoni* and other above-named *Laminariæ* of the digitata group. It may be remarked that the number and size of the muciferous lacunæ are subject to rather considerable variations. In general they are smaller and fewer in individuals from higher than from lower latitudes. In specimens from Spitzbergen they are sometimes exceedingly few and very difficult to distinguish from the adjoining cells, in specimens from the north coast of Norway, on the contrary, they are not seldom very numerous and sometimes very large in proportion to the size of the cells in the intermediate layer. They are then situate sometimes nearer to the cortical layer, sometimes farther inwards in the intermediate layer, and in the latter case they are often in a great part of their periphery surrounded with cells differing in size and shape from the other cells in the intermediate layer of the lamina.

Habitat. This species is usually sublitoral in the Polar Sea, but on the north coast of Norway it even ascends into the litoral zone, living then in rock-pools. However, it is to be found more abundant and fully developed only at or immediately below low-water mark and from this line it descends down to a depth of ten fathoms. But the greatest and densest masses are met with in the upper part of the sublitoral zone. In the Arctic Sea proper it belongs to the middle and lower part of the sublitoral zone, never rising here to low-water mark.

It is most vigorously developed on a bottom of solid rock, but occurs also on gravelly ground. According to FOSLIE it likes steep slopes and localities where there are colonies of *Mytilus edulis*. It is gregarious and lives on exposed as well as sheltered coasts, even in the interior of deep bays. It is certainly known to bear zoosporangia during the summer months at Spitzbergen; cp. J. G. AG. Spetsb. Alg. Till. p. 30 under

[1]) It is seen, however, by my description of the form that I have never meant to attribute to it any higher degree of independence, as GOBI appears to suppose. Cp. GOBI Algenfl. Weiss. Meer. and KJELLM. Kariska hafvets Alg. p. 27, Algenv. Murm. Meer. p. 77.

L. nigripes. Whether such organs are produced even in winter or not, I am unable to determine, because in my notes I have confounded this species with *L. nigripes.* At Finmarken and Novaya Zemlya I have found only sterile specimens.

Geogr. Distrib. In the Polar Sea this species seems to be confined to the Atlantic province and the Spitzbergen province. I did not see it anywhere in the eastern part of the Kara Sea and in the Siberian Sea. It has not been recorded from the American Arctic Sea and the statements of its occurrence in Baffin Bay are uncertain; cp. J. G. AG. Grönl. Lam. och Fuc. p. 11 and 18. Its maximum of frequency is on the north coast of Norway. The most northern point where it has been taken is the North Cape of Spitzbergen Lat. N. 80° 31'.

Localities: The Norwegian Polar Sea: Nordlanden, Tromsö amt, for instance, at Tromsö and Carlsö; Finmarken: Maasö, Gjesvær, Öxfjord, Talvik a. s. o.,' common and abundant everywhere.

The Greenland Sea: the east coast of Greenland at Sabine Isle (?), Beeren Eiland, the coasts of Spitzbergen, at the last-named place common and often plentiful.

The Murman Sea: common and often abundant on the west coast of Novaya Zemlya and Waygats; Jugor Shar.

The White Sea: common and abundant.

The Kara Sea: the east coast of Novaya Zemlya at Uddebay.

Baffin Bay: the west coast of Greenland at Godhavn and Whale Sound (?).

Of the forms mentioned f. *complanata* is known from the coasts of Novaya Zemlya. Here as on the coasts of Spitzbergen there occurred partly a form apparently most nearly related to f. *valida*, partly f. *ensifolia* and f. *ovata.* As for the distribution of the forms on the north coast of Norway the reader is referred to the above-cited work of FOSLIE.

Laminaria stenophylla HARV. (J. G. AG.)

Lam. p. 18. Laminaria digitata var. stenophylla HARV. Phyc. Brit. t. 338.

Descr. Laminaria stenophylla J. G. AG. l. c.

Fig. Laminaria digitata var. stenophylla HARV. l. c.

Syn. Laminaria digitata var. stenophylla KLEEN, Nordl. Alg. p. 33.

Remark on the species. J. G. AGARDH l. c. has strongly pointed out the claims of this *Laminaria* to be regarded as a separate species, setting forth its differences from the *L. digitata* occurring on the coast of Sweden and south-western Norway. Such a view seems to me very well justified already on account of the reasons brought forward by AGARDH. Recent investigations have, however, brought into light some facts that appear to prove that the *Laminaria* of Scotland and Ireland, the specific difference of which from *L. Clustoni* I have above pointed out and which has been set down by HARVEY as a f. *stenophylla* of *L. digitata*, is not identical either with *L. Clustoni* or with that species which has gone under the name of *L. digitata* in later Swedish algological treatises. For it is stated by CLUSTON that the *Laminaria* alluded to does not change its lamina regularly, as *L. Clustoni*, »but the great distinction in this part.... is that the Cuvy« (= *L. Clustoni*) »annually throws off the old leaf and acquires a new one, while this has never been observed in the Tangle» (= *L. stenophylla*). I think no one is entitled to doubt that this statement of CLUSTON is founded on complete researches continued through the whole year. Nevertheless it proceeds from J. E. ARESCHOUG'S investigations — according to oral communications kindly given — that the Scandinavian *L. digitata* changes its lamina periodically in the same manner as *L. Clustoni* and that at the fall of the lamina the limit between the young and the old one is quite as distinctly marked as in the last-named species. The same observation has been made by FOSLIE and set forth in his work Digit. Lam. p. 6. He mentions also that there are quite as distinct yearly rings to be seen in *L. digitata* as in *L. Clustoni*, which confirms my observations in respect to *L. digitata* f. *complanata*. Cp. KJELLM. Kariska hafvets Algv. tab. 1, fig. 14. Such rings, according to LE JOLIS, are not to be found in *L. flexicaulis*, which I have taken to be identical with *L. digitata* var. *stenophylla* HARV. — For these reasons I am opinion that, even after *L. Clustoni* has been segregated from *L. digitata*, there still remains in the last-named old collective species not one but two species, the one perennial, annually throwing off its lamina regularly as *L. Clustoni* and possessing in its stipe thickening-layers of the same kind as in this species, and the other possibly biennial, as LE JOLIS supposes, fast increasing in size, with a stipe that »ne présente pas d'anneaux concentriques» and a lamina that »se développe d'une manière continue et indéfinie; et pour ce motif peut atteindre de grandes dimensions». The former species is *L. digitata* of J. G. AGARDH and J. E. ARESCHOUG, the latter is *L. digitata* EDM., *L. digitata* var. *stenophylla* HARV., *L. stenophylla* J. G. AG. It possibly is the former of these species, *L. digitata* J. G. AG., ARESCH., that LE JOLIS refers to l. c. p. 553, when he says: »Cependant on trouve quelquefois dans le Lam. flexicaulis un état de végétation qui le rapproche du Lam. Clustoni. Alors un certain arrêt a lieu dans la croissance de la plante et un léger rétrécissement s'est manifesté dans une partie de la fronde.... c'est dans ce cas encore qu'on remarque dans le stipe du Lam. flexicaulis de traces d'anneaux colores...» In such a case *L. flexicaulis* LE JOL. would include not only *L. stenophylla* J. G. AG., but also *L. digitata* J. G. AG., ARESCH.

It is another question whether this *L. stenophylla* is really to be found within the Polar Sea. I have never myself seen any specimen taken here. My recording it among the arctic algæ is founded only on KLEEN'S statement that he has collected it at Nordlanden. I had so much the less reason to doubt its occurrence here, because the species is apparently northern and moreover is stated by J. G. AGARDH to have been collected by himself at Throndhjem. Neverthless I am not quite sure whether KLEEN has not mistaken also young individuals of L. *digitata* with cuneate base growing in rock-pools for L. *digitata* var. *stenophylla*. I have myself met with specimens at Nordlanden which resemble HARVEY'S figure of *L. stenophylla* in the form of the lamina, but are undoubtedly young specimens of *L. digitata*.

Habitat. It is said by KLEEN to live at low-water mark as well as in rock-pools within the litoral zone.

Geogr. Distrib. Reported from the Norwegian Polar Sea.

Locality: Nordlanden: common and abundant according to KLEEN.

Gen. **Chorda** (STACKH.) LAMOUR.

Ess. p. 46; STACKH. Ner. Brit. p. XVI; ex parte.

Chorda filum (L.) STACKH.

l. c. Fucus filum L. Spec. Pl. p. 1162.

f. *typica.*

Descr. Chorda filum ARESCH. Obs. Phyc. 3, p. 13.
Fig. » » HARV. Phyc. Brit. t. 107.
Exsicc. » » ARESCH. Alg. Scand. exsicc. N:o 92.

f. *subtomentosa* ARESCH.

Obs. Phyc. 3, p. 13.

Descr. Chorda filum β subtomentosa ARESCH. l. c.
Exsicc. » » var. tomentosa ARESCH. Alg. Scand. exsicc. N:o 168.

f. *crassipes* nob.

f. thallo 20—40 cm. alto, diametro 1—2 mm., flavescenti-olivacea apicem versus valde, at basim versus obsolete attenuato, parte basali partem apicalem crassitudine multo superante. Tab. 26, fig. 16.

Syn. Chorda filum ARESCH. Phyc. Scand. p. 365 et l. c.
» » J. G. AG. Spetsb. Alg. Till. p. 28; Grönl. Alg. p. 110.
» » CROALL, Fl. Disc. p. 457.
» » GOBI, Algenfl. Weiss. Meer. p. 74.
» » HARV. Fl. West. Esk. p. 49.
» » KJELLM. Spetsb. Thall. 2, p. 27; Algenv. Murm. Meer. p. 41.
» » KLEEN, Nordl. Alg. p. 34.
» » POST. et RUPR. Ill. Alg. p. II.
Fucus filum GUNN. Fl. Norv. 2, p. 10.
» » SCORESBY, Account. 1, p. 132.
» » WG. Fl. Lapp. p. 505.
Scytosiphon filum J. G. AG. Spetsb. Alg. Progr. p. 2; Bidr. p. 11.

Remark on f. crassipes. This form differs pretty much in habit from typical *Chorda filum.* It tapers rather imperceptibly towards the callus and on that account is, immediately above it, very much thicker that at the extremity and almost as thick as at the middle, while the thallus of the typical form is almost as much attenuated towards the base as towards the tip. Besides, it is smaller and of a lighter and clearer colour than common *Ch. filum.* In inner structure, however, it accords very nearly with it. I have not observed any hairs in f. *crassipes.* Its being a deep-water form explains its lighter colour as well as the absence of hairs.

Habitat. In the Arctic Sea proper this species is always sublitoral, in the Norwegian Polar Sea it is usually so; but it occurs here also in rock-pools within the litoral zone. Generally it lives at inconsiderable depths on gravelly bottom, particularly when the ground is composed of broken *Lithothamnia,* one to three fathoms below low-water mark. I have dredged f. *crassipes* in the lower part of the sublitoral region in 15—20 fathoms. The present species is gregarious and flourishes both in exposed and sheltered localities. Though it certainly has its typical appearance when growing in rock-pools, it is always low-sized at such places, 1—2 feet long. In deeper water the typical form even at Finmarken attains a considerable length, even two metres and a half. The typical form bears zoosporangia in August on the north coast of Norway. On the north coast of Spitzbergen I have collected specimens of f. *subtomentosa* with zoosporangia in course of development in the month of August, and on the west coast of Novaya Zemlya specimens with ripe zoosporangia were taken in September.

Geogr. Distrib. This species seems to be widely disseminated in the Arctic Sea, but it is only on the coast of Norway that it is met with in any considerable number. Its maximum of frequency is situate south of the Polar Circle. It is not known from the Kara and Siberian Seas. The northernmost point where it is certainly found is Fairhaven on the north-west coast of Spitzbergen Lat. N. 79° 49'.

Localities: The Norwegian Polar Sea: Nordlanden common and abundant; Tromsö amt, common and abundant about the town of Tromsö; Finmarken at Maasö, Gjesvær, Öxfjord, and Talvik, pretty common and plentiful.

The Greenland Sea: the west coast of Spitzbergen, local and scarce.

The Murman Sea: the coast of Russian Lapland; the west coast of Novaya Zemlya, local, scarce.

The White Sea: rather rare.

The American Arctic Sea: the north coast of Western Eskimaux-land.

Baffin Bay: the west coast of Greenland: Tessarmiut, Unartok, Neuherrnhut, Godthaab, Kapiselik, Godhavn, Sakkak, and Waygats Strait.

The typical form is known from the north coast of Norway and the west coast of Greenland; f. *crassipes* from the north coast of Norway (Maasö). At Spitzbergen and Novaya Zemlya I have found a form that resembles most nearly f. *subtomentosa.*

Chorda tomentosa LYNGB.

Hydr. Dan. p. 74.

Descr. Chorda tomentosa ARESCH. Obs. Phyc. 3, p. 14.
Fig. » » LYNGB. l. c. t. 19, fig. A.
Exsicc. » » ARESCH. Alg. Scand. exsicc. N:o 93.

Habitat. With regard to the habitat of this species, the reader is referred to ARESCH. Obs. Phyc. l. c.

Locality: It is said by ARESCHOUG to have been taken in the Norwegian Polar Sea at East Finmarken in Engelsvigen.

Fam. ENCOELIEÆ (KÜTZ.)

Phyc. gener. p. 336; lim. mut.

Gen. Stilophora J. G. AG.

Nov. p. 16.

Stilophora Lyngbyei J. G. AG.
Symb. 1, p. 6.

Descr. Stilophora Lyngbyei J. G. AG. Spec. Alg. 1. p. 84.
Fig. » » HARV. Phyc. Brit. t. 237.
Exsicc. » » ARESCH. Alg. Scand. exsicc. N:o 91.
Syn. Stilophora Lyngbyei KLEEN, Nordl. Alg. p. 33.

Habitat. It has been found within the sublitoral zone at a depth of two or three fathoms on sandy bottom. In the month of August it was provided with zoosporangia.

Geogr. Distrib. Known only from the Norwegian Polar Sea. Here it is rare.

Locality: Nordlanden at Fleinvær, the most northern place where this species has been found to occur.

Gen. Asperococcus LAMOUR.

Ess. p. 277.

Asperococcus echinatus (MERT.) GREV.
Alg. Brit. p. 50. Conferva echinata MERT. in ROTH, Cat. Bot. 3, p. 170.

Descr. Asperococcus echinatus J. G. AG. Spec. Alg. 1, p. 76.
Fig. » » HARV. Phyc. Brit. t. 194.
Exsicc. » » ARESCH. Alg. Scand. exsicc. N:o 267.
Syn. Asperococcus echinatus KLEEN, Nordl. Alg. p. 33.

Habitat. It grows scattered in rock-pools within the litoral zone, both on exposed and sheltered coasts, attached to other algæ, as *Corallina officinalis*, *Furcellaria fastigiata* a. o. In July, August, and the beginning of September it has been found with zoosporangia on the north coast of Norway.

Geogr. Distrib. Known only from the Norwegian Polar Sea, where it is beyond its proper area of distribution. The northernmost point where it has been collected is Öxfjord in Norwegian Finmarken about Lat. N. 70°.

Localities: The Norwegian Polar Sea: Nordlanden rather rare; Finmarken dwarfed and scarce at Öxfjord.

Asperococcus bullosus LAMOUR.

Ess. p. 277.

Descr. Asperococcus bullosus J. G. AG. Spetsb. Alg. 1, p. 77.
Fig. » Turneri HARV. Phyc. Brit. t. 11.
Exsicc. » bullosus ARESCH. Alg. Scand. exsicc. N:o 89.
Syn. Asperococcus Turneri DICKIE, Alg. Sutherl. 1, p. 141.

Habitat. Nothing is known about the habitat of this species in the Arctic Sea.

Locality: It is stated by DICKIE l. c. to have been found in Baffin Bay at Fiskernæs on the west coast of Greenland.

Gen. Ralfsia BERK.

Engl. Bot. Suppl. t. 2866.

Ralfsia deusta (AG.) J. G. AG.

Ralfsia (?) deusta J. G. AG. Spec. Alg. 1, p. 63. Zonaria deusta AG. Syn. Alg. p. 40.

Descr. Ralfsia (?) deusta J. G. AG. l. c.
Fig. Fucus fungularis Fl. Dan. t. 420.
Syn. Padina deusta POST. et RUPR. Ill. Alg. p. II.
Ralfsia deusta ARESCH. Phyc. Scand. p 361.
» » DICKIE, Alg. Cumberl. p. 237; ex parte.
» » GOBI, Algenfl. Weiss. Meer. p. 73.
» » KJELLM. Algenv. Murm. Meer, p. 40.
» » Nyl. et Sæl. Herb. Fenn. p. 73.

Habitat. According to my observations this species is sublitoral in the Polar Sea, growing scattered in 2—5 fathoms, on exposed coasts, attached to stones, muscles, and stout stems of *Laminariæ*.

Geogr. Distrib. The present species has been found at several, widely distant places in the Polar Sea, but it seems never to occur there in any considerable numbers. The most northerly locality where it is certainly known to occur is Matotshin Shar on the west coast of Novaya Zemlya Lat. N. 73° 15'.

Localities: *The Norwegian Polar Sea:* Nordlanden according to specimens in KLEEN'S herbarium. KLEEN does not record the species from here. The specimens in his collections are young or little developed and have probably been considered by him to belong to *Ralfsia verrucosa*.

The Murman Sea: the coast of Russian Lapland, where the species seems to be rather abundant; the west coast of Novaya Zemlya from Matotshin Shar to Rogatshew Bay, pretty common and plentiful.

Baffin Bay: Cumberland Sound; the west coast of Greenland: Nenese and Nanortalik. Besides there are specimens in the herbarium of the Copenhague Museum without any locality noted.

Ralfsia verrucosa (ARESCH.) J. G. AG.

Spec. Alg. 1, p. 62. Cruoria verrucosa ARESCH. Alg. Pugill. p. 264.

Descr. Ralfsia verrucosa J. G. AG. l. c.
Fig. » » KÜTZ. Tab. Phyc. 9, t. 77.

Syn. Ralfsia verrucosa KLEEN, Nordl. Alg. p. 33; ex parte.

Habitat. It is litoral, attached to stones or shells. It prefers sheltered bays, but grows also on exposed coasts. In the Polar Sea it lives scattered in small numbers and has only been found sterile here.

Geogr. Distrib. Known only from the Norwegian Polar Sea. Its northernmost place of occurrence is the south coast of Magerö about Lat. N, 71°.

Localities: The Norwegian Polar Sea: Nordlanden, common according to KLEEN; Finmarken scarce on the south coast of Magerö.

Fam. CHORDARIACEÆ (AG.) FARL.

New. Engl. Alg. p. 83; Ag. Syst. Alg. p. XXXVI; lim. mut.

Gen. Chordaria (AG.) J. G. AG.

Alg. Syst. 2, p. 62; AG. Syn. Alg. p. XII; lim. mut.

Chordaria flagelliformis MÜLL.

Fl. Dan. t. 650.

f. *typica*.

Descr. Chordaria flagelliformis J. G. AG. Spec. Alg. 1, p. 66; excl. var.
Fig. » » HARV. Phyc. Brit. t. 111.
Exsicc. » » ARESCH. Alg. Scand. exsicc. N:o 97.

f. *chordæformis* KJELLM.

Spetsb. Thall. 2, p. 28.

Descr. et *Fig.* Chordaria flagelliformis f. chordæformis KJELLM. l. c. et t. 1, fig. 13—15.

f. *ramusculifera* KJELLM.

l. c. p. 29.

Descr. et *Fig.* Chordaria flagelliformis f. ramusculifera KJELLM. l. c. et t. 1, fig. 10—12.

Syn. Chordaria divaricata GOBI, Algenfl. Weiss. Meer. p. 70, huic formæ proxima.

f. *subsimplex* KJELLM.

Descr. et *Fig.* Chordaria flagelliformis f. subsimplex KJELLM. l. c. et t. 1, fig. 16—18.

Syn. Chordaria divaricata GOBI, Algenfl. Weiss. Meer. p. 69; sec. spec.
» flagelliformis ARESCH. Phyc. Scand. p. 366.
» » J. G. AG. Spetsb. Alg. Progr. p. 2; Bidr. p. 11; Grönl. Alg. p. 110.
» » CROALL, Fl. Disc. p. 458.
» » DICKIE, Alg. Cumberl. p. 237; Alg. Sutherl. 1, p. 141; Alg. Nares p. 7.
» » GOBI, l. c. p. 70.

Syn. Chordaria flagelliformis KJELLM. Vinteralgv. p. 64; Spetsb. Thall. 2, p. 28; Algenv. Murm. Meer. p. 41.
» » KLEEN, Nordl. Alg. p. 34.
» » LINDBL. Bot. Not. p. 157.
» » Nyl. et Sæl. Herb. Fenn. p. 73.
» » POST. et RUPR. Ill. Alg. p. II.
» » ZELLER, Zweite d. Polarf. p. 84.
Fucus flagelliformis WG. Fl. Lapp. p. 505.

Remark on the synonymy. The alga set down by GOBI l. c. as *Ch. divaricata*, judging by the specimen kindly communicated to me by GOBI, is not of that species, but a *Ch. flagelliformis.* It is most nearly related with that form of this species which I have called f. *ramusculifera*, though by its slenderness and loose consistency it approaches also f. *subsimplex.* I have found a form very nearly coinciding with that collected by GOBI, in the eastern part of the Murman Sea, and in my account of the marine vegetation of that sea I have recorded it as an intermediate form between the two just-mentioned ones.

Habitat. From the Norwegian Polar Sea only the typical form of the present species is known. It occurs here sometimes litoral, sometimes sublitoral, fastened partly to other algæ, as *Halosaccion ramentaceum, Fucus serratus* a. o., partly to stones. In other parts of the Polar Sea I have always found it within the sublitoral zone, but in the upper part of it. It is met with both on exposed and sheltered coasts and in the Norwegian Polar Sea is occasionally gregarious. The typical form on the north coast of Norway bears zoosporangia during the whole summer, at least to the middle of September. At Spitzbergen I have found this form provided with such organs in August, and on the west coast of Novaya Zemlya at the end of June and the beginning of July. The form *chordæformis* has been taken with zoosporangia at Spitzbergen in January, February, May, July, August, and December, at the west coast of Novaya Zemlya during the earlier part of July; f. *ramusculifera* on the north coast of Spitzbergen in July and August, f. *subsimplex* at Spitzbergen at the end of August and the commencement of September.

Geogr. Distrib. The present species has been found in the Arctic Sea proper at several places at very different longitudes, but nowhere in any considerable numbers. Its maximum of frequency is situate to the south of that region. The most northern point where it has been taken is Discovery Bay on the west coast of Greenland, Lat. N. 81° 41′.

Localities: The Norwegian Polar Sea: (f. *typica*) Nordlanden common and abundant; Finmarken common and pretty abundant, as at Maasö, Gjesvær, the south coast of Magerö, and Talvik.

The Greenland Sea: the east coast of Greenland at Sabine Isle (?); the west and north coasts of Spitzbergen, local and scarce. All the forms recorded have been found at Spitzbergen.

The Murman Sea: the coast of Russian Lapland (f. *typica*), probably common; the west coast of Novaya Zemlya (f. *typica*, f. *chordæformis*, and a form intermediate between f. *ramusculifera* and f. *subsimplex*) rather generally dispersed, but scarce.

The White Sea: f. *ramusculifera.*

The Siberian Sea: (f. *typica* and f. *ramusculifera*) rare and so dwarfed as to be almost unrecognizable at Irkaypi and Pitlekay.

Baffin Bay: Cumberland Sound; the west coast of Greenland probably common; found at Nanortalik, Fiskernæs, Sukkertoppen, Hunde Islands, Godhavn, Egedesminde, Rittenbenk, Melville Bay, Whale Islands, Discovery Bay.

Gen. **Castagnea** (DERB. et SOL.) J. G. AG.

Alg. Syst. 2, p. 33; DERB. et SOL. Mem. phys. Alg. p. 56, sec. J. G. AG. l. c.

Castagnea divaricata (AG.) J. G. AG.

l. c. p. 37. Chordaria divaricata AG. Syn. Ag. p. 12.

Descr. Castagnea divaricata J. G. AG. l. c.
Fig. Mesogloia divaricata KÜTZ. Tab. Phyc. 8, t. 8.
Exsicc. Chordaria divaricata ARESCH. Alg. Scand. exsicc. N:o 98.
Syn. Chordaria divaricata KLEEN, Nordl. Alg. p. 39.

Locality. It is uncertain whether this species really lives in the Polar Sea. A specimen furnished with sporangia was found in August washed ashore at Bodö in Nordlanden, accordingly at the southern limit of the Polar Sea. Most probably it has grown in the neighbourhood, and the species may thus be regarded as belonging to the Flora of the Polar Sea.

Gen. **Eudesme** J. G. AG.

Alg. Syst. 2, p. 29.

Eudesme virescens (CARM.) J. G. AG.

l. c. 31. Mesogloia virescens CARM. in HOOK. Brit. Fl. 2, p. 387.

Descr. Eudesme virescens J. G. AG. l. c.
Mesogloia virescens » Spec. Alg. 1, p. 56.
Fig. » » HARV. Ner. Am. 1, t. 10, fig. *b.*
Exsicc. Castagnea virescens ARESCH. Alg. Scand. exsicc. N:o 315.
Syn. Castagnea virescens GOBI, Algenfl. Weiss. Meer. p. 72.
» » KLEEN, Nordl. Alg. p. 35.
» Zosteræ » » » » »

Habitat. It grows sometimes in the upper part of the sublitoral zone, usually within the litoral zone, partly on stones or algæ above low-water mark, partly in rock-pools below that line. It occurs scattered both on exposed and on sheltered coasts. It is found with zoosporangia in the Norwegian Polar Sea during the whole summer to the middle of September.

Geogr. Distrib. Known only from the Norwegian Polar Sea, the most western part of the Murman Sea, and the White Sea. Already in the most northern part of the Norwegian Polar Sea it becomes rare. The northernmost point where it has been collected is Gjesvær on the north coast of Norway about Lat. N. 71°.

Localities: The Norwegian Polar Sea: Nordlanden common, according to KLEEN; Finmarken, local and very scarce at Gjesvær and Öxfjord.

The Murman Sea: the coast af Russian Lapland.

The White Sea: at Solowetzki Isles.

Gen. **Mesogloia** (AG.) J. G. AG.

Alg. Syst. 2, p. 27; AG. Syn. Alg. 1, p. XXXVII. lim. mut.

Mesogloia vermicularis AG.

l. c. p. 126.

Descr. Mesogloia vermicularis J. G. AG. Spec. Alg. 1, p. 58.
Fig. » » HARV. Phyc. Brit. t. 31.
Exsicc. » » ARESCH. Alg. Scand. exsicc. N:o 99.
Syn. Mesogloia vermicularis GOBI, Algenfl. Weiss. Meer. p. 72.
» » KLEEN, Nordl. Alg. p. 34.

Habitat. This species grows scattered at a depth of 4—5 fathoms on stony and shelly bottom. It has been found with zoosporangia at Nordlanden in August.

Geogr. Distrib. Known only from the Norwegian Polar Sea and the most western part of the Murman Sea. The most northern point where it is known to occur is at Tri-Ostrowa on the Murman coast about Lat. N. 68°.

Localities: The Norwegian Polar Sea: found at Nordlanden in one locality: at Fleinvær.

The Murman Sea: the coast of Russian Lapland at Tri-Ostrowa.

Fam. MYRIONEMATEÆ THUR.

in LE JOL. Liste Alg. Cherb. p. 15 et p. 23.

Gen. **Leathesia** (GRAY.) J. G. AG.

Alg. Syst. 2, p. 40; GRAY. Brit. Pl. 1, p. 301; char. mut. sec. J. G. AG. Spec. Alg. 1, p. 50.

Leathesia difformis (L.) ARESCH.

Phyc. Scand. p. 376. Tremella difformis L. Fl. Suec. p. 429, sec. Fr. Fl. Scand. p. 316.

Descr. Leathesia marina J. G. Ag. Spec. Alg. 1, p. 52.
Fig. » tuberiformis HARV. Phyc. Brit. t. 324.
Exsicc. » difformis ARESCH. Alg. Scand. exsicc. N:o 101.
Syn. Leathesia difformis KLEEN, Nordl. Alg. p. 35.

Habitat. This species lives somewhat gregarious in rock-pools within the litoral zone, on exposed as well as sheltered coasts, seldom attached to stones, usually to litoral algæ as *Corallina officinalis, Polysiphonia nigrescens, Ahnfeltia plicata, Cladophora rupestris.* On the coast of Norway it has zoosporangia during the latter part of the summer, in August and September.

Geogr. Distrib. Known only from the Norwegian Polar Sea. Its northernmost point is Öxfjord in Finmarken about Lat. N. 70°.

Localities: The Norwegian Polar Sea: Nordlanden common according to KLEEN; Finmarken local and scarce at Öxfjord.

Gen. **Elachista** DUBY.

Mem. Cor. 1, p. 19, sec. J. G. AG. Spec. Alg. 1, p. 7.

Elachista fucicola (VELL.) ARESCH.

Alg. Pugill. p. 235. Conferva fucicola VELL. Mar. Plant. N:o 4, sec. ARESCH. l. c.

Descr. Elachista fucicola ARESCH, Phyc. Scand. p. 377.

Fig. » » » » » t. 9, fig. c.

Exsicc. » » » Alg. Scand. exsicc. N:o 102.

Syn. Conferva fucicola WG. Fl. Lapp. p. 514.

» » POST et RUPR. Ill. Alg. p. II (?).

Elachista fucicola ARESCH. Phyc. Scand. p. 377.

» » CROALL, Fl. Disc. p. 458.

» » DICKIE, Alg. Cumberl. p. 237; Alg. Sutherl. 1, p. 141.

» » GOBI, Algenfl. Weiss. Meer. p. 72.

» » KJELLM. Spetsb. Thall. 2, p. 31; Algenv. Murm. Meer. p. 42; Kariska hafvets algv. p. 27.

» » KLEEN, Nordl. Alg. p. 35.

Habitat. In the Norwegian Polar Sea the present species is litoral; in other parts of the Polar Sea I have always found it sublitoral. It is always attached to other algæ, usually *Fuci*, but often other species, as *Rhodomela lycopodioides, Halosaccion ramentaceum, Rhodymenia palmata, Chondrus crispus, Gigartina mamillosa, Porphyra laciniata, Chætomorpha melagonium.* It flourishes both on exposed coasts and in sheltered localities, for instance, in the interior of deep bays. On the north coast of Norway it occasionally occurs gregarious and contributes rather essentially to mark the character of the vegetation on ranges of comparatively large extent. On the coast of Spitzbergen I have found that form of the species which has been named f. *globosa.* It bears sporangia through all the summer on the coast of Norway. At Spitzbergen it has been found with such organs in February, March, July, and August, on the west coast of Novaya Zemlya in July.

Geogr. Distrib. With the exception of the American Arctic Sea, this species is known in all parts of the Arctic Sea. Its most northern point is Musselbay on the north coast of Spitzbergen Lat. N. 79° 53'.

Localities: The Norwegian Polar Sea: Nordlanden common and abundant; Finmarken common and pretty plentiful, as at Maasö, Gjesvær, Öxfjord, and Talvik.

The Greenland Sea: Local and scarce on the north and west coasts of Spitzbergen.

The Murman Sea: the coast of Russian Lapland; the west coast of Novaya Zemlya local and scarce.

The Kara Sea: the east coast of Novaya Zemlya at Uddebay scarce.

The Siberian Sea: at Pitlekay scarce.

Baffin Bay: Cumberland Sound; the west coast of Greenland at Fiskernæs, Jakobshavn, and Whale Islands. Specimens have also been brought home by BERGGREN from the west coast of Greenland.

Elachista lubrica RUPR.

Alg. Och. p. 388.

Descr. Elachista lubrica ARESCH. Obs. Phyc. 3, p. 18.
Exsicc. » » » Alg. Scand. exsicc. N:o 217.
Syn. Elachista flaccida DICKIE, Alg. Sutherl. 1, p. 141 (?).
» lubrica J. G. AG. Grönl. Alg. p. 110.
» » ARESCH. l. c.
» » GOBI, Algenfl. Weiss. Meer. p. 73.
» » KJELLM. Vinteralgv. p. 65; Spetsb. Thall. 2, p. 31; Algenv. Murm. Meer. p. 42.
» » RUPR. l. c. p. 389.

Habitat. The present species agrees in habitat with the preceding one. It is usually fastened to *Halosaccion ramentaceum*, but occurs epiphytic also on other species, as *Rhodomela lycopodioides, Polysiphonia arctica, Rhodymenia palmata, Fucus edentatus, Desmarestia aculeata, Chætopteris plumosa.* In Norway I have found it only at exposed points, but at Spitzbergen it lives also in the interior of deep bays. I have taken a form *globosa* in May and June at Spitzbergen. It appears to bear zoosporangia all the year round. At Spitzbergen I have seen specimens furnished with such organs in all the months of the year with the exception of June and September, in which months I had no opportunity of examining it. In January, March, April, and December it was very plentifully provided with zoosporangia. On the coast of Norway it has been collected with zoosporangia in July and August, on the west coast of Novaya Zemlya in June and July.

Geogr. Distrib. The present species has been found hitherto only in those parts of the Polar Sea that are situate north of the Atlantic. It attains its maximum of frequency on the north coast of Norway, but it is abundant also in the eastern part of the Greenland Sea. The most northern point where it is known to occur is Musselbay on the north coast of Spitzbergen Lat. N. 79° 53'.

Localities: The Norwegian Polar Sea: Nordlanden at Röst; Finmarken, common and abundant at Maasö and Gjesvær.

The Greenland Sea: the north and west coasts af Spitzbergen, pretty common and abundant.

The Murman Sea: the coast of Russian Lapland; the west coast of Novaya Zemlya local and scarce.

The White Sea: at Solowetzki Isles.

Baffin Bay: the west coast of Greenland at Sukkertoppen.

Gen. **Myrionema** GREV.

Crypt. Fl. N:o 300, sec. J. G. AG. Spec. Alg. 1, p. 47.

Myrionema strangulans GREV.

l. c.

Descr. et *Fig.* Myrionema strangulans HARV. Phyc. Brit. t. 280.
Exsicc. » vulgare KJELLM. in ARESCH. Alg. Scand. exsicc. N:o 415.

Syn. Myrionema strangulans DICKIE, Alg. Sutherl. 1, p. 141.
» vulgare KLEEN, Nordl. Alg. p. 35.

Habitat. Attached to litoral algæ, chiefly *Ulvaceæ*, but also others, as *Polysiphonia urceolata, Dumontia filiformis,* and *Cladophoræ.* It occurs both at exposed and sheltered points and is gregarious. On account of its smallness it can, however, very little influence the character of the vegetation. On the coast of Norway it has been found with plenty of zoosporangia in July and August.

Geogr. Distrib. It is known with certainty from the Norwegian Polar Sea; but it appears to occur also in Baffin Bay. The most northern place where it is certainly known to live is Gjesvær on the north coast of Norway about Lat. N. 71°.

Localities: The Norwegian Polar Sea: Nordlanden common and abundant; Finmarken local but plentiful at Gjesvær.

Baffin Bay: Cp. DICKIE, l. c.

Fam. LITHODERMATEÆ nob.

Gen. **Lithoderma** ARESCH.

Obs. Phyc. 3, p. 22.

Lithoderma fatiscens ARESCH.

l. c. p. 23.

Descr. Lithoderma fatiscens ARESCH. l. c.
Fig. » » tab. nostra 26, fig. 6—7.

Syn. Lithoderma fatiscens KJELLM. Spetsb. Thall. 2, p. 43; Algenv. Murm. Meer, p. 49; Kariska hafvets algv. p. 28.
Ralfsia fatiscens GOBI, Algenfl. Weiss. Meer. p. 74.
» spec. KJELLM. Vinteralgv. p. 64.

Remark. GOBI's refusal to acknowledge as valid ARESCHOUG's genus of *Lithoderma* is apparently occasioned by his having misunderstood ARESCHOUG's description of the so called unilocular sporangia (cp. Bot. Zeit. 1877, p. 532). Fig. 6 in tab. 26 shows these organs in the genus *Lithoderma* to differ most essentially in appearance, arrangement, and origin from the same organs in the species of the genus *Ralfsia*. The difference is so essential that the alga in question cannot even, in accordance with the modern principles of the systematic arrangement of the *Phæozoosporaceæ*, be referred to the same family as the *Ralfsiæ*. It does not with regard to those organs agree with the *Encœlieæ*, with which family *Ralfsia* must be placed, but rather with the *Punctariaceæ*. It differs, however, so considerably from these in the morphological and anatomical character of the frond, that it cannot be referred to that family either. The genus *Lithoderma* is distinguished from all *Phæozoosporaceæ* proper that I know, by the so called multilocular sporangia (gametangia) being arranged in specific stands issuing from the surface-cells of the frond, and on this ground I have thought it best to make as yet a separate family; (tab. 26, fig. 7). The structure and development of the frond are also different in *Lithoderma* and *Ralfsia*. The former is in this respect analogous to *Melobesia*, the latter to *Lithophyllum* among the *Corallinaceæ*.

Habitat. The present species is sublitoral, covering small stones on gravelly bottom in 5—15 fathoms water. It usually occurs gregarious, characterizing the vegetation of extensive areas. It is most often met with on exposed coasts. I have found it in the Polar Sea with zoosporangia multilocularia (gametangia) during the latter part of September and in December.

Geogr. Distrib. It is probably circumpolar, though it is not known at present in the American Arctic Sea. I have found it most abundant in the eastern part of the Greenland Sea and in the Murman Sea. Its northernmost point is Treurenberg Bay on the north coast of Spitzbergen Lat. N. 79° 56'.

Localities: The Norwegian Polar Sea: Finmarken, local and scarce at Maasö and Gjesvær.

The Greenland Sea: common and rather plentiful on the coasts of Spitzbergen.

The Murman Sea: common and abundant on the west coast of Novaya Zemlya and Waygats.

The White Sea: at Solowetzki Isles.

The Kara Sea: pretty abundant at Uddebay.

The Siberian Sea: Lat. N. 76° 8', Long. O. 90° 25'; Irkaypi, Pitlekay, and the Tshutsh-villages east of this point, at no place common or abundant.

Baffin Bay: the west coast of Greenland, according to specimens collected by Prof. TH. M. FRIES.

Lithoderma lignicola nob.

L. thallo crustas elongatas, plus minus confluentes formante; filis verticalibus ex articulis 20 vel pluribus, crassitudine longioribus vel æquilongis contextis. Tab. 26, fig. 8—11.

Description. This species forms thin, uneven, blackish-brown when dried, elongated, almost linear, more or less confluent crusts on old submersed wood (fig. 8). The basal layer of these crusts is composed of firmly united, branching rows of cells, pretty distinctly parallel to one another, never flabellate as in *L. fatiscens*. These cells have thick walls; in optical longitudinal section they are square, rectangular, elliptic or irregularly 4—6-angular (fig. 9). The thickening layer is formed of vertical cell-rows issuing from the basal layer, terete-sixangular in periphery, simple or scarcely branching, which are rather loosely united upwards so as to become separated from one other by stronger pression. They are at least 250 μ. in length and in full-grown individuals are composed of at least 20 cells which are square or rectangular in optical longitudinal section, and more high than thick. Their length is 10—15 μ., thickness 8—10 μ. (fig. 10—11). In *L. fatiscens* these cells are less high than thick. Reproductive organs of the species are not known at present. By the shape of the crust, the direction of the horizontal cell-rows, and the shape of their cells the present species is distinguished from the known *L. fatiscens*.

Habitat. It has been found litoral, growing in sheltered localities on old wood in company with *Chætophora pellicula* and *Callothrix Harveyi*.

Locality: *The Norwegian Polar Sea:* Collected at Talvik in the interior of Altenfjord in Finmarken in the middle of September.

Fam. SCYTOSIPHONEÆ Thur.

in Le Jol. Liste Alg. Cherb. p. 14 et 20.

Gen. Ilea (Fr.) Aresch.

Phyc. Scand. p. 353; Fr. Fl. Scan. p. 319; ex parte.

Ilea fascia (Müll.) Fr.

l. c. p. 321. Fucus fascia Müll. Fl. Dan. t. 768.

f. *typica*.

Descr. Laminaria fascia J. G. Ag. Spec. Alg. 1, p. 129.
Fig. Fucus fascia Müll. l. c.
Exsicc. Ilea fascia Aresch. Alg. Scand. exsicc. N:o 96; spec. thallo angusto.

f. *cæspitosa* (J. G. Ag.) Farl.

New Engl. Alg. p. 62. Laminaria cæspitosa J. G. Ag. Nov. p. 14.

Descr. Laminaria cæspitosa J. G. Ag. Spec. Alg. 1, p. 130.
Fig. Phyllitis cæspitosa Born. et Thur. Etud. Phycol. t. 4.
Exsicc. Ilea fascia Aresch. Alg. Scand. exsicc. N:o 96; spec. thallo latiore.

Syn. Ilea fascia Kleen, Nordl. Alg. p. 39.
Laminaria fascia J. G. Ag. Grönl. Alg. p. 110.
» » Ashm. Alg. Hayes, p. 96.
» » Dickie, Alg. Cumberl. p. 237; Alg. Sutherl. 1, p. 140; 2, p. 191.
Phyllitis fascia Gobi, Algenfl. Weiss. Meer. p. 69.

Remark on the definition of the forms. The algæ named by J. G. AGARDH *Laminaria fascia* and *L. cæspitosa*, although very different in their typical forms, are yet connected with each other by transitions so that it is impossible to draw any distinct limit between them. In the Polar Sea f. *cæspitosa* predominates, and at certain places, for instance, at Finmarken on exposed coasts, becomes very large-sized. At Gjesvær individuals of half a metre in length and 3—4,5 cm. in breadth were common. In the interior of Altenfjord the plant was smaller and narrower, possessing nearly the same aspect as the broader of the two forms distributed in Alg. Scand. exsicc. under N:o 96. The same form exists at Greenland. Farther southwards on the coast of Norway, at Nordlanden, it becomes still narrower, assuming the appearance of the typical form. However, it occurs here also in forms that are intermediate between the typical one and f. *cæspitosa*.

Habitat. The present species belongs to the lower part of the litoral zone. It is usually attached to stones and occurs in the most exposed localities as well as in the interior of deep bays, generally growing in large masses so as to determine essentially the character of the vegetation. I have found specimens with zoosporangia at Finmarken at the beginning of August and I have seen such specimens from Greenland, which were probably collected at the end of September or the commencement of October.

Geogr. Distrib. This species goes north of the 78:th latitude on the west coast of Greenland, in case the alga determined by ASHMEAD as *Laminaria fascia* is really identical with the present one. In the eastern hemisphere it hardly penetrates into the Arctic Sea proper. For it is known neither from the Greenland Sea, nor from the eastern Murman Sea, nor from the Siberian Sea. Its northernmost point here is Gjesvær about Lat. N. 71°. In certain localities of Finmarken it was abundant.

Localities: The Norwegian Polar Sea: Nordlanden (f. *typica* and transitions to f. *cæspitosa*) common and abundant; Finmarken (f. *cæspitosa*): Gjesvær local but abundant, Talvik local and rather scarce.

The Murman Sea: the coast of Russian Lapland.

The White Sea: Solowetzki Isles.

The American Arctic Sea: Union Bay.

Baffin Bay: Cumberland Sound, the west coast of Greenland at Julianeshaab, Godthaab, Sukkertoppen, Hunde Islands, and between the 78:th and 82:d latitude.

Gen. **Scytosiphon** (AG.) THUR.

in LE JOL. Liste Alg. Cherb. p. 20; AG. Spec. Alg. 1, p. 160; char. mut.

Scytosiphon lomentarius (LYNGB.) J. G. AG.

Spec. Alg. 1, p. 126. Chorda lomentaria LYNGB. Hydr. Dan. p. 74.

Descr. Scytosiphon lomentarius J. G. AG. l. c.
Fig. Chorda lomentaria HARV. Phyc. Brit. t. 285.
Exsicc. Chorda lomentaria ARESCH. Alg. Scand. exsicc. N:o 94.

Syn. Chorda lomentaria ARESCH. Phyc. Scand. p. 365.
» » DICKIE, Alg. Cumberl. p. 237.
» » Nyl. et Sæl. Herb. Fenn. p. 73.
Fucus lomentaria SOMMERF. Suppl. p. 184.
Scytosiphon lomentarius GOBI, Algenfl. Weiss. Meer. p. 68.
» » KLEEN, Nordl. Alg. p. 39.

Habitat. This species grows scattered both on exposed and sheltered coasts, attached to rocks within the litoral zone, partly in its lowest part, partly higher up, in the latter case usually in rock-pools. At Nordlanden it is fully developed in June, bearing then so called zoosporangia. Farther northward I have not found specimens with propagative organs until somewhat later in the year, at the end of July.

Geogr. Distrib. It belongs only to the most southern parts of the Polar Sea. Its maximum of frequency is in the southern portion of the Norwegian Polar Sea. The northernmost point where it has been found is Maasö on the north coast of Norway about Lat. N. 71°.

Localities: The Norwegian Polar Sea: Nordlanden common and abundant; Finmarken local and very scarce (in July and August) at Maasö.

The Murman Sea: the coast of Russian Lapland at Sviatoi-Noss.

Baffin Bay: Cumberland Sound; the west coast of Greenland at Godthaab. In the herbarium of the Copenhague Museum there are specimens collected by WORMSKIOLD at Greenland, without any notes as to their special locality.

Scytosiphon attenuatus nob.

Sc. laxe cæspitans, thallo cylindraceo vel cylindraceo-claviformi, basim versus valde et longe attenuato, 5—8 cm. alto, diametro usque 1,5 mm., fusco-olivaceo, opaco, pilis parce vestito; zoosporangiis multilocularibus (gametangiis) conicis, 30—40 μ. longis, 15—20 μ. crassis, superne liberis cum cellulis obovoideis vel breviter claviformibus subhyalinis, magnis, usque 120 μ. longis, 55 μ. crassis stratum subcontiguum formantibus. Tab. 26 fig. 1—5.

Syn. Coilonema chordaria f. simpliciuscula KJELLM. Spetsb. Thall. 2, p. 40.

Description. This plant forms thin clusters. It is attached by a callus radicalis. The most vigorously developed individuals that I have seen have a cylindrical-claviform frond, evenly and considerably tapering towards the base. Their length is 5—8 cm. The diameter is 1,5 mm. in the upper part. Other specimens are much narrower, cylindrical with a more narrow base; fig. 1—2. The alga is light-brown with an olive-brown tinge, and not shining. The frond is solid in its lowest part at the base, but hollow in the greatest part of its length. Its sterile part as well as that which bears zoosporangia (gametangia) is covered with thin, rather long hairs of the structure common in the *Phæosporaceæ;* fig. 5. The outermost layer of the wall of the frond is composed of cells that are squarish or rectangular in longitudinal section and richly provided with endochrome. The inner part of the wall is formed of cells with little or no endochrome, the outer ones being smaller, the inner ones becoming wider and longer; fig. 3—4. The cortical cells produce zoosporangia (gametangia) and so called paraphyses. The former are narrowly or thickly conical, obtuse, issuing from a broad base, with free tips. They are usually simple, sometimes branched, downward composed of two or more rows of cells, upward of only one row. They form, together with the numerous paraphyses, an almost coherent layer, which is not however so dense as in *Sc. lomentarius* nor surrounded, as in that species, with a common tegument without distinct structure, a so called cuticula, which is dissolved in the formation of zoospores (gamets). The paraphyses are large cells, several times larger than the zoosporangia (gametangia), nearly obovate, sometimes almost club-shaped. They are always poor in endochrome, sometimes apparently altogether destitute of it. In arrangement and number they vary in the present species as in the preceding in different individuals and in different parts of the same individual, being sometimes very few, sometimes very numerous. The present species is essentially different from the preceding and easily distinguished from it by the shape of the frond and, above all, by the shape, size, and disposition of the zoosporangia (gametangia).

In my account of the marine Flora of Spitzbergen I have determined this plant as *Coilonena chordaria* f. *simpliciuscula*. The specimens at my disposal were so badly preserved that I failed to get a clear view of the structure. Seen from the surface, the part provided with zoosporangia much resembles a *Coilonema*. The paraphyses appear as void zoosporangia and the extremities of the zoosporangia as rounded cortical cells. I interpreted my preparations in this manner. By a suitable treatment of some of the least injured specimens I have since found that this interpretation was wrong and that the present plant is plainly a *Scytosiphon*, though of another species than the common *Sc. lomentarius*.

Habitat. It grows on the upper part of the sublitoral zone, at a depth of 3—4 fathoms, attached to other algæ, both on exposed and sheltered coasts. Specimens taken in July on the east coast of Spitzbergen bear reproductive organs.

Locality: The Greenland Sea: in Icefjord at Goose Isles and in Smeerenberg Bay.

PUNCTARIACEÆ (THUR.) KJELLM.

Pl. Scand. p. 9; THUR. in LE JOL. Liste Alg. Cherb. p. 14; lim. mut.

Gen. **Punctaria** GREV.

Alg. Brit. p. XLII.

Punctaria plantaginea (ROTH) GREV.

l. c. p. 53. Ulva plantaginea ROTH. Cat. Bot. 3, p. 243.

f. *typica*.

Descr. Punctaria plantaginea J. G. AG. Spec. Alg. 1, p. 73.
Fig. » » HARV. Phyc. Brit. t. 128.
Exsicc. » » ARESCH. Alg. Scand. exsicc. N:o 170.

f. *linearis* Foslie.

Arct. Havalg. p. 9.

Descr. Punctaria plantaginea β. linearis Foslie, l. c.

Syn. Punctaria plantaginea Dickie, Alg. Cumberl. p. 237.

» » Kjellm. Spetsb. Thall. 2, p. 42; Algenv. Murm. Meer. p. 48.

» » Kleen, Nordl. Alg. p. 39.

Habitat. The typical form of this species grows in the Norwegian Polar Sea, sometimes on the upper part of the sublitoral zone in 2—5 fathoms, sometimes in rock-pools in the litoral zone. In other parts of the Polar Sea I have always found it sublitoral within the formation of *Laminariaceæ*. It is usually attached to small stones or shells. It seems to prefer sheltered localities and always occurs in small number at one and the same place. At Spitzbergen and Novaya Zemlya it has been collected with zoosporangia in July, at Nordlanden in June. At Finmarken it was found in a state of dissolution in the middle of September. The form *linearis* grows according to Foslie at a depth of 2—4 fathoms, attached to small stones and shells. Specimens taken in August were fully developed and provided with zoosporangia.

Geogr. Distrib. Known only from those parts of the Polar Sea which lie north of the Atlantic. The typical form has not been found abundant anywhere here. But f. *linearis* was plentiful in the locality where it was met with. The northernmost point of this species is Skansbay at Spitzbergen Lat. N. 78° 31'.

Localities: The Norwegian Polar Sea: Nordlanden (f. *typica*) pretty commonly dispersed, but scarce; Finmarken (f. *typica*) local and scarce at Öxfjord; (f. *linearis*) abundant at Russemark in Porsangerfjord.

The Greenland Sea: the west coast of Spitzbergen, rare and very scarce.

The Murman Sea: rare and very scarce on the west coast of Novaya Zemlya.

Baffin Bay: Cumberland Sound, probably rare, according to Dickie l. c.

Fam. DESMARESTIACEÆ (Thur.) Kjellm.

Pl. Scand. p. 10; Thur. in Le Jol. Liste Alg. Cherb. p. 10 et 21; lim. mut.

Gen. **Desmarestia** (Lamour.) Grev.

Alg. Brit. p. XXXIX; Lamour. Ess. p. 43; spec. excl.

Desmarestia aculeata (L.) Lamour.

l. c. p. 45. Fucus aculeatus L. Spec. Pl. Ed. 2, p. 1632.

Descr. Desmarestia aculeata J. G. Ag. Spec. Alg. 1, p. 167.

Fig. » » Harv. Phyc. Brit. t. 49.

Exsicc. » » Aresch. Alg. Scand. exsicc. N:o 87.

Syn. Desmarestia aculeata J. G. Ag. Progr. p. 2; Bidr. p. 11; Till. p. 28; Grönl. Alg. p. 110.

» » Aresch. Phyc. Scand. p, 347.

» » Ashm. Alg. Hayes, p. 96.

» » Croall, Fl. Disc. p. 457.

Syn. Desmarestia aculeata DICKIE, Alg. Sutherl. 1, p. 140; 2, p. 191; Alg. Nares, p. 6; Alg. Cumberl. p. 236.
» » EATON, List, p. 44.
» » GOBI, Algenfl. Weiss. Meer, p 67.
» » KLEEN, Nordl. Alg. p. 39.
» » KJELLM. Vinteralgv. p. 65; Spetsb. Thall. 2, p. 42; Algenv. Murm. Meer. p. 48; Kariska hafvets algv. p. 29.
» » Nyl. et Sæl. Herb. Fenn. p. 73.
» » POST. et RUPR. Ill. Alg. p. II.
» » SCHÜBELER, in Heugl. Reise, p. 317.
» » ZELLER, Zweite d. Polarf. p. 84.
» » WITTR. in Heugl. Reise, p. 284.
» inanis POST. et RUPR. l. c.
Desmia aculeata LYNGB. Hydr. Dan. p. 34.
Fucus aculeatus Pall. Reise 3, p. 34.
» » β. WG. Fl. Lapp. p. 502.
» muscoides GUNN. Fl. Norv. 2, p. 139; Cfr. Act. Nidros, p, 83, t. 7.
» virgatus GUNN. Fl. Norv. 1, p. 45.
Sporochnus aculeatus LINDBL. Bot. Not. p. 157.
» » SCHRENK, Ural. Reise, p. 547.
» » Cfr. MARTENS, Voyage Spitzb. p. 79—80.

Habitat. This species is both litoral and sublitoral on the coasts of Norway. In other parts of the Polar Sea I have only found it sublitoral, usually in 2—15 fathoms, on stony or rocky bottom, as an element of the formation of *Laminariaceæ*. It is usually attached to stones. It prefers exposed coasts, but is not wanting in sheltered localities either. Sometimes it occurs gregarious. It has not been found with reproductive organs.

Geogr. Distrib. It is known from all parts of the Arctic Sea, with the exception of the Siberian Sea. The maximum of frequency is in the Spitzbergen province. The most northern place where it has been collected is Discovery Bay in Smith Sound, Lat. N. 81° 41'.

Localities: The Norwegian Polar Sea: Nordlanden common; Tromsö amt about the town of Tromsö and at Carlsö; Finmarken at Maasö, Gjesvær, and Talvik, at all these places local but pretty abundant.

The Greenland Sea: Sabine Island on the east coast of Greenland; it is the most common and abundant *Phæozoosporacea* of Spitzbergen after *Chætopteris plumosa.*

The Murman Sea: the coast of Russian Lapland and Samoyede-land; Kolgujew Isle; the west coast of Novaya Zemlya and Waygats common and abundant; the coast of the continent at Jugor Shar.

The White Sea: one of the commonest algæ.

The Kara Sea: Kara Bay; Uddebay.

Baffin Bay: Cumberland Sound; the west coast of Greenland common according to J. G. AGARDH. Known localities: Nanortalik, Smallesund, Fiskernæs, Neuherrnhut, Hunde Islands, Jakobshavn, Godhavn, Lat. N. 73° 20', Whale Islands, Besselsbay, Discovery Bay, Smith Sound between the 78:th and 82:d latitude.

Gen. Dichloria Grev.

Alg. Brit. p. XL.

Dichloria viridis (Müll.) Grev.

l. c. p. 39. Fucus viridis Müll. Fl. Dan. t. 886.
Descr. Dichloria viridis J. G. Ag. Spec. Alg. 1, p. 164.
Fig. Desmarestia viridis Harv. Phyc. Brit. t. 312.
Exsicc. » » Aresch. Alg. Scand. exsicc. N:o 88.

Syn. Desmarestia viridis Aresch Phyc. Scand. p. 348.
» » Dickie, Alg. Sutherl. 1, p. 140.
» » Kjellm. Vinteralgv. p. 65.
» » Post. et Rupr. Ill. Alg. p. II.
Dichloria viridis J. G. Ag. Spetsb. Alg. Till. p. 27; Grönl. Alg. p. 110.
» » Gobi, Algenfl. Weiss. Meer. p. 67.
» » Kjellm. Spetsb. Thall. 2, p. 42; Algenv. Murm. Meer. p. 48.
» » Kleen, Nordl. Alg. p. 39.
Dictyosiphon foeniculaceus Zeller, Zweite d. Polarf. p. 86; sec. spec.
Fucus viridis Wg. Fl. Lapp. p. 503.

Habitat. In the Norwegian Polar Sea this species lives both within the litoral and the sublitoral zone. In the latter case it is usually found among the formation of *Laminariaceæ*, but it descends deeper down than thisdoes. In the Arctic Sea proper it is never litoral; on the contrary it is often elitoral, growing at greater depths than any other alga. At Spitzbergen it occurs at a depth of 5—150 fathoms. It is usually attached to stones, sometimes to larger algæ, and prefers exposed localities both near the coast and far out in the open sea. It grows scattered. On the coast of Spitzbergen it attains a high degree of luxuriancy. Here there are not seldom found bushing specimens, richly branched, half a metre in length. Only sterile individuals have been met with.

Geogr. Distrib. Known only from the Polar Sea to the north of the Atlantic. Its maximum of frequency is on the coasts of Spitzbergen. The most northern point where it has been taken is Treurenberg Bay on the north coast of Spitzbergen, Lat. N. 79° 56′.

Localities: *The Norwegian Polar Sea:* Nordlanden at several places, but rather scarce; Tromsö amt about the town of Tromsö; Finmarken: Maasö, Gjesvær, the south coast of Magerö, Öxfjord, Talvik, rather common and abundant.

The Greenland Sea: the east coast of Greenland; along the north and west coasts of Spitzbergen common and abundant.

The Murman Sea: the coast of Cisuralian Samoyede-land; the west coast of Novaya Zemlya common and abundant.

The White Sea: rare.

Baffin Bay: the west coast of Greenland: Hunde Islands, Godhavn, Jakobshavn.

Gen. **Phloeospora** ARESCH.

Bot. Not. 1873, p. 163.

Phloeospora subarticulata ARESCH.

l. c. p. 164.

Descr. Phloeospora subarticulata ARESCH. Bot. Not. 1876, p. 33.
Exsicc. Dictyosiphon foeniculaceus var. subarticulatus ARESCH. Alg. Scand. exsicc. N:o 104.
Syn. Phloeospora subarticulata GOBI, Algenfl. Weiss. Meer. p. 64.
» » KJELLM. Spetsb. Thall. 2, p. 40; Algenv. Murm. Meer. p. 45.

Habitat. This species grows sublitoral in the Spitzbergen province, scattered, attached to stones, both on exposed coasts and in sheltered localities. At Spitzbergen I have found in July some specimens with plenty of zoosporangia.

Geogr. Distrib. The present alga is of rare occurrence in the Polar Sea and has hitherto been found only in the eastern part of that region which lies north of the Atlantic. Even here it is everywhere scarce. The most northern point where it is known to occur is Fairhaven on the north-west coast of Spitzbergen, Lat. N. 79° 49'.

Localities: The Norwegian Polar Sea: Nordlanden according to specimens in KLEEN's herbarium; not recorded from here in Nordl. Alg.

The Greenland Sea: at two places on the north-west and west coast of Spitzbergen, local and scarce.

The Murman Sea: at N. Gusinnoi Cape on the west coast of Novaya Zemlya, rare.

The White Sea: Solowetzki Isles, probably scarce.

Phloeospora tortilis (RUPR.) ARESCH.

Bot. Not. 1876, p. 34. Scytosiphon tortilis RUPR. Alg. Och. p. 373.

Descr. Phloeospora tortilis ARESCH. l. c.
Fig. Dictyosiphon tortilis GOBI, Brauntange t. 2, fig. 12—16.
Phloeospora tortilis KJELLM. Spets. Thall. 2, t. 1, fig. 21.
Exsicc. » » ARESCH. Alg. Scand. Exsicc. N:o 413.
Syn. Dictyosiphon spec. KJELLM. Vinteralgv. p. 65.
Phloeospora Lofotensis FOSLIE, Arct. Havalg. p. 8.
» tortilis GOBI, Algenfl. Weiss. Meer. p. 64.
» » KJELLM. Spetsb. Thall. 2, p. 40; Algenv. Murm. Meer. p. 45; Kariska hafvets algv. p. 29.

Habitat. This species grows in salt or brackish water, sometimes of very slight salinity, within the litoral or sublitoral zone, always at little depth (2—5 fathoms). It flourishes both on exposed and sheltered coasts. When young, it is attached to stones, but later it occurs in greater or less masses of indefinite shape lying loose on the bottom. It is somewhat gregarious and is occasionally found in such numbers as to influence essentially the character of the vegetation. In the Arctic Sea it increases vigorously in a vegatative manner by branches and branchsystems being detached and individualized.

On the north coast of Spitzbergen this kind of propagation was especially lively during the winter. The plant also develops zoospores here in the same season; however, such organs are chiefly produced during the latter part of the summer, August and September. FOSLIE has collected specimens with zoosporangia at the end of September at Nordlanden.

Geogr. Distrib. The present species is probably circumpolar. However, it has not been observed as yet in the Siberian and American Arctic Seas. The maximum of frequency is on the coasts of Spitzbergen. The most northern point where it has been found is Musselbay on the north coast of Spitzbergen Lat. N. 79° 53'.

Localities: *The Norwegian Polar Sea:* Nordlanden; Lofoden abundant; Finmarken at Talvik local and scarce.

The Greenland Sea: common and abundant on the north and west coasts of Spitzbergen.

The Murman Sea: the west coast of Novaya Zemlya, Jugor Shar, local and scarce.

The White Sea: Solowetski Isles.

The Kara Sea: Uddebay scarce; Cape Palander rather abundant; Actinia Bay scarce.

Baffin Bay: the west coast of Greenland: Neuherrnhut.

Phloeospora pumila KJELLM.

Algenv. Murm. Meer. p. 45.

Descr. Phloeospora pumila KJELLM. l. c.
Fig. » » » t. 1, fig. 16—22.

Remark on the species. In my description of this species I have expressly stated that I could not determine with certainty how far it is related to *Phl. tortilis*, whether it is to be regarded as a distinct species or as a dwarfed form of the latter alga produced by external conditions. I have not since that time acquired any new facts for deciding this question. But in the mean time GOBI has brought forward a remarkable view with regard to the present alga. He believes he has found that it is »nichts weiter als vegetative Sprosse der Phl. tortilis» (GOBI, Algenfl. Weiss. Meer. p. 65). When I described *Phl. pumila*, I was perfectly well acquainted with the peculiar manner of vegetative propagation in *Phl. tortilis*, having given a detailed account of it in Spetsb. Thall. 2, p. 41. I am quite willing to admit that the axes and systems of axes that detach themselves from older individuals of *Phl. tortilis* and then develop independently, are very similar to *Phl. pumila*. But GOBI, in identifying these formations with *Phl. pumila*, has overlooked my express statement that *Phl. pumila* forms distinct tufts or small mats and that these are attached to a substratum by unmistakable rhizoids. I have seen plenty of such formations as are mentioned by GOBI, but I have always found them more or less firmly united or entangled to indefinite masses lying loose on the bottom, never composing distinct tufts or mats, covering or fastened to a substratum, as in the case in *Phl. pumila*. In this alga a considerable number of rhizoids combined into a cushion-shaped plexus issue from the lower part of the frond. Such a plexus is delineated in tab. 1, fig. 16 in Algenv. Murm. Meer. The rhizoids are usually unbranched, cylindrical, more or less curved, monosiphonic, but sometimes polysiphonic, and in this case club-shaped with the upper joints almost spherical; tab. 26, fig. 17.

I have not yet found any decisive reason to abandon my opinion that *Phl. pumila* is another species than *Phl. tortilis* although only feebly distinguished from it. My not having found it with zoosporangia is a matter of no consequence with regard to its claim to be considered a separate species, because such organs may either be developed at another season than when I happened to meet with this plant or replaced perhaps by vegetative propagation in the manner observed in this species. On a renewed examination of the specimens once collected I have failed to detect any hairs.

Habitat. This alga belongs to those few species which are litoral within the Arctic Sea proper. It has been found growing in sheltered places in rock-pools with sandy bottom below low-water mark, forming sometimes scattered tufts, sometimes mats of rather considerable extent. Specimens with zoosporangia as yet unknown.

Geogr. Distrib. Known as yet only from the eastern part of the Murman Sea.

Localities: The Murman Sea: Matotshin Shar to Besimannaja Bay on the west coast of Novaya Zemlya.

Gen. **Coilonema** ARESCH.

Alg. Scand. exsicc. N:o 323.

Coilonema Ekmani ARESCH.

Obs. Phyc. 3, p. 33.

Descr. Dictyosiphon (Coilonema) Ekmani l. c.

Syn. Lithosiphon Lomentariæ KLEEN, Nordl. Alg. p. 40.

Habitat. This species is a litoral alga, attached to Scytosiphon lomentarius. Specimens taken at the end of June on the north-west coast of Norway are profusely furnished with mature zoosporangia.

Geogr. Distrib. Known only from the most southern part of the Polar Sea.

Locality: *The Norwegian Polar Sea*: Bodö in Nordlanden.

Coilonema chordaria ARESCH.

f. *bahusiensis* ARESCH.

Bot. Not. 1873, p. 170.

Descr. Dictyosiphon (Coilonema) chordaria ARESCH. Obs. Phyc. 3, p. 32.

Fig. » chordaria ARESCH. Phyc. Scand. t. 8, B.

Syn. Dictyosiphon (Coilonema) Finmarkicum FOSLIE, Arct. Havalg. p. 6.

Habitat. A litoral alga usually occurring in rock-pools and attached to stones. It is to be met with on exposed as well as sheltered coasts. In exposed localities it grows larger and more luxuriant. I have collected specimens at Gjesvær in Finmarken of the length of 30 cm. In the interior of Altenfjord it was in general low but bushy. Though being often pretty gregarious, it occupies only inconsiderable spaces. In August and September it is richly provided with zoosporangia on the north coast of Norway.

Geogr. Distrib. Known only from the Atlantic region of the Polar Sea, at several places abundant. The most northern point where it has been taken is Gjesvær about Lat. N. 71°.

Localities: *The Norwegian Polar Sea*: Finmarken at Gjesvær, Talvik, and Sværholt, at the two first-named places rather abundant, but local.

Gen. **Dictyosiphon** (GREV.) ARESCH.

Bot. Not. 1873, p. 164; GREV. Alg. Brit. p. 55; char. mut.

Dictyosiphon corymbosus nob.

L. fronde fusco-flavescente, solido; axi primario distincto, ramis subcorymbosis, elongatis, simplicibus vel parce ramulosis; zoosporangiis sæpe confertis, a superficie thalli visis vulgo ellipsoideis.

f. *abbreviata* nob.

f. ramis vix semipedalibus, crassitudinem setæ excedentibus.
Tab. 26, fig. 12—15.

f. *elongata* nob.

f. ramis usque ultra pedalibus quam in præcedente tenuioribus.
Syn. Dictyosiphon hippuroides KJELLM. Algenv. Murm. Meer. p. 46; ex parte.

Description. This alga is attached by a callus and the frond becomes from half a foot (f. *abbreviata*) to more than one foot (f. *elongata*) long. It resembles *D. fœniculaceus* in colour and by its branching reminds one much of *Coilonema chordaria* f. *bahusiensis*. The main axis is distinct and beset in its whole length with rather numerous secondary axes of the first order, from half a foot to one foot long, usually unbranched, sometimes bearing one or two branches of the second order. Branches of a higher order than the second are rare. The secondary axes of the first order are almost corymbose. The principal axis is markedly attenuated towards the base, the secondary axes not at all or exceedingly little, by which marks the present species is distinguished in habit from species of *Coilonema*. The branches taper perceptibly towards the tip, more in f. *elongata*, less in f. *abbreviata*. With regard to structure this species is most nearly allied to *D. hippuroides*. The cortical layer is composed of small angular cells which in optical longitudinal section are squarish or irregularly four-sided, in the lower part of the frond rectangular. The latter are arranged in rather regular, longitudinal rows. The endochrome of the cortical cells is less plentiful and lighter in colour than in *D. hippuroides*. The thick central layer is composed of elongated cells of varying width with comparatively thin walls. In full-grown individuals there is to be found in the centre of the frond a small number of fine cell-rows resembling those of *D. hippuroides*: fig. 13. Hairs are rarely found. The zoosporangia in this as in other species of *Dictyosiphon* are generated beneath the cortical layer and at first covered by it. Seen from the surface, they are ellipsoidical or rounded-ellipsoidical in shape, usually with their longer axis in the direction of the length of the frond; occasionally they are spherical. They occur in dense masses in f. *abbreviata*, less numerous in f. *elongata*. When fully developed, their long axis (in the direction of the length of the frond) attains 60 μ.; fig. 14—15.

It seems to me that the species *D. hippuroides* would become too vague, if a plant so different from its typical form as that now described should be included within it, as I have formerly done l. c.

Habitat. The present species occurs on the upper part of the sublitoral zone, attached to stones, in 2—5 fathoms, f. *abbreviata* in exposed localities, f. *elongata* in sheltered ones. The former was found gregarious in considerable numbers. I have collected specimens with zoosporangia of f. *abbreviata* on the west coast of Novaya Zemlya in July, f. *elongata* on the north coast of Norway in September.

Geogr. Distrib. Known from the Norwegian Polar Sea and the Murman Sea. The northernmost point is Gribowa Bay on the west coast of Novaya Zemlya Lat. N. 73°.

Localities: *The Norwegian Polar Sea*: Finmarken at Talvik, f. *elongata*, local and scarce.

The Murman Sea: Gribowa Bay f. *abbreviata*.

Dictyosiphon hippuroides (LYNGB.) KÜTZ.

Tab. Phyc. 6, p. 19. Scytosiphon hippuroides LYNGB. Hydr. Dan. p. 63.

f. *typica*.

Descr. Dictyosiphon hippuroides ARESCH. Obs. Phyc. 3, p. 26.
Fig. » » Kütz. l. c. t. 52.
Exsicc. » » ARESCH. Alg. Scand. exsicc. N:o 105, 320, 321.

f. *fragilis* HARV. (nob).

Dictyosiphon fragilis HARV. in KÜTZ. Spec. Alg. p. 485.
Descr. Dictyosiphon fragilis KÜTZ. l. c.
Fig. » » » Tab. Phyc. 6, t 52.
Syn. Dictyosiphon foeniculaceus α ARESCH. Phyc. Scand. p. 369.
» hippuroides GOBI, Algenfl. Weiss. Meer. p. 66.
» » KJELLM. Spetsb. Thall. 2, p. 38; Algenv. Murm. Meer. p. 46; ex parte.
» » KLEEN, Nordl. Alg. p. 34.

Remark on the forms. This species, though less multiform in the Polar Sea than the following one, still occurs in several marked forms, of which especially one, besides the typical, seems to be characteristic. The typical form I hold to be that which ARESCHOUG has distributed under N:o 105 in Alg. Scand. Exsicc. It is the commonest. Besides this, there exists another in the Norwegian Polar Sea, in which I believe to recognize *D. fragilis* described and figured by KÜTZING. By its looser consistency and its dense, coarse branches increasing in thickness upwards, it differs considerably from

f. *typica*, and resembles rather much a *Coilonema*. The form in question is not independant, approaching sometimes so nearly to the typical that it is difficult to draw a limit between them. The form distributed in ARESCH. Alg. Scand. Exsicc. under N:o 321 is the most common in the Arctic Sea proper.

Habitat. This species is chiefly litoral in the Norwegian Polar Sea, occasionally sublitoral as in the other parts of the Arctic Sea. It does not descend to any considerable depth, growing usually epiphytical on other algæ, mostly *Chordaria*, not seldom attached to stones. It occurs on the coast of Norway in rather considerable masses and flourishes at exposed as well as sheltered points. On the coast of Norway it bears zoosporangia during all the summer, at least to the middle of September. On the coast of Spitzbergen I have collected it with such organs both in summer (August) and winter (December).

Geogr. Distrib. Known only from the Polar Sea north of the Atlantic. It is most abundant and most richly developed in the Norwegian Polar Sea. The most northern point is Musselbay on the north coast of Spitzbergen, Lat. N. 79° 53'.

Localities: The Norwegian Polar Sea: Nordlanden, common and abundant; Finmarken: Gjesvær (f. *fragilis*) abundant, Magerö Sound and Öxfjord (f. *typica*) common and abundant.

The Greenland Sea: local and scarce on the north-west and west coasts of Spitzbergen.

The Murman Sea: the coast of Russian Lapland and Samoyede-land; on the west coast of Novaya Zemlya rather common, but not abundant.

The White Sea: probably common and abundant (cp. GOBI l. c. p. 11).

Baffin Bay: the west coast of Greenland: Lichtenau.

Dictyosiphon foeniculaceus (HUDS.) GREV.

Alg. Brit. p. 56. Conferva foeniculacea HUDS. Fl. Angl. p. 164.

f. *typica.*

Descr. Dictyosiphon foeniculaceus ARESCH. Obs. Phyc. 3, p. 30.
Fig. » » » Phyc. Scand. t. 7.
Exsicc. » » » Alg. Scand. exsicc. N:o 103 et 319.

f. *flaccida* ARESCH.

Subspec. Dictyosiphon flaccidus ARESCH. Bot. Not. 1873, p. 169.

Descr. Dictyosiphon foeniculaceus var. flaccidus ARESCH. Obs. Phyc. 3, p. 31.

Syn. Chordaria flagelliformis var. J. G. AG. Grönl. Alg. p. 110; sec. spec.
Dictyosiphon foeniculaceus β ARESCH. Phyc. Scand. p. 370.
» » CROALL, Fl. Disc. p. 458.
» » DICKIE, Alg. Sutherl. 1, p. 141; Alg. Cumberl. p. 237; Alg. Nares, p. 7.
» » GOBI, Algenfl. Weiss. Meer, p. 66.
» » HARV. Fl. West-Esk. p. 49.
» » KJELLM. Spetsb. Thall. 2, p. 38, excl. subspec. 2; Algenv. Murm. Meer. p. 47.
» » KLEEN, Nordl. Alg. p. 34.

Syn. Scytosiphon foeniculaceus LYNGB. Hydr. Dan. p. 63; ex parte.
» » Nyl. et Sæl. p. 73.
» » POST. et RUPR. Ill. Alg. p. II.

Habitat. This species is litoral in the Norwegian Polar Sea, epiphytic on *Fucaceæ*, sublitoral in other parts of the Arctic Sea, chiefly attached to *Chordaria flagelliformis* or stones. It grows rather scattered both on exposed and on sheltered coasts. On the north coast of Spitzbergen there occurred through the whole winter an intermediate form between the typical one and f. *flaccida*. It retained its characteristic appearance and was in development all the time. At Nordlanden it bears zoosporangia during the summer, at Finmarken in August and September, at Spitzbergen in July and August, on the west coast of Novaya Zemlya in July.

Geogr. Distrib. Known from the Polar Sea north of the Atlantic. The maximum of frequency is on the north coast of Norway. The most northerly point where it has been found is Rawlingsbay in Smith Sound Lat. N. 80° 20′.

Localities: *The Norwegian Polar Sea:* Nordlanden, common and abundant; Finmarken, common and abundant: Magerö Sound, Maasö, Gjesvær, Öxfjord, and Talvik.

The Greenland Sea: the north and west coasts of Spitzbergen rather common, but not abundant.

The Murman Sea: the coast of Russian Lapland, the west coast of Novaya Zemlya local, but rather plentiful.

The White Sea: probably common and plentiful (cp. GOBI l. c. p. 11).

Baffin Bay: Cumberland Sound not rare; the west coast of Greenland at Tessarmiut, Lichtenau, Neuherrnhut, Godthaab, Holstenborg, Egedesminde, Hunde Islands, Jakobshavn, Claushavn, Disco Isle, Rittenbenk, Sakkak, Rawlingsbay, probably commonly disseminated along the whole coast (cp. CROALL l. c.).

Both the mentioned forms have the same extent of distribution, but according to my experience f. *flaccida* and forms most closely related with it are the most common in the North.

Dictyosiphon hispidus KJELLM.

Algeuv. Murm. Meer. p. 47.

Descr. et *Fig.* Dictyosiphon foeniculaceus subspec. hispidus KJELLM. Spetsb. Thall. 2, p. 39 et t. 2, fig. 1.
Syn. Enteromorpha ramulosa ZELLER, Zweite d. Polarf. p. 84; sec. spec.

Remark on the species. GOBI supposes this alga to be a *D. foeniculaceus* f. *flaccida* somewhat more richly branching. It certainly reminds one of f. *flaccida* by its soft, flexible, very tubulose frond, but it differs from it partly by smaller zoosporangia, partly by its peculiar branching. In the case of a genus with so feebly marked forms as *Dictyosiphon* it must however in a certain degree be arbitrary whether a given form should be considered a variety or a species. The chief reason why I think it is more differentiated than f. *flaccida* and other forms of *Dictyosiphon* is the fact that it is found with its characteristic habit in widely distant parts of the Arctic Sea.

Habitat. It grows sublitoral in 2—5 fathoms, attached to stones, scattered, both on exposed and sheltered coasts. At the beginning of August it was provided with zoosporangia in Jugor Shar.

Geogr. Distrib. Known from the Spitzbergen province. Nowhere found in great numbers. The most northern point where it certainly occurs is Treurenberg Bay on the north coast of Spitzbergen Lat. N. 79° 56′.

Localities: The Greenland Sea: the east coast of Greenland at Sabine Island (?) (cp. ZELLER l. c. p. 87); the north and west coasts of Spitzbergen local, scarce.

The Murman Sea: the west mouth of Jugor Shar, scarce.

Gen. **Lithosiphon** HARV.

Man. Ed. 2, p. 43.

Lithosiphon Laminariæ (LYNGB.) HARV.

l. c. Bangia Laminariæ LYNGB. Hydr. Dan. p. 84.

Descr. et *Fig.* Lithosiphon Laminariæ HARV. Phyc. Brit. t. 295.
Exsicc. » » ARESCH. Alg. Scand. exsicc. N:o 21.

Syn. Bangia Laminariæ POST. et RUPR. Ill. Alg. p. II.
Lithosiphon » KLEEN, Nordl. Alg. p. 39.

Habitat. Having never myself found this species in the Polar Sea, I am unacquainted by my own observations as to its habitat within that region. It occurs according to KLEEN at Nordlanden in the latter part of the summer, epiphytic on *Alaria esculenta.*

Geogr. Distrib. Known with certainty only from the Norwegian Polar Sea. It is reported however also from the eastern Murman Sea.

Localities: The Norwegian Polar Sea: Nordlanden according to KLEEN.

The Murman Sea: the coast of Novaya Zemlya according to POST. et RUPR. (cp. KJELLM. Algenv. Murm. Meer. p. 49).

Fam. AGLAOZONIACEÆ THUR.

in LE JOL. Liste Alg. Cherb. p. 14.

Gen. **Aglaozonia** ZANARD.

Sagg. p. 38.

Aglaozonia parvula (GREV.) ZANARD.

l. c. p. 38. Zonaria parvula GREV. Crypt. Fl. t. 360.

Descr. Zonaria parvula J. G. AG. Spec. Alg. 1, p. 107.
Fig. » » HARV. Phyc. Brit. t. 341.
Exsicc. Padinella parvula ARESCH. Alg. Scand. exsicc. N:r 22.

Syn. Padinella parvula KLEEN, Nordl. Alg. p. 39.

Habitat. This species lives at a depth of several fathoms, attached to shells.

Geogr. Distrib. Found only within the Norwegian Polar Sea.

Locality: Nordlanden at Fleinvær.

Fam. SPHACELARIACEÆ J. G. AG.

Alg. Med. p. 27.

Gen. **Cladostephus** (AG.) J. G. AG.

Spec. Alg. 1, p. 41; AG. Syn. Alg. p. XXV; spec. excl.

Cladostephus spongiosus (LIGHTF.) AG.

l. c. p. XXVI. Conferva spongiosa LIGHTF. Fl. Scot. p. 983

Descr. Cladostephus spongiosus J. G. AG. Spec. Alg. 1, p. 43.

Fig. » » HARV. Phyc. Brit. 138.

Exsicc. » » ARESCH. Alg. Scand. exsicc. N:o 172.

Syn. Cladostephus spongiosus ARESCH. Phyc. Scand. p. 368.

» » KLEEN, Nordl. Alg. p. 35.

Habitat. This alga occurs in the litoral zone in rock-pools. It prefers exposed localities and is somewhat gregarious, attached to stones. I do not know at what season it develops reproductive organs in the Arctic Sea. There are certainly to be found in the herbarium of the Copenhague Museum specimens from Greenland with gametangia (zoosporangia multilocularia), but the time when they were collected is not noted. Only sterile individuals are known from the Norwegian Polar Sea, where this species has been collected in summer.

Geogr. Distrib. The southern part of the Norwegian Polar Sea and Baffin Bay. Its proper range is certainly beyond the limits of the Arctic Sea. The most northern point where it has been taken is Westfjord in Nordlanden.

Localities: The Norwegian Polar Sea: Nordlanden scarce on the shores of Westfjord. *Baffin Bay:* the coast of Greenland; the special locality is not noted.

Gen. **Stupocaulon** KÜTZ.

Phyc. gener. p. 293.

Stupocaulon scoparium (L.) KÜTZ.

l. c. Conferva scoparia L. Spec. Pl. Ed. 2, p. 1635.

Descr. Stupocaulon scoparium KÜTZ. Spec. Alg. p. 466.

Fig. » » Tab. Phyc. 5, t. 96.

Syn. Stupocaulon scoparium ZELLER, Zweite d. Polarf. p. 84.

Locality: This species is said by ZELLER l. c. to have been brought home from Greenland by the second German Polar expedition, probably from Sabine Island on the east coast. I have not seen any specimens.

Gen. **Chætopteris** KÜTZ.

Phyc. gener. p. 293.

Chætopteris plumosa (LYNGB.) KÜTZ.

l. c. Sphacelaria plumosa LYNGB. Hydr. Dan. p. 103.

Descr. Chætopteris plumosa J. G. AG. Spec. Alg. 1, p. 41.

Fig. Sphacelaria plumosa HARV. Phyc. Brit. t. 87.

Chætopteris plumosa ARESCH. Obs. Phyc. 3, t. 2, fig. 4.

» » KJELLM. Spetsb. Thall. 2, t. 2, fig. 2—3.

Exsicc. Sphacelaria plumosa ARESCH. Alg. Scand. exsicc. N:o 107.

Chætopteris plumosa KJELLM. in ARESCH. Alg. Scand. exsicc. N:o 408.

Syn. Chætopteris plumosa J. G. AG. Grönl. Alg. p. 110.

» » DICKIE, Alg. Sutherl. 1, p. 141; 2, p. 191; Alg. Cumberl. p. 238; Alg. Nares, p. 7.

» » GOBI, Algenfl. Weiss. Meer. p. 63.

» » HARV. Fl. West-Esk. p 49.

» » KJELLM. Vinteralgv. p. 65; Spetsb. Thall. 2, p. 32; Algenv. Murm. Meer. p. 42, Kariska hafvets algv. p. 27.

» » KLEEN, Nordl. Alg. p. 35.

» » RUPR. Alg. Och. p. 378.

Conferva pennata WG. Fl. Lapp. p. 512; ex parte.

Sphacelaria plumosa J. G. AG. Spetsb. Alg. Progr. 2; Bidr. p. 11.

» » CROALL, Fl. Disc. p. 458.

» » EATON, List. p. 44.

» » LYNGB. l. c.

» » POST. et RUPR. Ill. Alg. p. II.

Habitat. In the Norwegian Polar Sea this species usually grows in rock-pools within the litoral zone, occasionally attached to stones within the sublitoral. In other parts of the Polar Sea it is almost without exception sublitoral, keeping however chiefly in the upper part of this zone. It is usually found in 2—5 fathoms. It is a common element of the formation of *Laminariaceæ*, but occurs also within other formations. It prefers gravelly and stony bottom. At Spitzbergen it is not seldom gregarious, occurring in rather considerable masses. It flourishes both in exposed and sheltered localities, and is even found far off from the coast. KLEEN has collected specimens with gametangia in August at Nordlanden. At Spitzbergen it bears reproductive organs, zoosporangia and gametangia, from November to May, most plentifully from the middle of November to the beginning of March. After the end of March individuals with such organs were rare. In other parts of the Arctic Sea I have only had an opportunity of examining specimens of this alga in summer and autumn, and at these seasons I have always found it sterile.

Geogr. Distrib. This species is circumpolar, common in the Arctic Sea, and very abundant at certain places. According to my observations its maximum of frequency and luxuriancy is in the eastern part of the Greenland Sea on the coasts of Spitzbergen. Its northernmost point is in Smith Sound Lat. N. 82° 27'.

Localities: The Norwegian Polar Sea: Nordlanden common; Finmarken, local and pretty scarce at Gjesvær, the south coast of Magerö, Öxfjord, and Talvik.

The Greenland Sea: common and abundant on the coasts of Spitzbergen.

The Murman Sea: the coast of Samoyede-land, Kolgujew Isle; the west coast of Novaya Zemlya and Waygats rather common, but not abundant.

The White Sea: rather rare.

The Kara Sea: Uddebay pretty plentiful; Cape Palander and Actinia Bay scarce.

The Siberian Sea: Irkaypi rather abundant and luxuriant; Koljutshin Isle, Pitlekay, and the coast east of this point, rather common and not scarce.

The American Arctic Sea: Western Eskimaux-land.

Baffin Bay: Cumberland Sound; the west coast of Greenland at Nanortalik, Lichtenau, Kakortok, Smallesund, Holstenborg, Hunde Islands, Godhavn, Lat. N. 73° 20'; Whale Island, Floeberg Beach.

Gen. **Sphacelaria** (LYNGB.) J. G. AG.

Spec. Alg. 1, p. 29; LYNGB. Hydr. Dan. p. 103; spec. excl.

Sphacelaria cirrhosa (ROTH) AG.

Syst. Alg. p. 164. Conferva cirrhosa ROTH, Cat. Bot. 2, p. 214.

Descr. Sphacelaria cirrhosa J. G. AG. Spec. Alg, p. 1, 34.
Fig. » » HARV. Phyc. Brit. t. 178.
Exsicc. » » ARESCH. Alg. Scand. exsicc. N:o 108—109.

Syn. Conferva pennata WG. Fl. Lapp. p. 512; ex parte sec. herb.
Sphacelaria cirrhata CROALL, Fl. Disc. p. 458 (?).
» cirrhosa DICKIE, Alg. Cumberl. p. 238.
» » KLEEN, Nordl. Alg. p. 36.
» pennata LYNGB. Hydr. Dan. p. 105 (?).

Habitat. Either litoral or sublitoral, attached to other algæ or to stones, growing scattered both on exposed and sheltered coasts. KLEEN has collected it with zoosporangia (and gametangia?) at Nordlanden in August.

Geogr. Distrib. Known from the Norwegian Polar Sea and Baffin Bay. It is uncertain, however, whether the *Sphacelaria cirrhosa* reported from the west coast of Greenland is really the present species and not *Sph. arctica.* Its maximum of frequency is at Nordlanden, this being also the most northern place where it occurs.

Localities: The Norwegian Polar Sea: Nordlanden common.

Baffin Bay: Cumberland Sound; the coast of Greenland according to CROALL and LYNGBYE l. c.

Sphacelaria arctica HARV.

sec. J. G. AG. Grönl. Alg. p. 110; Cfr. DICKIE, Alg. Cumberl. p. 238.

Descr. et *Fig.* Sphacelaria arctica KJELLM. Spetsb. Thall. 2, p. 34 et t. 2, fig. 4—6.

Syn. Conferva pennata WG. Fl. Lapp. p. 512; ex parte sec. herb.
Spacelaria arctica J. G. AG. Grönl. Alg. p. 110.
» » DICKIE, Alg. Cumberl. p. 238.
» » GOBI, Algenfl. Weiss. Meer. p. 62.
» » KJELLM. Vinteralgv. p. 65; Spetsb. Thall. 2, p. 34; Algenv. Murm. Meer. p. 43; Kariska hafvets algv. p. 28.
» » KLEEN, Nordl. Alg. p. 36.
» cirrhosa POST. et RUPR. Ill. Alg. p. II.
» heteronema POST. et RUPR. l. c. Cfr. GOBI l. c. p. 62, in adnot.

Habitat. This species is litoral in the Norwegian Polar Sea, but in other parts of the Arctic Sea it occurs on the upper part of the sublitoral zone, usually within the formation of *Laminariaceæ*, in 2—5 fathoms water. It is attached to other algæ or stones, and grows scattered. Apparently it prefers rather exposed localities. At Spitzbergen I have found it with zoosporangia in December, January, and April, and with gametangia in February, March, and April. In other parts of the Arctic Sea I have met only with sterile specimens.

Geogr. Distrib. This alga is extensively distributed in the Arctic Sea, probably circumpolar. The northernmost point where it occurs is Low Island on the north coast of Spitzbergen Lat. N. 80° 20′. Its maximum of frequency is within the Spitzbergen province. In the greatest luxuriancy it has been found in the Kara Sea.

Localities: The Norwegian Polar Sea: Nordlanden not rare; Finmarken local and scarce, at Gjesvær, the south coast of Magerö, and Öxfjord.

The Greenland Sea: the north and west coasts of Spitzbergen common and rather abundant.

The Murman Sea: the coast of Cisuralian Samoyede-land, Kolgujew Isle, the west coast of Novaya Zemlya and Waygats rather common, even abundant in some places.

The White Sea: cp. GOBI l. c.

The Kara Sea: Uddebay rather abundant and very luxuriant; Cape Palander scarce; Actinia Bay rather abundant; Cape Tshelyuskin rare.

The Siberian Sea: Koljutshin Isle; Pitlekay, and the coast east of this point, rather common and abundant.

Baffin Bay: Cumberland Sound, not rare; the west coast of Greenland at Neuherrnhut and Godhavn.

Sphacelaria olivacea (DILLW.) AG.

Spec. Alg. 2, p. 30. Conferva olivacea DILLW. Brit. Conf. p. 57.

Descr. Sphacelaria olivacea J. G. AG. Spec. Alg. 1, p. 30.
Fig. » radicans HARV. Phyc. Brit. t. 189.
Exsicc. » olivacea KJELLM. in ARESCH. Alg. Scand. exsicc. N:o 410.
Syn. Sphacelaria olivacea KLEEN, Nordl. Alg. p. 36.

Habitat. A litoral alga, growing gregarious on stones or poles, usually in sheltered localities. From the Polar Sea, where it has as yet been collected only in summer, I have not seen any but sterile specimens. It probably bears reproductive organs in winter here as farther to the south, for inst. on the coast of Sweden.

Geogr. Distrib. It belongs only to the most southern part of the Polar Sea. At certain places on the coast of Finmarken it is still plentiful. The northernmost point where it has been found is Talvik in Finmarken about Lat. N. 70°.

Localities: The Norwegian Polar Sea: Nordlanden at Bodö and on both sides of Westfjord, rather common; Finmarken at Talvik abundant.

Baffin Bay: the coast of Greenland according to J. G. AG. Spec. Alg. p. 31, and specimens in the herbarium of the Copenhague Museum with no special locality noted.

Fam. ECTOCARPACEÆ (AG.) THUR.

in LE JOL. Liste Alg. Cherb. p. 14 et 21; AG. Syst. Alg. p. XXX; lim. mut.

Gen. **Isthmoplea** KJELLM.

Algenv. Murm. Meer. p. 30.

Isthmoplea sphærophora (HARV.) KJELLM.

I. c. Ectocarpus sphærophorus HARV. Engl. Fl. 5, p. 326.

Descr. Capsicarpella sphærophora KJELLM. Skand. Ect. och Tilopt. p. 20.
Fig. » » » » » » » t. 1. fig. 2.
Exsicc. » » ARESCH. Alg. Scand. exsicc. N:o 414.

Syn. Capsicarpella sphærophora KLEEN, Nordl. Alg. p. 36.
Isthmoplea sphærophora GOBI, Algenfl. Weiss. Meer. p. 58.

Habitat. Litoral, attached to other algæ, as *Rhodomela lycopodioides*, *Polysiphonia fastigiata*, *Gigartina mamillosa*, *Ptilota plumosa*, and *Pt. elegans*. In the Polar Sea it does not occur anywhere in greater masses. Apparently it prefers exposed localities, neither KLEEN nor myself having met with it in the interior of deep bays. It bears reproductive organs in July and August on the coast of Norway.

Geogr. Distrib. Found in the Norwegian Polar Sea and the White Sea. It is not properly an arctic alga, decreasing both in frequency and luxuriancy towards the North. Its most northern point is Gjesvær in Finmarken about Lat. N. 71°.

Localities: The Norwegian Polar Sea: Nordlanden common and abundant; Tromsö amt at the town of Tromsö; Finmarken local and rather scarce at Maasö and Gjesvær.

The White Sea: cp. GOBI l. c. p. 12.

Gen. **Ectocarpus** (LYNGB.) KJELLM.

Skand. Ect. och Tilopt. p. 34; LYNGB. Hydr. Dan. p. 130; char. mut.

Ectocarpus confervoides (ROTH) LE JOL.

Liste Alg. Cherb. p. 75. Ceramium confervoides ROTH, Cat. Bot. 1, p. 151.

f. *arcta* KÜTZ. (KJELLM.)

Ectocarpus arctus KÜTZ. Phyc. gener. p. 289.

Descr. Ectocarpus confervoides f. arcta KJELLM. Skand. Ect. och Tilopt. p. 71.
Fig. Corticularia arcta KÜTZ. Tab. Phyc. 5, t. 80.
Exsicc. Ectocarpus pseudosiliculosus CROUAN, Exsicc. N:o 27.

f. *siliculosa* DILLW. (KJELLM.)

Conferva siliculosa DILLW. Conf. p. 69.

Descr. Ectocarpus confervoides f. siliculosa KJELLM. l. c. p. 73.
Fig. » siliculosus LYNGB. Hydr. Dan. t. 43, fig. c.
Exsicc. » » ARESCH. Alg. Scand. exsicc. N:o 176.

f. *spalatina* KÜTZ. (KJELLM.)

Ectocarpus spalatinus KÜTZ. Phyc. gener. p. 288.

Descr. Ectocarpus confervoides f. spalatina KJELLM. l. c. p. 76.
Fig. » spalatinus KÜTZ. Tab. Phyc. 5, t. 63.

f. *typica* nob.

Descr. Ectocarpus confervoides s. s. KJELLM. l. c. p. 77.
Fig. » patens KÜTZ. Tab. Phyc. 5, t. 67.
Exsicc. » litoralis var. ARESCH. Alg. Scand. exsicc. N:o 111.

f. *penicillata* AG.

Syst. Alg. p. 162.

Descr. Ectocarpus confervoides f. penicillata KJELLM. l. c. p. 80.
Fig. Corticularia Nægeliana KÜTZ. Tab. Phyc. 5, t. 81.
Exsicc. Ectocarpus siliculosus ARESCH. Alg. Scand. exsicc. N:o 112.

f. *hiemalis* CROUAN (KJELLM.).

Ectocarpus hiemalis CROUAN, Alg. Finist. N:o 26; saltem ex parte.

Descr. Ectocarpus confervoides f. hiemalis KJELLM. l. c. p. 83.

Syn. Conferva litoralis WG. Fl. Lapp. p. 513; ex parte.
» siliculosa SOMMERF. Suppl. p. 193.
Ectocarpus confervoides KJELLM. Spetsb. Thall. 2, p. 35; Algenv. Murm. Meer. p. 44.
» » KLEEN, Nordl. Alg. p. 37.
» Nægelianus, GOBI, Algenfl. Weiss. Meer. p. 61.
» siliculosus J. G. AG. Spetsb. Alg. Progr. p. 2; Bidr. p. 11.
» » CROALL, Fl. Disc. p. 458.
» » DICKIE, Alg. Nares. p. 7.

Remark on the synonymy. I have been blamed by GOBI because I have identified *Corticularia Nægeliana* KÜTZ. with that form of *Ectocarpus* which ARESCHOUG has distributed under the name of *E. siliculosus* in Alg. Scand. exsicc. N:o 112. Cp. GOBI Algenfl. Weiss. Meer. p. 61 note 3. As far as I can judge by dried specimens, there is nothing to prevent such an identification. In branching and in the shape of the gametangia ARESCHOUG'S alga resembles the plant figured by KÜTZING so closely that they must be regarded as belonging to the same type. The form passes on the one side into *E. arctus* or *E. furcatus* KÜTZ., on the other side into *E. siliculosus* KÜTZ.

and the forms distributed in Alg. Scand. exsicc. N:o 176 and 111. Undoubtedly *E. siliculosus*, f. *penicillata* AG. belongs also to the same group of forms as *Corticularia Nægeliana*.

Habitat. The present species as here understood is usually litoral in the Norwegian Polar Sea, always sublitoral in the other parts of the Arctic Sea, as far as my observations go. It is attached sometimes to other algæ, sometimes to stones. Certain forms of it sometimes occur gregarious in large masses on the coast of Norway. The species lives both on exposed and sheltered coasts. It bears reproductive organs, usually gametangia, in summer.

Geogr. Distrib. It belongs properly to the Atlantic region of the Arctic Sea and attains here its maximum of frequency, occurring in several different forms. Within the proper Arctic region I have always found it very scarce, usually dwarfed, always in the typical form or such forms as are most nearly allied to this. Its most northerly point is Besselsbay in Smith Sound Lat. N. 81° 7'.

Localities: *The Norwegian Polar Sea*: Nordlanden, all the forms recorded, ff. *typica*, *penicillata*, and *hiemalis* being most common; Tromsö amt near the town of Tromsö, f. *typica;* Finmarken: Maasö, Gjesvær, Öxfjord, and Talvik, ff. *arcta*, *typica*, and *penicillata*, the two last-named being most common and abundant.

The Greenland Sea: Skansbay at Spitzbergen, f. *typica*, scarce.

The Murman Sea: the coast of Russian Lapland and Cisuralian Samoyede-land, f. *penicillata* or forms most nearly approaching this; the west coast of Novaya Zemlya, f. *typica*, rare.

The White Sea: cp. GOBI l. c.

Baffin Bay: Found floating off Holstenborg. Said to have been collected at Besselsbay by the English expedition under Nares.

Ectocarpus pygmæus ARESCH.

in KJELLM. Skand. Ect. och Tilopt. p. 85.

Descr. Ectocarpus pygmæus KJELLM. l. c.

Syn. Ectocarpus pygmæus KLEEN, Nordl. Alg. p. 38.

Habitat. Found growing on shelly bottom at a depth of some fathoms. It has gametangia in July and August on the coast of Norway.

Geogr. Distrib. Known only from the Norwegian Polar Sea.

Locality: Nordlanden at Fleinvær and two or three points at Lofoten. Cp. KLEEN, l. c.

Ectocarpus draparnaldioides CROUAN.

Alg. Finist. N:o 24.

Descr. Ectocarpus draparnaldioides KJELLM. Skand. Ect. och Tilopt. p. 37.

Exsicc. » » CROUAN l. c.

Syn. Ectocarpus draparnaldioides KLEEN, Nordl. Alg. p. 38.

Habitat. Sublitoral, attached to *Laminaria digitata.* It bears gametangia at Nordlanden in July and August.

Locality: *The Norwegian Polar Sea*: Nordlanden at Lofoten.

Ectocarpus fasciculatus HARV.

Man. p. 40; ex parte.

Descr. Ectocarpus fasciculatus KJELLM. Skand. Ect. och Tilopt. p. 89.
Fig. » » HARV. Phyc. Brit. t. 273.
Exsicc. » » ARESCH. Alg. Scand. exsicc. N:o 114.
Syn. Ectocarpus fasciculatus KLEEN, Nordl. Alg. p. 38.

Habitat. This species grows in the litoral zone, scattered, on exposed coasts, usually attached to other algæ, as species of *Monostroma*, occasionally to stones. It bears gametangia in the Norwegian Polar Sea in July and August.

Geogr. Distrib. It belongs to the Atlantic region of the Arctic Sea, being rare in the northern part, common in the southern. Its northernmost point is Gjesvær in Finmarken about Lat. N. 71°.

Localities: *The Norwegian Polar Sea*: Nordlanden common and abundant; Finmarken local and scarce at Maasö and Gjesvær.

Ectocarpus tomentosus (HUDS.) LYNGB.

Hydr. Dan. p. 132. Conferva tomentosa HUDS. Fl. Ang. p. 594.

Descr. Ectocarpus tomentosus KJELLM. Skand. Ect. och Tilopt. p. 63.
Fig. » » HARV. Phyc. Brit. t. 182.
Exsicc. » » ARESCH. Alg. Scand. exsicc. N:o 110.
Syn. Ectocarpus tomentosus KLEEN, Nordl. Alg. p. 37.

Habitat. Litoral, attached to other algæ, usually *Fucaceæ.* It is often gregarious and lives chiefly in localities exposed to a heavy surge. In July and August it has been found with gametangia.

Geogr. Distrib. Known only from the southern part of the Atlantic region of the Polar Sea.

Locality: *The Norwegian Polar Sea*: Nordlanden common and abundant.

Ectocarpus ovatus KJELLM.

Spetsb. Thall. 2, p. 35.

Descr. et *Fig.* Ectocarpus polycarpus KJELLM. Skand. Ect. och Tilopt. p. 93, et t. 1, fig. 5.
Syn. Ectocarpus ovatus KJELLM. Spetsb. Thall. 2, p. 35.
» polycarpus KLEEN, Nordl. Alg. p. 38.

Habitat. This species is litoral in the Norwegian Polar Sea, attached to *Corallina officinalis*, sublitoral and fastened to other algæ in the Greenland Sea. It grows scattered. It has been collected with gametangia at Spitzbergen in July, at Finmarken in September, at Nordlanden in August.

Geogr. Distrib. It is nowhere common within the Polar Sea and has only a restricted range. The most northerly place where it has been found is Skansbay on the west coast of Spitzbergen Lat. N. 78° 31'.

Localities: *The Norwegian Polar Sea*: Nordlanden at Fleinvær; Finmarken scarce at Öxfjord.

The Greenland Sea: the west coast of Spitzbergen at Skansbay, local and scarce.

Ectocarpus Lebelii ARESCH.

? f. borealis KJELLM.

Skand. Ect. och Tilopl. p. 57.

Descr. Ectocarpus Lebelii (?) f. borealis KJELLM. l. c.

Syn. Ectocarpus Lebelii (?) f. borealis KLEEN, Nordl. Alg. p. 37.

Habitat. This species has been found only once. It grew on *Scytosiphon lomentarius* and, when collected in July, was furnished with gametangia.

Locality: *The Norwegian Polar Sea*: Gjesvær at Nordlanden.

Ectocarpus terminalis KÜTZ.

Phyc. gener. p. 236.

Descr. et *Fig.* Ectocarpus terminalis KJELLM. Skand. Ect. och Tilopl. p. 54 et t. 2, fig. 7.

Syn. Ectocarpus terminalis KLEEN, Nordl. Alg. p. 37.

Habitat. This is a litoral alga, epiphytic on *Callithamnia* and *Cladophora*, growing scattered but not seldom in rather considerable numbers, both on exposed and sheltered coasts. It has been taken with gametangia on the arctic coast of Norway, in July, August, and September.

Geogr. Distrib. Known only from the Atlantic region of the Polar Sea, where it is rather common and plentiful. Its northernmost point is Gjesvær in Finmarken about Lat. N. 71°.

Localities: *The Norwegian Polar Sea*: Nordlanden not rare; Finmarken, rather common and abundant at Gjesvær and Öxfjord.

Ectocarpus reptans CROUAN.

Flor. p. 161.

Descr. et *Fig.* Ectocarpus reptans KJELLM. Skand. Ect. och Tilopl. p. 52 et t. 2, fig. 8.

Syn. Ectocarpus reptans KLEEN, Nordl. Alg. p. 36.

Habitat. In this respect it agrees with the preceding species, together with which it is often found growing. Taken with gametangia in July and August.

Geogr. Distrib. Known only from the arctic coast of Norway. Its northernmost point is Gjesvær about Lat. N. 71°.

Locality: *The Norwegian Polar Sea*: Nordlanden pretty common; Finmarken local and scarce at Gjesvær.

Gen. Pylaiella BORY.

Dict. Class. 4, p. 393.

Pylaiella litoralis (L.) KJELLM.

Skand. Ect. och Tilopt. p. 99. Conferva litoralis L. Spec. Plant. p. 1165; ex parte.

Descr. Pylaiella litoralis KJELLM. l. c.
Fig. Ectocarpus litoralis HARV. Phyc. Brit. t. 197.
» » KÜTZ. Tab. Phyc. 5, t. 76.
» compactus » » » » » »
Exsicc. » firmus ARESCH. Alg. Scand. exsicc. N:o 24.
» » f. vernalis ARESCH. Alg. Scand. exsicc. N:o 173.
» » var. rupincola » » » » » 113.
Syn. Conferva litoralis GUNN. Fl. Norv. 2, p. 106 (?).
» » WG. Fl. Lapp. p. 513; ex parte.
Ectocarpus crinitus CROALL, Fl. Disc. p. 458 (?).
» firmus WITTR. in Heugl. Reise, p. 284.
» litoralis J. G. AG. Spetsb. Alg. Progr. p. 2; Bidr. p. 11; Till. p. 28.
» » DICKIE, Alg. Sutherl. 1, p. 141; Alg. Cumberl. p. 238.
» » NYL. et SÆL. Herb. Fenn. p. 75.
» » POST. et RUPR. Ill. Alg. p. II.
» ochraceus, ZELLER, Zweite d. Polarf. p. 84.
Pylaiella flexilis RUPR. Alg. Och. p. 385.
» litoralis GOBI, Algenfl. Weiss. Meer. p. 59.
» » KJELLM. Vinteralgv. p. 65; Spetsb. Thall. 2, p. 36; ex parte; Algenv. Murm. Meer. p. 44; Kariska hafvets algv. p. 28.
» » KLEEN, Nordl. Alg. p. 38.
» Nordlandica RUPR. l. c. p. 386.
» pyrrhogon » » » 385.
» saxatilis » » » 386.

Remark on the species. The present alga occurs, in the Polar Sea as well as in the Atlantic, in a number of forms differing from one other in size, mode of growth, colour, and branching. As I have not succeeded in drawing any limits between them, I am obliged to place them all under one name. The figures quoted and the specimens distributed in ARESHOUG's Alg. Scand. exsicc., which I have cited, will show some of the forms that I understand by the present name.

Habitat. This species is usually litoral, occasionally sublitoral in the Norwegian Polar Sea, in the other parts of the Polar Sea it is almost constantly sublitoral. It descends to a depth of several fathoms, growing epiphytic on other algæ or fastened to stones, often gregarious in considerable masses, both on exposed and sheltered coasts.

On the north coast of Norway it has been found with zoosporangia and gametangia from July to September; on the coast of Spitzbergen I have seen specimens with reproductive organs in all the months of the year with the exception of May and October. They were however rare during the winter. On the west coast of Novaya Zemlya I have collected individuals with such organs in July, in the Kara Sea in July and August.

Geogr. Distrib. The present species is probably circumpolar. It is however not known as yet from the American Arctic Sea. The maximum of frequency is in the Norwegian Polar Sea. The most northerly point is Low Island on the north coast of Spitzbergen Lat. N. 80° 20'.

Localities: The Norwegian Polar Sea: Nordlanden common and abundant; Tromsö amt common and abundant about the town of Tromsö, at Renö and Carlsö; Finmarken common and abundant at Maasö, Gjesvær, the south coast of Magerö, Öxfjord, and Talvik.

The Greenland Sea: the east coast of Greenland at Sabine Island; on the coasts of Spitzbergen common, but not abundant; Beeren Eiland.

The Murman Sea: the coast of Russian Lapland and Cisuralian Samoyede-land; the west coast of Novaya Zemlya and Waygats rather common but not abundant.

The White Sea: common and abundant.

The Kara Sea: Uddebay rather abundant; Actinia Bay and Cape Tshelyuskin scarce.

The Siberian Sea: at Pitlekay rather abundant.

Baffin Bay: Cumberland Sound common; the west coast of Greenland: Tessarmiut, Nanortalik, Lichtenau, Kakortok, Fiskernæs, Hunde Islands, Lat. N. 73° (floating).

Pylaiella varia nob.

P. thallo racemose-ramoso; ramis sub angulo fere recto egredientibus duplicis generis, longioribus et brevissimis; his e singula bis denis cellulis constructis, omnibus vel saltem nonnulis, vulgo divisione vario modo peracta, in zoosporangia vario modo disposita mutatis. Tab. 27, fig. 1—12.

Syn. Ectocarpus Landsburgii DICKIE, Alg. Sutherl. 1, p. 142.
» Vidovichii WITTR. in Heugl. Reise, 3, p. 284.
Pylaiella litoralis KJELLM. Spetsb. Thall. 2, p. 36; ex parte.

Description. In the collections af algæ brought home by HEUGLIN from the coast of Spitzbergen there was found a peculiar *Ectocarpacea*, which was determined as *Ectocarpus Vidovichii.* Having myself met with the same alga on the coast of Spitzbergen, I found that it belonged to the genus *Pylaiella.* I held it to be an abnormally developed form of the multiform *P. litoralis.* Both the specimens of HEUGLIN and those collected by myself were in a stage of development that prevented their being certainly determined. I have later found numerous, fully developed individuals of the same alga on the north coast of Norway. These differ so much in many respects from the common forms of *Pylaiella* that I cannot but hold them to belong to another species. I propose to give this new species the name of *varia*, by which I mean to denote the great variations exhibited in the development of the zoosporangia. The alga forms loosely complicated mats of a dark olive-brown, lying free on the bottom or hanging on larger algæ. I have not found attached specimens. The frond is repeatedly racemosely branched with distinct main axis with branches of at least four orders. The branches are of two kinds: long branches with many cells and short ones with from one to ten cells. The former are few in number and issue partly alone partly in pairs opposite to each other. The short branches are numerous; by these the present species is easily recognized from *P. litoralis*. In long parts of the frond such a branch issues from every cell. They are always isolated, and issue at a right or nearly right angle. The long branches are somewhat attenuated towards the tip and generally end in some long hair-cells. The short branches are cylindrical or slightly claviform, with an apical cell rich in endochrome, which cell is finally transformed into a zoosporangium.

The cells of the frond are usually short, cylindrical or slightly tun-shaped, equally or even twice as long as thick. Those cells which give rise to a long branch are commonly short. If a short branch issues from a cell which is more long than thick, as is often the case, the branch is almost always placed at the middle of the longer wall. The thickness of the principal axis amounts to about 50 μ. With the exception of the hair-cells, all the cells are rich in granular, equally distributed endochrome (fig. 1).

The development of the zoosporangia and their arrangement dependent thereon are subject to very great variations. I have represented the most common of these modes of development fig. 2—12. The zoosporangia are sometimes arranged as in *P. litoralis* (fig. 2—3). A modification of this type is exhibited by fig. 4, in which all the cells, not only the ultimate ones, are transformed into zoosporangia. Sometimes it is only the apical cell that becomes a zoosporangium, the branch be composed of one or more cells (fig. 9). It also happens often that a greater or less number of cells are divided by longitudinal or oblique walls, and that it is the secondary cells, produced by this division, that are developed into zoosporangia (fig. 5, 7, 8, 10, 12). In this case the division of the cells and the development of the zoosporangia takes place sometimes in such a manner that the zoosporangia become arranged in whorls (fig. 11). I have not observed the bursting forth of the zoosporangia.

Habitat. This alga is found sublitoral in 2—3 fathoms on exposed coasts. It occurs sometimes at Norway in considerable masses, and has been collected here with almost ripe zoosporangia at the beginning of August.

Geogr. Distrib. It belongs both to the Atlantic and the arctic region of the Polar Sea and appears to have a wide range in the latter. It is most abundant however in the Norwegian Polar Sea. Its most northern point is Musselbay on the north coast of Spitzbergen Lat. N. 79° 53'.

Localities: The Norwegian Polar Sea: Finmarken pretty common and plentiful at Maasö.

The Greenland Sea: Dunö and Musselbay on the coast of Spitzbergen.

The Siberian Sea: rather abundant in Actinia Bay.

Baffin Bay: Hunde Island on the west coast of Greenland, in case *Ectocarpus Landsburgii* DICKIE belongs to the present species, as is most probable.

Pylaiella nana nob.

P. thallo maculam minutam, diametro 1 mm. in aliis algis formante, e filis repentibus, in membranam fere confertis, fila verticalia, cylindrica emittentibus contexto; zoosporangiis seriatis, gametangiisque sæpe ramosis, cylindricis vel elongato-conicis fila verticalia terminantibus. Tab. 27, fig. 13—17.

Description. In its vegetative system this alga much resembles an *Ectocarpus terminalis* or *E. reptans.* As in these species the frond is formed of branching cell-rows creeping on other algæ, densely compressed, throwing out at right angles unbranched or very little branched cell-rows. These vertical cell-rows are of three kinds. Some of them are almost quite cylindrical, composed downwards of cells rich in endochrome, upwards of cells that are poor or altogether wanting in endochrome. Other rows end in organs that resemble the gametangia of *Ectocarpi* and undoubtedly are of that nature. Those of the third kind are usually slightly claviform, ending in a simple or branching row of spherical or short tun-shaped cells, which possess a dense, very profuse endochrome. I am not quite clear as to the nature of these rows of cells, but they resemble the rows of young zoosporangia in a *Pylaiella* so much that I cannot but regard them as yet as organs of that kind..

The procumbent filaments are composed of cells that in longitudinal section are rectangular, or square, or from five-angular to polygonal, rich in endochrome and with rather thick walls, like the other cells of the frond. The vertical cell-rows issue from the inner part of the disk formed of those densely compact filaments. Those vertical rows which are sterile are about 750 μ. in length and 15 μ. in thickness, usually unbranched, composed of cylindrical cells that are from one to one and a half time as long as thick.

The gametangia are almost always branching, though not much, and elongated-conical, 80 μ. long, 20—25 μ. thick. The gamets pass out through a hole formed in the top of every gametangium (fig. 14). The chains of zoosporangia are usually unbranched, composed of from three to more zoosporangia (fig. 15—16), sometimes rather much branching. Every branch is formed of one or more zoosporangia, which are different in shape, usually almost spherical, short cylindrical or tun-shaped. Their walls are thicker than those of the vegetative cells. I have not seen zoosporangia with fully developed zoospores (fig. 15—17).

If this alga is really a *Pylaiella*, it is sharply distinguished from all hitherto known species of this genus by its smallness, by the structure of its vegetative system, and the shape of its gametangia.

Habitat. It grows scattered within the litoral zone in exposed places, epiphytic on *Cladophoræ*, and has been found with gametangia and sporangia in August.

Locality: The Norwegian Polar Sea: Gjesvær in Finmarken.

Gen. **Myriotrichia** Harv.

in Hook. Journ. Bot. 1, p. 300.

Myriotrichia filiformis Harv.

Man. p. 44.

Descr. Myriotrichia filiformis J. G. Ag. Spec. Alg. 1, p. 14.
Fig. » » Harv. Phyc. Brit. t. 156.

Syn. Myriotrichia filiformis Kleen, Nordl. Alg. p. 39.

Habitat. Epiphytic on litoral algæ, as *Corallina officinalis, Asperococcus echinatus, Scytosiphon lomentarius, Dictyosiphon hippuroides* a. o., which are sometimes completely covered by it. I have met with it only on exposed coasts. It bears zoosporangia in July and August on the arctic coast of Norway.

Geogr. Distrib. It belongs to the Atlantic region of the Polar Sea. Its northernmost point is Öxfjord in Finmarken about Lat. N. 70°.

Localities: The Norwegian Polar Sea: Nordlanden at Gjesvær; Finmarken local, but rather abundant at Öxfjord.

(?) Gen. **Gleothamnion** Cienk. [1])

Bericht. p. 25.

Gleothamnion palmelloides Cienk.

l. c.

Descr. et *Fig.* Gleothamnion palmelloides Cienk, l. c. p. 25 et t. 1, fig. 12—16, t. 2, fig. 17—19.

Habitat. Cp. Cienk. l. c.
Locality: The White Sea.

Series **CHLOROPHYLLOPHYCEÆ** (Rabenh.) Wittr.

Pithoph. p. 42; Rabenh. Fl. Eur. Alg. 3, p. 1 lim. mut.

Fam. CHÆTOPHORACEÆ (Harv.) Wittr.

Pl. Scand. p. 15. Chætophoroideæ Harv. Man. p. 10; ex parte.

Gen. **Chætophora** Schrank.

Bair. Fl. Cfr. Wittr. Gotl. och Öl. Alg. p. 25.

Chætophora maritima Kjellm.

Spetsb. Thall. 2, p. 51.

Descr. et *Fig.* Chætophora maritima Kjellm. l. c. et t. 5, fig. 15—16.

Syn. Chætophora maritima Kjellm. l. c. et Algenv. Murm. Meer. p. 53.

---

[1]) By the sign of interrogation I have meant to denote that I am uncertain if this genus belongs to the family of *Ectocarpaceæ*.

Habitat. The present species grows litoral, on exposed coasts, forming in company with *Calothrix scopulorum* a thin layer on stones and rocks above low-water mark. It is gregarious and occurs sometimes in considerable masses. As yet found only sterile.

Geogr. Distrib. Known only from the Polar Sea north of the Atlantic. Its northernmost point is Fairhaven on the north-west coast of Spitzbergen. Lat. N. 79° 49'.

Localities: The Greenland Sea: the north-west and west coast of Spitzbergen local, but at certain places abundant.

The Murman Sea: the west coast of Novaya Zemlya local and scarce.

Chætophora pellicula nob.

Ch. crustam membranaceam, 200—300 μ. crassam, e viride flavescentem formans, crusta e filis repentibus dense confertis, fila adscendentia plus minus ramosa, pilifera, muco uberiore cohibita emittentibus; cellulis vegetativis forma varia, 10—20 μ. longis, 5—10 μ. crassis, membrana crassa; cellulis zoosporigenis subcylindricis, 15—20 μ. longis, 8—12 μ. crassis. Tab. 31, fig. 4—7.

Description. This alga forms a thin slimy membrane 200—300 μ. in thickness of a light green or yellowish green colour. This membrane is composed of branching cell-rows imbedded in slime, the leading axes and some secondary axes of which are densely compressed and horizontally expanded on the substratum; other secondary axes rise upwards, issuing at a greater or smaller angle. The branching of the cell-rows varies considerably, being sometimes very scarce, sometimes so profuse that almost every cell puts forth a branch. The branches of the decumbent cell-rows are unilateral, those of the rising ones issue from many sides (fig. 4—5). The hairs are long and rather numerous. The vegetative cells are very variable in form, sometimes almost spherical, sometimes square, rectangular, elliptic, irregularly three-, four-, or five-angular in optical longitudinal section, 5—10 μ. thick by 10—12 μ. long. Their membrane is thick, the endochrome abundant. The zoosporogenic cells are cylindrical, somewhat bulging, 15—20 μ. long, 8—12 μ. thick. The opening is about in the middle of the longer wall (fig. 7).

Besides by zoospores, the present species is propagated by resting cells, produced by the transformation of vegetative cells. These rest either in the parent plant or in other plants together with which it occurs. Thus cells of that kind were found very numerous in the frond of *Lithoderma lignicola* growing on *Ch. pellicula.* After becoming free, they increase considerably in size, their contents are augmented and their membrane thickened. Their further development is unknown to me.

This species is closely related to *Ch. maritima*, but it differs essentially from it in branching and by the crustaceous form of the frond.

Habitat. It has been found in sheltered localities within the litoral zone, on old decaying wood, growing together with *Lithoderma lignicola* and *Calothrix Harveyi.* Specimens collected at the beginning of September bore scarce zoosporangia.

Locality: The Norwegian Polar Sea: Finmarken at Talvik.

Fam. ULVACEÆ (Ag.) Wittr.

Pl. Scand. p. 17; Ag. Syst. Alg. p. XXX; lim. mut.

Gen. **Enteromorpha** (Link) Harv.

Man. p. 173; Link. Epist. p. 5; ex parte.

Enteromorpha clathrata (Roth) Grev.

Alg. Brit. p. 181. Conferva clathrata Roth, Cat. Bot. 3, p. 175.

Descr. Enteromorpha clathrata Ahln. Enterom. p. 43.

f. *Agardhiana* Le Jol.

Liste Alg. Cherb. p. 49.

Descr. Ulva clathrata α Agardhiana Le Jol. l. c.

Syn. Enteromorpha clathrata Dickie, Alg. Cumberl. p. 239 (?); Alg. Nares p. (7).
» » Kleen, Nordl. Alg. p. 40.

Habitat. It is usually litoral, sometimes sublitoral according to Kleen, attached to stones. In general it is gregarious, but does not occur in any greater numbers together. I have found it only on sheltered coasts. It would seem however from the account of Kleen as if it should live at Nordlanden even in exposed localities. Only sterile specimens are known from the Polar Sea.

Geogr. Distrib. This species is known to occur in the Norwegian Polar Sea and reported from Baffin Bay. In other parts of the Arctic waters it has not been met with. It seems to me probable that Dickie's statement of its occurrence in Baffin Bay in the high North depends on fine, richly branching forms of *E. compressa* having been somehow confounded with *E. clathrata.* However, not having myself seen the specimens determined by Dickie, I cannot decidedly dispute the correctness of his statement. The present species is rather abundant, but not common, in the Norwegian Polar Sea. Its northernmost point would be Port Sheridan in Smith Sound Lat. N. 82° 27'.

Localities: The Norwegian Polar Sea: Nordlanden common and abundant; Finmarken pretty plentiful at Öxfjord and Talvik. I have not observed it elsewhere.

Baffin Bay: Cumberland Sound, Hayes Sound, Buchenau Strait, and Port Sheridan according to Dickie.

Enteromorpha intestinalis (L.) Link.

Epist. p. 5. Ulva intestinalis L. Spec. Pl. p. 1163.

f. *genuina* Ahln.

Enterom. p. 18.

Descr. Enteromorpha intestinalis α genuina Ahln. l. c.
Fig. » » Kütz. Tab. Phyc. 6, t. 31.
Exsicc. » » Aresch. Alg. Scand. exsicc. N:o 122.

f. *attenuata* AHLN.

l. c. p. 20.

Descr. Enteromorpha intestinalis *b* attenuata AHLN. l. c.
Exsicc. » » var. » WITTR. et NORDST. Alg. exsicc. N:o 136.
» » f. longissima ARESCH. Alg. Scand. exsicc. N:o 327.

f. *cornucopiæ* LYNGB.

Scytosiphon intestinalis *γ* cornucopiæ LYNGB. Hydr. Dan. p. 67.
Descr. Enteromorpha intestinalis *c* cornucopiæ AHLN. l. c. p. 21.
Exsicc. » » var. » WITTR. et NORDST. Alg. exsicc. N:o 137.
Syn. Enteromorpha intestinalis ARESCH. Phyc. Scand. p. 415.
» » DICKIE, Alg. Sutherl. 1, p. 143; ex parte; Alg. Cumberl. p. 239.
» » KLEEN, Nordl. Alg. p. 40.
Ulva compressa WG. Fl. Lapp. p. 508; ex parte.
» intestinalis GUNN. Fl. Norv. 2, p. 120.
» » SOMMERF. Suppl. p. 185.

Remark on the forms of this species. The present species is certainly very rich in forms within the Polar Sea, but it appears to me that the forms occurring here may be grouped round the three types distinguished by AHLNER. Of f. *genuina* I have seen two variations, the one most nearly coinciding with N:o 122 in ARESCH. Alg. Scand. exsicc., though broader than this and more abruptly constricted at the base, the other surely identical with *E. intestinalis, ε. mesenteriæformis* KÜTZ, (Spec. Alg. p. 478), a large form more than half a metre long and even 3 cm. in diameter. It tapers gradually, but rather much, towards the base and is of a pretty dark grass-green colour. The form *attenuata* is more variable than the typical form. The most characteristic form in my collections from Finmarken is a long form of a pale yellow-green colour, which resembles N:o 36 *a* in WITTR. et NORDST. Alg. exsicc. To the group of *attenuata* I have referred also a dark green form, more than a foot long by one inch broad, which, according to AHLNER who has had the kindness to examine my collections of arctic *Enteromorphæ*, differs not inconsiderably from f. *attenuata*, but is nevertheless most closely allied to this form. Specimens of f. *cornucopiæ* from the Arctic Sea resemble N:o 137 *a* in WITTR. et NORDST. Alg. exsicc. but are considerably larger and broader. Many of those are 4—5 cm. in diameter upwards.

Habitat. The present species is litoral, growing gregarious between tide-marks, chiefly in rock-pools which are filled with water at low-water. It is attached to other algæ or to stones, flourishes on exposed as well as sheltered coasts, and even enters river-mouths where the water is only little salt. Reproductive organs are developed in July, August, and September on the coast of Finmarken.

Geogr. Distrib. This species exists with certainty in the Norwegian Polar Sea. I have also seen an Enteromorpha from Baffin Bay which I think ought to be referred to this species. In the White Sea typical *E. intestinalis* is not to be found, according to GOBI. It is possible however that f. *attenuata* exists here; but without having access to a greater number of specimens I dare not decide this question. The *E. intestinalis* reported from other parts of the Arctic Sea appears to me to belong to other species, especially *E. compressa.* Its maximum of frequency is certainly in the Norwegian Polar Sea. The most northerly place where it is as yet known to live is Gjesvær about Lat. N. 71°.

Localities: The Norwegian Polar Sea: Nordlanden common and abundant; Tromsö amt common and abundant, occurring here as at Nordlanden in all the three above-mentioned forms; Finmarken common and abundant at Maasö (f. *genuina* and f. *attenuata*), Öxfjord (f. *attenuata*), Talvik (f. *attenuata*).

Baffin Bay: Cumberland Sound according to DICKIE, the west coast of Greenland at Frideriksdal, Tessarmiut and Nanortalik. It is said by DICKIE to have been taken at Hunde Islands and Cape Bowen.

Enteromorpha compressa (L.) LINK.

Epist. p. 5. Ulva compressa L. Spec. Pl. p. 1163; char. emend. Cfr. AHLN. Enterom. p. 31.

f. *typica.*

Descr. Enteromorpha compressa AHLN. Enterom. p. 31.
Fig. » » KÜTZ. Tab. Phyc. 6, t. 38.
Exsicc. » » HOHENACK. Alg. Mar. N:o 259.

f. *capillacea* KÜTZ.

Enteromorpha compressa β capillacea KÜTZ. Spec. Alg. p. 480.
Descr. Enteromorpha compressa *b* capillacea AHLN. l. c. p. 32.
Exsicc. » intestinalis *a* capillacea HOHENACK. l. c. N:o 258.

f. *racemosa* AHLN.

l. c. p. 33.

α. *Ahlnerii* nob.

Descr. Enteromorpha compressa *c* racemosa AHLN. l. c.
Exsicc. » ramulosa ARESCH. Alg. Scand. exsicc. N:o 226.

β. *abbreviata* nob.

f. frondis axi primario vix unciali, ramis confertis.

γ. *elongata* nob.

f. frondis axi primario pedali et ultra, capillaceo, ramis distantibus.

f. *prolifera* AG.

Ulva compressa β prolifera AG. Spec. Alg. 1, p. 421.
Descr. Enteromorpha compressa *d* prolifera AHLN. l. c. p. 35.
Fig. » percursa HARV. Phyc. Brit. t. 352.
Exsicc. » complanata var. crinita RABENH. Alg. Eur. N:o 911.
Syn. Enteromorpha clathrata J. G. AG. Grönl. Alg. p. 110; sec. spec.
» compressa ASHM. Alg. Hayes, p. 96.
» » CROALL, Fl. Disc. p. 461.
» » DICKIE, Alg. Cumberl. p. 239; Alg. Walker, p. 86; Alg. Sutherl. 2, p. 193.
» » KLEEN, Nordl. Alg. p. 40.
» » POST. et RUPR. Ill. Alg. p. II; saltem ex parte.

Syn. Enteromorpha intestinalis J. G. AG. Spetsb. Alg. Progr. p. 2; Bidr. p. 11; Till. p. 28.
» » f. compressa KJELLM. Spetsb. Thall. 2, p. 43; Algenv. Murm. Meer. p. 49.
Scytosiphon compressus β crispatus LYNGB. Hydr. Dan. p. 64.
Ulva compressa GUNN. Fl. Norv. 2, p. 120 (?).
» » SCHRENK, Ural. Reise, p. 547.
» » SOMMERF. Suppl. p. 186.
» » WG. Fl. Lapp. p. 508; ex parte.
» Enteromorpha GOBI, Algenfl. Weiss. Meer. p. 80; saltem ex parte.

Habitat. This is one of the few species that are litoral in the whole Polar Sea. It is attached to stones or algæ and occurs both on exposed and sheltered coasts, sometimes gregarious in considerable numbers. I have found it with reproductive organs at Spitzbergen at the end of July, on the coast of Norway in August and September.

Geogr. Distrib. The present species has been found circumpolar in some form or other. The northernmost point where it is certainly known to occur is Fairhaven on the north-west coast of Spitzbergen Lat. N. 79° 49'.

Localities: The Norwegian Polar Sea: Nordlanden, f. *typica* and f. *capillacea*, common and abundant; Tromsö amt about the town of Tromsö, f. *typica* and f. *prolifera*; Finmarken at Maasö, Gjesvær, the southern coast of Magerö, Talvik, f. *typica* and f. *prolifera* scarce, f. *capillacea* and f. *racemosa* rather abundant.

The Greenland Sea: f. *typica* and a transition between it and f. *racemosa*, pretty common but not abundant on the north-west and north coasts af Spitzbergen.

The Murman Sea: on the west coast of Novaya Zemlya rather local and scarce in a form most nearly related to f. *racemosa:* Kolgujew Isle and the coast of Cisuralian Samoyede-land (?).

The White Sea: at several places according to GOBI.

The Siberian Sea: Actinia Bay, in a low, almost typical form.

The American Arctic Sea: Port Kennedy, Beachey Island, Assistance Bay, Baring Bay.

Baffin Bay: Cumberland Sound not rare; the west coast of Greenland at Frideriksdal, Nanortalik, Lichtenau, Julianeshaab, Smallesund, Ameralik, Jakobshavn, Sakkak, Rittenbenk, ff. *capillacea*, *prolifera*, and *racemosa;* Smith Sound between the 78:th and 82:d latitude.

Enteromorpha complanata KÜTZ.

Spec. Alg. p. 480; Cfr. AHLN. Enterom. p. 25.

f. *prolifera* nob.

f. thalli axi primario elongato, complanato, apicem versus dilatato, secundum totam longitudinem ramos crebros, elongatos, axem primarium æmulantes emittente.

Description. A leading axis is plainly to be distinguished in the frond, attaining a length of even 30 cm., downwards narrow, almost terete, upwards thickening and flattened, about 1 cm. broad in its uppermost part. It bears along its whole length a great number of simple branches of the same shape as the main axis.

This form much resembles *Phycoceris racemosa* KÜTZ. (Tab. Phyc. 6, tab. 26) in habit. In structure it coincides very closely with *E. complanata* var. *subsimplex* ARESCH. (Ahln. Enterom p. 29), but it has somewhat narrower and less regular cells with somewhat more abundant endochrome.

Habitat. A litoral alga, attached to stones, growing scattered both on exposed and sheltered coasts. On the north coast of Norway it bears zoospores (gamets?) in August and September.

Geogr. Distrib. Known only from the Atlantic region of the Polar Sea. Its northernmost point is Maasö about Lat. N. 71°.

Localities: The Norwegian Polar Sea: Finmarken at Maasö and Talvik, loca and scarce.

Enteromorpha minima NÄG.

in KÜTZ. Spec. Alg. p. 482; Cfr. AHLN. Enterom. p. 48.

f. *glacialis* KJELLM.

in WITTR. et NORDST. Alg. exsicc. N:o 43.

Descr. Enteromorpha minima f. glacialis KJELLM. Algenv. Murm. Meer. p. 50.
Exsicc. » » » » » in WITTR. et NORDST. l. c.

Habitat. This species, wherever hitherto found, formed dense, rather extensive mats within the litoral zone on flat rocks that sloped towards the sea and were moistened at low tide by snow- and ice-water dropping down.

Locality: The Murman Sea: Besimannaja Bay on the west coast of Novaya Zemlya.

Enteromorpha tubulosa KÜTZ.

Tab. Phyc. 6, p. 11. E. intestinalis γ tubulosa KÜTZ. Spec. Alg. p. 478; Cfr. AHLN. Enterom. p. 49.

f. *pilifera* KÜTZ. (AHLN.).

l. c. p. 50; Enteromorpha pilifera KÜTZ. Tab. Phyc. 6, p. 11.

Descr. Enteromorpha tubulosa *b* pilifera AHLN. l. c.
Fig. » pilifera KÜTZ. l. c. t. 30.

Locality: E mari groenlandico according to WORMSKIOLD in the herbarium of the Copenhagen Museum.

Enteromorpha micrococca KÜTZ.

Tab. Phyc. 6, p. 11.

f. *typica* nob.

Descr. Enteromorpha micrococca AHLN. Enterom. p. 46.
Fig. » » KÜTZ. l. c. tab. 30.
» » AHLN. l. c. t. 1, fig. 7.

f. *subsalsa* nob.

f. strata valde intricata, forma indefinita, e viride flavescentia vel fere albida, in fundo libera expansa formans; thalli axi primario distincto, latiusculo, compresso secundum totam longitudinem ramos longiores et breviores, simplices vel ramulosos, patentes, uncinatos, curvatos, axi primario graciliores emittente; structura formæ typicæ persimilis. Tab. 31, fig. 1—3.

Exsicc. Enteromorpha clathrata var. uncinata KJELLM. in WITTR. et NORDST. Alg. exsicc. N:o 131.

Syn. Enteromorpha clathrata J. G. AG. Spetsb. Alg. Progr. p. 3; Bidr. p. 11 (?).

» » f. uncinata KJELLM. Spetsb. Thall. 2, p. 44; Algenv. Murm. Meer. p. 50. Ulva micrococca GOBI, Algenfl. Weiss. Meer. p. 81.

Remark on f. subsalsa. I have already mentioned, in my treatise on the marine vegetation of the Murman Sea, that the alga recorded here under the name of *E. clathrata* agreed less, with regard to its structure, with the species of that name than with *E. micrococca* KÜTZ. My determination of the alga in question was based on the opinion pronounced by LE JOLIS, that the morphological characters are of more importance than the anatomical ones in determining the species of the genus *Enteromorpha.* I must now relinquish this view, on account of the results acquired by recent investigations.

The alga in question closely agrees with *E. micrococca* in structure. It is certainly different from it in habit by the branching of the frond, but the degree of branching varying very much and even typical *E. micrococca* being occasionally branched, I have thought best to regard the arctic plant here referred to as a form of the last-named species. It is distinguished from its typical form by being always more richly branching and by forming large, irregular, intertwisted masses lying loose on the bottom.

Habitat. The principal form of this species grows scattered within the litoral zone, in sheltered localities, attached to stones. The form *subsalsa* is known only from lagoons with brackish water, often occurring in large masses so as to determine the character of the vegetation. It persists through the winter enclosed in ice, resuming its development when the ice has melted. Of the typical form I have observed specimens with zoospores in September.

Geogr. Distrib. Known both from the Atlantic and arctic region of the Polar Sea. It is apparently widely distributed in the latter region and it has here its maximum of frequency. The northernmost point where it is known to occur is Treurenberg Bay on the north coast of Spitzbergen Lat. N. 79° 56′.

Localities: The Norwegian Polar Sea: (f. *typica*) scarce and local at Öxfjord and Talvik.

The Greenland Sea: abundant in lagoons at Musselbay and Treurenberg Bay on the north coast of Spitzbergen.

The Murman Sea: the coast of Russian Lapland, f. *typica;* the west coast of Novaya Zemlya, f. *subsalsa*, rather plentiful in lagoons at Besimannaja Bay, and Karmakul Bay.

The Siberian Sea: f. *subsalsa* abundant at Pitlekay.

Baffin Bay: f. *subsalsa* on the west coast of Greenland at Tessarmiut.

Gen. **Ulva** (L.) WITTR.

Monostr. p. 9; L. Syst. Nat. Ed. 10, p. 1346.

Ulva crassa KJELLM.

Spetsb. Thall. 2, p. 44.

Descr. et *Fig.* Ulva crassa KJELLM. l. c. et t. 3.

Syn. Ulva crassa KJELLM. l. c. et Algenv. Murm. Meer. p. 51.
» latissima KJELLM. Vinteralgv. p, 65.

Remark on this species. Since I have had the opportunity of examining a greater number of forms of Ulva from different parts of the Scandinavian coast and especially from the north coast of Norway, I have found that *U. crassa* is not such an independent form as I believed at the time when I described it. It is nearly related to *U. lactuca*, but it differs from that species by greater thickness, the greater height of the cells, the smaller transverse diameter of the cell-rooms and by more abundant endochrome. Until it has been fully demonstrated that these characteristics cannot be employed as specific differences in the genus of *Ulva*, I think *U. crassa* ought to be retained as a separate species.

Habitat. It grows sublitoral in 3—5 fathoms water, attached to other algæ and to stones, always scattered and in sheltered localities, On the coast of Spitzbergen it occurs also in winter. I found young individuals at Musselbay in November, January, and April and an older specimen with zoospores on the 23 December. I have besides taken specimens bearing zoospores in July and August on the coast of Spitzbergen, and in July on the west coast of Novaya Zemlya.

Geogr. Distrib. This species belongs to the arctic region of the Polar Sea, but judging from one very fragmentary specimen occurs also on the north coast of Norway. Greater numbers of it have not been met with anywhere. Its most northern point is Musselbay on the north coast of Spitzbergen Lat. N. 79° 53'.

Localities: The Norwegian Polar Sea: Finmarken at Öxfjord, only one incomplete specimen found.

The Greenland Sea: on the north and west coasts of Spitzbergen, rather commonly dispersed, but always scarce.

The Murman Sea: on the west coast of Novaya Zemlya local and very scarce.

Ulva lactuca L.

Spec. Pl. 2, p. 1163.

Descr. et *Fig.* Ulva lactuca BORN. et THUR. Etud. Phycol. p. 5 et t. 2—3.

Syn. Ulva latissima Nordl. Alg. p. 40.
» » SCHÜBELER, in Heugl. Reise, p. 317 (?).
» linza ARESCH. Phyc. Scand. p. 410 (?).
» » KLEEN, l. c. (?).

Habitat. In the Norwegian Polar Sea this species is litoral, attached to stones, growing scattered in sheltered localities. I have collected individuals with zoospores (gamets) in September on the coast of Finmarken.

Geogr. Distrib. It is known with certainty from the Norwegian Polar Sea, in the northern part of which it is scarce, and from Baffin Bay. As I do not know if *U. latissima* SCHÜBEL. l. c. is to be referred to this species, I cannot decide whether it occurs in the Murman Sea. I never observed it there myself. The most northerly place where it is as yet certainly known to grow is Öxfjord on the coast of Finmarken about Lat. N. 70°.

Localities: *The Norwegian Polar Sea:* Nordlanden probably common and rather abundant; Finmarken local and scarce at Öxfjord and Talvik.

The Murman Sea: (?).

Baffin Bay: the west coast of Greenland near Godhavn according to specimens in the herbarium of the Copenhague Museum.

Gen. **Monostroma** (THUR.) WITTR.

Monostr. p. 15; THUR. Note s. Ulv. p. 29, sec. WITTR. l. c.; lim. mut.

Monostroma latissimum (KÜTZ.) WITTR.

Monostr. p. 33. Ulva latissima KÜTZ. Phyc. gener. p. 296.

Descr. Monostroma latissimum WITTR. l. c.

Fig. » » » » t. 1, fig. 4.

Exsicc. » » WITTR. et NORDST. Alg. exsicc. N:o 145.

Syn. Monostroma arcticum KLEEN, Nordl. Alg. p. 41; sec. spec.; ex parte. (?).
» latissimum KLEEN, Nordl. Alg. p. 41 (?); ex parte(?).

Remark on the synonymy. KLEEN records *M. latissimum* in Nordl. Alg. The specimens of his collections which bear this name, do not however belong to the present species but to *M. saccodeum.* On the contrary that alga which is called *M. arcticum* in his collections, is in fact *M. latissimum.* Of *M. arcticum* no specimens are to be found there. It is certain, accordingly, that *M. latissimum* really exists at Nordlanden, but it is uncertain whether it is included under this name in Nordl. Alg. and whether KLEEN has really found *M. arcticum.* His statement of its occurrence in brackish water, directly contrary to my own observations, seems to show that he has confounded *M. arcticum* and *M. latissimum.* Still it is possible that *M. arcticum*, which is rather common in Finmarken at certain places, is really to be found in Nordlanden, and that KLEEN'S determination is correct, although the specimens collected have happened afterwards to be confounded.

Habitat. There is nothing known with certainty about the habitat of the present species in the Polar Sea. If *M. arcticum* KLEEN Nordl. Alg. should be exclusively the present species, this would be a brackish-water alga at Nordlanden, as it is often elsewhere, growing in the lower part of the sublitoral zone, attached to other algæ.

Geogr. Distrib. Known only from the Norwegian Polar Sea.

Locality: Nordlanden. I have collected it myself at Bodö. Possibly KLEEN's specimens are derived from the same place.

Monostroma undulatum WITTR.

Monostr. p. 46.

Descr. et *Fig.* Monostroma undulatum WITTR. l. c. et t. 3, fig. 9.

Syn. Monostroma undulatum KLEEN, Nordl. Alg. p. 41.
Ulva lactuca SOMMERF. Suppl. p. 185; sec. syn.
» » WG. Fl. Lapp. p. 507; sec. spec.

Habitat. Found on the upper part of the sublitoral zone, attached to *Corallina officinalis.*

Geogr. Distrib. Known only from the Norwegian Polar Sea.

Locality: Nordlanden at Kjerring Isle according to specimens in Wahlenberg's herbarium.

Monostroma lubricum KJELLM.

Spetsb. Thall. 2, p. 48.

Descr. et *Fig.* Monostroma lubricum KJELLM. l. c. et t. 4, fig. 8—9.

Syn. Monostroma lubricum GOBI, Algenfl. Weiss. Meer. p. 79.

The *habitat* of this species is unknown. Cp. KJELLM. l. c.

Geogr. Distrib. It belongs to the arctic region of the Polar Sea, where it is found at several places. Its most northern point is Fairhaven on the north-west coast of Spitzbergen Lat. N. 79° 49'.

Localities: The Greenland Sea: Fairhaven.

The Murman Sea: Sviatoi Noss on the coast of Russian Lapland.

Baffin Bay: the west coast of Greenland at Kakortok Bay according to specimens in the herbarium of Copenhagen Museum.

Monostroma cylindraceum nob.

M. thallo initio sacculum subcylindraceum, tenuem, flaccidum, flavescentem, bullosum, usque 10 cm. longum diametro 3—4 cm. formante, demum libero, rima longiore vel breviore exorta plus minus expanso, vix laciniato; parte monostromatica vegetativa 40—45 μ. crassa; corpore chlorophylloso lumen cellulare in sectione thalli transversa qvadratum vel rectangulare, 10—15 μ. altum non vel fere explente; parte zoosporifera e cellulis lumine cellulari in sectione transversa thalli elliptico vel circulari constructa. Tab. 30.

Description. The present alga is attached by a callus radicalis. The frond has at first the shape of an almost cylindrical sack, uneven and bullate on the surface, attaining a length of 10 cm. by 3—4 cm. in diameter, very pale green or yellowish green in colour, feebly or not at all shining, of loose consistency, and so slimy that it adheres strongly to the paper in drying. At last a reft is formed at the top, stretching more or less far downwards, sometimes nearly to the base, the frond expanding at the same time, but becoming only very slightly, if at all, lobed. The frond when bearing zoospores has a length of 15 cm. by an almost equal breadth in its upper expanded part; fig. 1.

Its lowest part is formed by club-shaped cells with very gelatinized membranes, by the shaft-ends of which it is attached. The club-heads are fusiform, fusiform-cylindrical, almost cylindrical, elongated-ovate, their cell-rooms being 5—10 μ. in cross section at their thickest part. In a transverse cut the club-heads are usually seen to occupy one side of the frond, the shafts being placed on the other. This part of the frond is 30—40 μ. thick; fig. 2—3.

At a distance of two millimetres from the callus the frond becomes monostromatic, being formed here as well as farther upwards, as far as it is sterile, by cells the rooms of which are four-angular in cross section with acute or rounded corners. They are sometimes square, sometimes rectangular, in the latter case usually having their greatest length in a parallal direction to the surface of the frond. The monostromatic part is 40—45 μ. thick; the rooms of the cells being 10—15 μ. high, the wall is accordingly of considerable thickness. The chlorophyl takes up sometimes the whole room of the cell, sometimes only a part of it. In the middle of the frond the vegetative cells, as seen from the surface, are 4—5-angular with thick walls. Their longest diameter is about 20—25 μ.; fig. 4—5. The part of the frond bearing zoospores is composed of cells the rooms of which are circular as seen from the surface, 10—17 μ. in diameter, with very thick walls. The rooms in cross section are either circular or circular-elliptical with their long axis, 17—22 μ., parallel to the surface of the frond; fig. 6—7.

Habitat. This species is litoral, attached to other algæ, especially *Halosaccion ramentaceum*, growing scattered on exposed coasts. Specimens with zoospores have been collected at the end of July and the beginning of August.

Geogr. Distrib. Known only from the Norwegian Polar Sea. Here it is scarce. Its northernmost point is Gjesvær about Lat. N. 71°.

Localities: Finmarken at Maasö and Gjesvær, local and scarce.

Monostroma saccodeum nob.

M. thallo callo radicali adnato, initio saccato, deinde membranaceo, in lacinias oblongas, lanceolatas vel ovatas, margine vel crispo plus minus decomposito-fisso; parte monostromatica inferne 30—40, superne 25—30 μ. crassa, e cellulis constructa a fronte visis lumina rotundata; semicircularia vel 3—5 angulata inter se membrana crassiuscula separata, in sectione transversa thalli visis lumina cellularia verticaliter elliptica 15—17 μ. alta, 8—10 lata præbentibus; corpore chlorophylloso lumen cellulare fere omnino explente. Tab. 28, fig. 1—10.

Syn. Monostroma latissimum KLEEN, Nordl. Alg. p. 41; saltem ex parte fide herb.

Description. The present species agrees with *M. Grevillei* in development, and is related to it even in habit, although in structure the two plants are sharply distinguished. The frond when young has the shape of an ellipsoidic or pyriformly cylindrical sack or bladder, even 4 cm. in length, of a light grass green colour, almost without gloss, with smooth wall, and attached by a callus radicalis; fig. 1—2. These bladders soon burst at the top, expand and become slit usually in their whole length. The segments when fully grown are even 10 cm. long, sometimes few in number, being then ovate or elliptical, sometimes numerous, being then usually lanceolate or oblong, in both cases often repeatedly and more or less deeply divided. The margin of the segments is generally wavy or crisp; fig. 3—4. When the frond grows older, it either detaches itself altogether from its substratum or else a number of segments are loosened that afterwards float on the surface of the water or lie loose on the bottom. These segments, after having become free, often increase considerably in size. There are to be found pieces which have a length of 15—20 cm. by a breadth of about 10 cm. When growing older the plant assumes a pale yellowish-green colour. It is loose and slimy in consistency so that it adheres strongly to the paper in drying. The lowest part of the frond nearest the callus is composed of claviform cells with very gelatinous walls, similar to those of several other species of *Monostroma*, as *M. arcticum* and *M. Grevillei*. Their rooms are 6—10 μ. thick in their longest diameter. The heads of the clubs occupy the middle of the frond, the shafts being placed towards the surface; fig. 5—6.

The rest of the frond is composed of cells with ellipsoidic or ovate rooms, the long axis of which is at right angles to the surface of the frond. The whole wall of the cell is covered by the chlorophyllous body; fig. 8. Downwards this part of the frond is 30—40 μ. thick, the cell-rooms are 15—17 μ. high and 8—10 μ. in diameter. At the upper margin of the frond the thickness is somewhat less, 25—30 μ., the rooms of the cells are somewhat lower and broader, almost round. The cells containing zoospores have somewhat thinner walls than the vegetative ones, partition-walls of more equal thickness, and somewhat larger rooms; fig. 9—10.

Habitat. This species grows within the litoral zone on exposed coasts, somewhat gregarious, but not in great numbers, attached to litoral algæ, as Corallina, Rhodomela lycopodioides, Halosaccion, Fucaceæ or to serpulids and stones. I have found at the beginning of August on the coast of Finmarken fully developed individuals with zoospores as well as very young ones scarcely one millimetre in length.

Geogr. Distrib. It is known as yet only from the north coast of Norway, the northernmost point where it has been met with being Gjesvær, about Lat. N. 71°.

Localities: Nordlanden according to specimens in KLEEN'S herbarium; Finmarken: Maasö local but rather abundant, Gjesvær local and scarce.

Monostroma angicava nob.

M. thallo callo radicali adnato, initio vesicam pyriformem constituente, deinde expanso, membranaceo, flaccido, lubrico, fusco-viride, demum pallescente, parce laciniato, margine plano, lacerato; parte monostromatica 45—60 μ. crassa, cellulis in sectione transversa thalli lumina cellularia verticaliter rectangularia, angulis rotundatis, 25—28 μ. alta, 8—10 μ. lata præbentibus, corpore chlorophylloso cellulas vegetativas non explente. Tab. 29.

Description. The present species is attached by a callus radicalis. It has when young the form of a pear-shaped bladder, attaining a length of 5 cm. by 4 cm. in diameter upwards. This bladder opening at the top and bursting down to the base, the frond expands and in its further growth becomes irregularly laciniate. A full-grown specimen with zoospores in my| collection is almost reniform in outline, with the margin lobed and slashed, but not crispy. The alga in its younger condition is dark green in colour with a brown tinge, but when older it becomes more pale. It has no gloss and is slimy and loose in consistency; fig. 1.

The lower part of the frond is composed of club-shaped cells extending in an oblique direction from one surface of the frond down to the other. The heads of the clubs are elongated, obovate, cylindrical or fusiform, often curved; fig. 2—3. The steril, monostromatic part of the frond is 45—50 μ. thick, composed of slender and high cells. A transverse section of the frond shows the rooms of the cells to be vertically rectangular with much rounded corners. Their height is about 25 μ. by 8—10 μ. in breadth. The endochrome does not fill the whole cell-room; fig. 4—5.

I have found the part of the frond bearing zoospores 55—60 μ. thick, with the cell-rooms 25—28 μ. high and usually 10 μ. broad; fig. 6—7. For other particulars the reader is referred to the figures cited.

The present species belongs to the same group of the genus *Monostroma* as the two preceding, *M. arcticum* and *M. Grevillei* a. o. It cannot be confounded with any species that has been before described, being easily distinguished from all by its numerous and high cell-rooms and the thick walls of the cells.

Habitat. At the place where it was found by me, it grew within the litoral region in an exposed locality on litoral algæ. At the beginning of August it was provided with zoospores.

Locality: It is known as yet only from the Norwegian Polar Sea, where I have found some few specimens at Maasö in Finmarken.

Monostroma Grevillei (THUR.) WITTR.

Monostr. p. 57. Enteromorpha Grevillei THUR. Note s. Ulv. p. 25.
Descr. Monostroma Grevillei WITTR. l. c.
Fig. » » » t. 4, fig. 14.
Exsicc. » » WITTR. et NORDST. Alg. exsicc. N:o 434.
Syn. Monostroma Grevillei GOBI, Algenfl. Weiss. Meer. p. 79.
» » KLEEN, Nordl. Alg. p. 41.

Habitat. This species lives rather gregarious, on exposed coasts, attached to stones within the litoral zone or lying loose on the bottom in the upper part of the sublitoral zone. Individuals collected by KLEEN at Nordlanden in June and at the beginning of July are furnished with zoospores.

Geogr. Distrib. It belongs only to that part of the Polar Sea which lies immediately to the north of the Atlantic. Its maximum of frequency is on the north-west coast of Norway. The most northerly place where it has been found is Hammerfest Lat. N. 70° 40′.

Localities: The Norwegian Polar Sea: Nordlanden: Tromsö amt at the town of Tromsö; Finmarken at Hammerfest.

The White Sea: at Solowetzki Isles.

Baffin Bay: the west coast of Greenland at Julianeshaab.

Monostroma arcticum Wittr.

Monostr. p. 44.

Descr. Monostroma arcticum Wittr. l. c.
Fig. » » » t. 2, fig. 8.
Exsicc. » » Kjellm. in Wittr. et Nordst. Alg. exsicc. N:o 144.

Syn. Monostroma arcticum Kleen, Nordl. Alg. p. 41. (?)

Addition to the description. This species agrees in development with the four preceding ones. The frond when very young forms an almost spherical, pear-shaped, or cylindrical bladder, which bursts and is split down to the base. Afterwards the frond expands and becomes laciniate. When fully developed it has a brownish colour.

Habitat. A litoral species attached to Balani or litoral algæ, as *Rhodomela lycopodioides, Halosaccion, Gigartina, Rhodymenia,* and old stems of *Fuci.* It prefers open coasts, occurring at places exposed to the unbroken violence of the sea, often gregarious in considerable numbers. Specimens with zoospores were common on the coast of Finmarken at the end of August.

Geogr. Distrib. Known from the north coast of Norway. It is plentiful at Finmarken. The most northern point where it has been collected is Gjesvær about Lat. N. 71°.

Localities: The Norwegian Polar Sea: Nordlanden at Bodö (?). (Cp. at *M. latissimum;*) Finmarken: Maasö and Gjesvær abundant.

Monostroma leptodermum Kjellm.

Algenv. Murm. Meer. p. 52.

Descr. et *Fig.* Monostroma leptodermum Kjellm. l. c. et t. 1, p. 23—24.

Habitat. Found only once, washed ashore. Known only as sterile.

Locality: The Murman Sea: Matotshin Shar at Novaya Zemlya.

Monostroma fuscum (Post. et Rupr.) Wittr.

Monostr. p. 53. Ulva fusca Post. et Rupr. Ill. Alg. p. 21.

Descr. Monostroma fuscum Wittr. l. c.
Fig. » » » t. 4, fig. 13.
Exsicc. Ulva sordida Aresch. Alg. Scand. exsicc. N:o 120.
Monostroma fuscum Kjellm. in Wittr. et Nordst. Alg. exsicc. N:o 143.

Syn. Monostroma fuscum Gobi, Algenfl. Weiss. Meer. p. 79.
» » Kjellm. Spetsb. Thall. 2, p. 49; Algenv. Murm. Meer. p. 52.
» » Kleen, Nordl. Alg. p. 41.
» » Wittr. l. c.
Ulva sordida Aresch. Phyc. Scand. p. 413.

Habitat. This species is always sublitoral in the Arctic Sea, growing scattered on exposed coasts, in 4—5 fathoms, attached to other algæ or stones. On the northern coast of Norway I have found it partly in the upper part of the sublitoral zone near low-water mark, partly at a considerable (10—15 fathoms) depth attached to stones, shells or muscles both in exposed and sheltered localities. It is often met with here floating free on the water in large quantities or washed ashore. Specimens with zoospores have been collected in August on the coast of Spitzbergen.

Geogr. Distrib. This species has its maximum of frequency in the Norwegian Polar Sea. In the other parts of the Polar Sea it is scarce and little spread. The most northern place where it is known to occur is Fairhaven on the north-west coast of Spitzbergen Lat. N. 79° 49'.

Localities: The Norwegian Polar Sea: Nordlanden abundant; Finmarken at Maasö, Gjesvær and Talvik, everywhere abundant and rather common.

The Greenland Sea: the north-west coast of Spitzbergen, rare.

The Murman Sea: the coast of Russian Lapland: the west coast of Novaya Zemlya: here only a couple of fragmentary specimens found washed ashore.

The White Sea: probably common.

Baffin Bay: the coast of Greenland according to specimens in the herbarium of the Copenhague Museum; the exact place of occurrence not stated.

Monostroma crispatum nob.

M. fronde callo radicali adnato, membranacea, obovata, obscure viridi nigrescente, margine lacerato et crispo, inferne 170, superne 50 μ. crassa; parte monostromatica e cellulis in sectione frondis transversa lumina qvadrangularia 35 μ. alta, 15—35 μ. lata præbentibus contexta. Tab. 28, fig. 11—13.

Description. This species belongs to the same section of the genus *Monostroma* as *M. fuscum* and *M. Blyttii*. I have found only three specimens, all of which agree most closely. The largest of them is figured in natural size (tab. 28, fig. 11). It is 4 cm. long and 1,5 cm. broad upwards where its transverse section is greatest. It is obovate in outline as the others. The margin is laciniate and very crisp, the colour is dark green, almost black green, becoming black in drying. Its attachment is a callus radicalis that is large in proportion to the plant. The stipe is short, but distinct (fig. 11). The greater part of the frond is formed of club-shaped cells with heads that are almost square in transverse section. The heads occupy the one side of the frond, the shafts the other. Upwards the frond is monostromatic, composed of cells square in transverse section, whose breadth is as great as or less than their height. The corners of the cell-rooms are slightly rounded. The endochrome does not cover the whole wall. The outer wall is comparatively thin. The frond is 50 μ. at the upper margin, the height of the cell-rooms here is 35 μ., the breadth 15—35 μ. The frond is still more than 150 μ. thick at the middle. The cells as seen from the surface are 4—6-angular with thin partition-walls.

Habitat. The specimens in my possession were dredged in the lowest part of the sublitoral zone in about 20 fathoms. They were attached to stones. All were sterile when taken in the middle of August.

Locality: *The Norwegian Polar Sea:* Finmarken at Maasö.

Monostroma Blyttii (ARESCH.) WITTR.

Monostr. p. 49. Ulva Blyttii ARESCH. in Fr. Sum. Veg. p. 129.

Descr. Monostroma Blyttii WITTR. l. c.
» » KLEEN, Nordl. Alg. p. 42.
Fig. » » KJELLM. Spetsb. Thall. 2, t. 4, fig. 1—7.
Exsicc. » » KJELLM. in ARESCH. Alg. Scand. exsicc. N:o 423 et WITTR. et NORDST. Alg. exsicc. N:o 44.

Syn. Monostroma Blyttii KJELLM. Spetsb. Thall. 2, p. 49; Algenv. Murm. Meer. p. 52.
» » KLEEN, Nordl. Alg. p. 42; excl. syn.
Ulva Blyttii ARESCH. Phyc. Scand. p. 412.
» fusca J. G. AG. Grönl. Alg. p. 110; fide spec.
» lactuca (?) J. G. AG. Spetsb. Alg. Progr. p. 2; Bidr. p. 11 (?).
» » (?) LINDBL. Bot. Not. p. 158 (?),
» latissima (?) ASHM. Alg. Hayes, p. 96 (?).
» » » CROALL, Fl. Disc. p. 461 (?).
» » » DICKIE, Alg. Sutherl. 1, p. 144; Alg. Cumberl. p. 239; Alg. Nares, p. 7 (?).
» rigida SOMMERF. Suppl. p. 185.

Remark on the synonymy. I have assumed that the *Ulva latissima* reported from Baffin Bay is the present species, because *U. latissima*, though said by CROALL and DICKIE to be common on the coast of Baffin Bay, is completely wanting in the collections from Greenland in the Copenhagen Museum, while of *Monostroma Blyttii* which is not reported from Baffin Bay either by DICKIE or CROALL there is to be found there a great quantity of very luxuriant specimens reminding one of *U. latissima*.

Habitat. In the Arctic Sea proper on the coasts of Spitzbergen and Novaya Zemlya this species is always sublitoral, growing scattered at a depth of 3—5 fathoms both on exposed and sheltered coasts, attached usually to small stones, sometimes to algæ. On the north coast of Norway it is, on the contrary, generally litoral, usually occurring in the upper part of this zone, especially in rock-pools with numbers of *Mytilus edulis*, both on exposed and sheltered coasts, in salt or somewhat brackish water, attached to stones or algæ. However, it is occasionally met with also here within the sublitoral zone, but then it lies loose on the bottom. Such specimens attain sometimes a considerable size. There is one in my collections of 30 cm. in length by 40 cm. in breadth. When it lives in such localities, viz. at the mouths of streamlets and rivers, where the water is somewhat brackish, it is yellowish green in colour and of looser consistency. It is generally gregarious on the coast of Scandinavia and contributes essentially to mark the aspect of the litoral vegetation.

Geogr. Distrib. It belongs to those parts of the Polar Sea which lie north of the Atlantic. On the coast of Norway it is very abundant, as seems to be the case also in Baffin Bay. The northernmost point where it is certainly known to occur is Fairhaven on the north-west coast of Spitzbergen Lat. N. 79° 49'.

Localities: The Norwegian Polar Sea: Nordlanden, rather common; Tromsö amt at Renö; Finmarken common and abundant at Maasö, Gjesvær, and the coast of Magerö, more scarce at Öxfjord and Talvik.

The Greenland Sea: Along the west and north-west coast of Spitzbergen, pretty common, but not abundant. It is the most common *Ulvacea* of Spitzbergen.

The Murman Sea: the west coast of Novaya Zemlya and Waygats rather common, but not abundant.

Baffin Bay: Cumberland Sound (?); probably common and abundant on the west coast of Greenland high towards the North. In case the alga called *Ulva latissima* by ASHMEAD is the present species, as I have supposed it to be, it would be found even north of 78:th Lat. I have seen specimens from Tessarmiut, Nanortalik, Julianeshaab, Godthaab, Sukkertoppen, and Holstenborg.

Gen. **Diplonema** novum nomen.

Typ. Ulva percursa LE JOL. Liste Alg. Cherb. p. 55.

Diplonema percursum (AG.)

Conferva percursa AG. Syn. Alg. p. 87.

f. *typica* nob.

f. thalli diametro longiore 20—25 μ. crasso; cellulis in sectione longitudinali rectangularibus, membranis tenuioribus.

Exsicc. Tetranema percursum ARESCH. Alg. Scand. exsicc. N:o 125.
Enteromorpha percursa WITTR. et NORDST. Alg. exsicc. N:o 140.

f. *crassiuscula* nob.

f. thallo diametro longiore 30—35 μ. crasso; cellulis in sectione longitudinali subquadratis, membranis crassioribus.

Syn. Enteromorpha confervoides J. G. AG. Spetsb. Alg. Bidr. p. 11.
» percursa CROALL, Fl. Disc. p. 463.
» » DICKIE, Alg. Sutherl. 1, p. 143.
» » KJELLM. Algenv. Murm. Meer. p. 51.
Scytosiphon compressus γ LINDBL. Bot. Not. p. 157.
» » γ confervoides SOMMERF. Spitsb. Fl. p. 232.
Ulva percursa SOMMERF. Suppl. p. 187.

Remark. I perfectly agree with ARESCHOUG that AGARDH'S *Conferva percursa* is to be regarded as the type of a separate genus, if *Ulva*, *Monostroma*, and *Enteromorpha* are considered as genera. But as the name given to the genus by ARESCHOUG ascribes to the alga in question a characteristic which it does not possess, namely that of being four-sized and formed of four cells in transverse section, I propose that that unsuitable name be exchanged for that of *Diplonema*. I have already mentioned (in Algenv.

Murm. Meer. p. 51) that all my examinations made on specimens from different localities, even on those distributed by ARESCHOUG in his Exsiccatæ under the name of *Tetranema percursum*, confirm the statements of LE JOLIS and KÜTZING that the plant is band-shaped, elliptic in transverse section, and formed of two cells (cp. KJELLM. l. c. tab. 1, fig. 25). BORY has already constituted a separate genus, *Percursaria*, for *Conferva percursa* (cp. LE JOLIS, Liste Alg. Cherb. p. 55). This denomination if accepted would oblige us to alter the old name of the species, as the combination of *Percursaria percursa* is impossible.

Swedish and arctic specimens on a more accurate comparison show rather remarkable differences. The former are narrower, their cells are longer in proportion to the breadth, with very thin walls while the cells are in a vegetative stage. The arctic individuals are about a third thicker than the others, often branching, composed of short cells with thick walls, that are square in transverse section or sometimes even rectangular with the long axis in the cross direction of the frond. I propose the name *crassiuscula* for this form.

Habitat. This species grows in lagoons and collections of brackish water in more or less considerable numbers both on exposed and sheltered coasts. I have found it with zoospores at the end of June at Novaya Zemlya.

Geogr. Distrib. It is known only from the Polar Sea north of the Atlantic. Its northernmost locality is Stans Foreland in the group of Isles at Spitzbergen, about Lat. N. 78°.

Locality: The Norwegian Polar Sea: Nordlanden according to SOMMERFELT and specimens in KLEEN'S collections (f. *typica*).

The Greenland Sea: the coast of Spitzbergen.

Baffin Bay: the west coast of Greenland: Jacobshavn, Disco Isle, Whale Islands. In the herbarium of the Copenhagen Museum there are found specimens without any exact note of the locality.

Gen. **Prasiola** (AG.) LAGERST.

Monogr. p. 9; AG. Spec. Alg. 1, p. 416; char. mut.

Prasiola stipitata SUHR.

in JESS. Monogr. p. 16, sec. LAGERST. l. c. p. 36.

Descr. Prasiola stipitata LAGERST. l. c.
Exsicc » » WITTR. et NORDST. Alg. exsicc. N:o 435.

Habitat. It grows on rocks at the sea-shore.

Locality: The Norwegian Polar Sea: Nordlanden at Lödingen according to specimens distributed by FOSLIE in WITTR. et NORDST. Alg. exsicc. under the abovementioned number.

Fam. CONFERVACEÆ (AG.) WITTR.

Pl. Scand. p. 16. Confervæ genuinæ AG. Syst. Alg. p. XXV; gen. excl.

Gen. Spongomorpha KÜTZ.

Phyc. gener. p. 273.

Spongomorpha spinescens KÜTZ.

Spec. Alg. p. 418.

Descr. Spongomorpha spinescens KÜTZ. l. c.
Fig. » » » Tab. Phyc. 4. t. 75.
Exsicc. » » KJELLM. in WITTR. et NORDST. Alg exsicc. N:o 115.

Addition to the description of this species. It is to be added to KÜTZING'S description of this species, that it forms very dense, usually perfectly spherical or flattened-spherical balls of light-green colour, composed of obpyramedical clusters with truncate tops. The numerous curved or spiral spine-shaped branches are characteristic of this species, connecting together with the rhizoids the particular branchsystems and clusters into dense masses.

Habitat. This is a litoral alga growing gregarious in great numbers on exposed coasts, attached to other algæ, most often to *Gigartina mamillosa*, sometimes to *Halosaccion ramentaceum.* I have collected specimens with zoospores in July on the coast of Finmarken.

Geogr. Distrib. It is known only from the Norwegian Polar Sea, where it was very abundant at one place and occurred commonly in suitable localities. The most northern point where it has been found is Gjesvær about Lat. N. 71°.

Locality: The Norwegian Polar Sea: Finmarken at Maasö, rather local and scarce; Gjesvær common and abundant.

Spongomorpha arcta (DILLW.) KÜTZ.

Spec. Alg. p. 417. Conferva arcta DILLW. Brit. Conf. Suppl. p. 67.

Descr. Conferva arcta ARESCH. Phyc. Scand. p. 426.
Fig. Cladophora arcta HARV. Phyc. Brit. t. 135.
Spongomorpha arcta KÜTZ. Tab. Phyc. 4, t. 74.
» centralis » » » 4, t. 80.
» cymosa » » » 4, t. 74.
Cladophora comosa » » » 3, t. 79.
» sacculifera » » » 3, t. 81.
» stricta » » » 3, t. 80.
» vaucheriæformis » » 3, t. 78.
Exsicc. Cladophora (Conf.) arcta ARESCH. Alg. Scand. exsicc. N:o 129, 334, 335.
Spongomorpha arcta WITTR. et NORDST. Alg. exsicc. N:o 114, 316, 413.
Syn. Cladophora arcta CROALL, Fl. Disc. p. 461.
» » DICKIE, Alg. Sutherl. 1, p. 143; Alg. Cumberl. p. 239.
» » GOBI, Algenfl. Weiss. Meer. p. 85.

Exsicc. Syn. Cladophora arcta Kjellm. Vinteralgv. p. 65; Spetsb. Thall. 2, p. 55; Algenv. Murm. Meer. p. 54.
» » Kleen, Nordl. Alg. p. 45.
» » Wittr. in Heugl. Reise, p. 284.
» polaris Harv. New Alg. p. 334.
Conferva arcta J. G. Ag. Spetsb. Alg. Progr. p. 2; Bidr. p. 11; Till. p. 38.
» centralis J. G. Ag. Grönl. Alg. p. 110.
» glomerata marina J. G. Ag. Spetsb. Alg. Bidr. p. 11.
» » β. » Lindbl. Bot Not. p. 158.

Remark on the determination of this species. This species is taken here in the wide sense given it by ARESCHOUG l. c. It then includes several forms that may well appear rather different and have indeed been described as separate species. Thus *Conferva Arcta* HARV. *C. centralis* LYNGB. belongs to it and so do perhaps all *Cladophoræ* of the group called *Comosæ* by KÜTZING (in Spec. Alg.), but at least *Cl. stricta, Cl. comosa, Cl. vaucheriformis*, and *Cl. sacculifera.* However these are all, according to ARESCHOUG and HARVEY, to be regarded as one and the same species at different stages of development and in different forms. I am obliged at present to subordinate my own opinion to that of these two great authorities.

Habitat. This species is litoral in the Norwegian Polar Sea, sublitoral, as far as I know, in the other parts of the Polar Sea, growing in 2—10 fathoms water, being in both cases sometimes epiphytic on other algæ, but usually attached to stones, poles a. s. o. It flourishes on exposed as well as sheltered coasts, both in salt and somewhat brackish water. On the north coast of Norway it is generally gregarious in great numbers, but in the Arctic Sea proper I have never found but isolated individuals. On one occasion I met with a form most nearly coinciding with *Conferva centralis* lying loose on the bottom within the sublitoral zone. The present alga occurs all the year round, even in winter, on the north coast of Spitzbergen. I never found any luxuriant specimens of it during the last-mentioned season, but they were nevertheless in full vigour and development. However it is most vigourously developed here in summer. I have collected specimens with zoospores in August on the coast of Spitzbergen and Finmarken.

Geogr. Distrib. This species has an extensive range in the Arctic Sea. Its maximum of frequency is on the north coast of Norway. Its northernmost certain locality is Low Island on the north coast of Spitzbergen Lat. N. 80° 20'.

Localities: The Norwegian Polar Sea: Nordlanden; Tromsö amt at Tromsö, Renö, and Karlsö; Finmarken: Maasö, Gjesvær, the south coast of Magerö, Talvik, common and abundant everywhere.

The Greenland Sea: Beeren Eiland; the north and west coast of Spitzbergen, commonly distributed, but in very small numbers.

The Murman Sea: at several places on the west coast of Novaya Zemlya and Waygats, but very scarce.

The White Sea: Solowetzki Isles.

The Kara Sea: scarce in Actinia Bay.

Baffin Bay: Cumberland Sound, common; the west coast of Greenland probably common. It is known from Tessarmiut, Nanortalik, Kakortok, Julianeshaab, Ameralik, Neuherrnhut, Godthaab, Holstenborg, Egedesminde, Godhavn, Claushavn, Sakkak, Rittenbenk, Whale Islands.

Spongomorpha Lanosa (ROTH) KÜTZ.

Spec. Alg. p. 420. Conferva lanosa Roth, Cat. Bot. 3, p. 291.

f. *typica*

Descr. Conferva uncialis *c* ARESCH. Phyc. Scand. p. 428.
Fig. Spongomorpha lanosa KÜTZ. Tab. Phyc. 4 t. 83.
Exsicc. Conferva lanosa ARESCH. Alg. Scand. exsicc. N:o 181 et 228.

f. *uncialis* FL. DAN. (THUR.).

in Le Jol. Liste Alg. Cherb. p. 63. Conferva uncialis Fl. Dan. t. 771.

Descr. Conferva uncialis *a* et *b* ARESCH. Phyc. Scand. p. 427 et 428.
Fig. Spongomorpha uncialis KÜTZ. Tab. Phyc. 4, t. 80.
Exsicc. Conferva uncialis ARESCH. Alg. Scand. exsicc. N:o 130.

Syn. Cladophora lanosa KJELLM., Algenv. Murm. Meer. 54.
» » DICKIE, Alg. Sutherl. 2, p. 192, Alg. Cumberl. p. 239.
» uncialis DICKIE, Alg. Sutherl. 1, p. 143.
» » KLEEN, Nordl. Alg. p. 45.
Conferva » SOMMERF. Suppl. p. 196.

Habitat. In the Norwegian Polar Sea this species is usually litoral, sometimes sublitoral. Those few individuals of the typical form which I have found within the proper Arctic Sea, grew in the deeper parts of the sublitoral zone. The typical form is epiphytic on other algæ, f. *uncialis* is attached to stones or to poles, in the former case partly in spots left bare at low water, partly in rock-pools between tides. Both the forms occur on exposed as well as sheltered coasts, usually gregarious, but not in any considerable number within the Arctic Sea. Specimens with zoospores have been collected in August and September on the coast of Norway, in the middle of Jyly on the west coast of Novaya Zemlya.

Geogr. Distrib. It is known from the Polar Sea north of the Atlantic, and attains its maximum of frequency on the north-west coast of Norway. The northernmost locality is Lat. N. 73° 20′ on the west coast of Greenland.

Localities: The Norwegian Polar Sea: both the forms common at Nordlanden; Finmarken: Gjesvær (f. *uncialis*) scarce.

The Murman Sea: (f. *typica*) rare on the east coast of Novaya Zemlya.

Baffin Bay: Cumberland Sound (f. *typica*) abundant; the west coast of Greenland, f. *uncialis* at Omenak and in Whale Sound, f. *typica* at Lat. N. 73° 20′

Gen. **Cladophora** Kütz.

Phyc. gener. p. 262.

Cladophora Rupestris (L.) Kütz.

Phyc. gener. p. 270. Conferva rupestris L. Spec. Pl. p. 1167.

Descr. et Fig. Cladophora rupestris Harv. Phyc. Brit. t. 180.
Exsicc. Conferva rupestris Aresch. Alg. Scand. exsicc. N:o 126.

Syn. Cladophora rupestris Dickie, Alg. Sutherl. 1, p. 143.
» » Gobi, Algenfl. Weiss. Meer. p. 84.
» » Kjellm. Algenv. Murm. Meer. p. 53.
Conferva » J. G. Ag. Grönl. Alg. p. 110.
» » Aresch. Phyc. Scand. p. 420.
» » Nyl. et Sæl. Herb. Fenn. p. 75.
» » Post. et Rupr. Ill. Alg. p. II.
» » Wg. Fl. Lapp. p. 512.

Habitat. This species is usually litoral, growing within the formation of *Fucaceæ*, sometimes sublitoral, attached to stones, more seldom epiphytic on algæ. It occurs both on exposed and on sheltered coasts, but is more luxuriant and more typically developed in unprotected localities. It is gregarious, but not in very great masses. Specimens with zoospores have been found on the coast of Finmarken at the end of August.

Geogr. Distrib. It is known from the Polar Sea north of the Atlantic and has its maximum of frequency in the Norwegian Polar Sea. The most northerly point where it has been taken is on the west coast of Novaya Zemlya.

Localities: The Norwegian Polar Sea: Nordlanden common and abundant; Tromsö amt at Tromsö and Renö; Finmarken rather common and abundant at Maasö, Gjesvær, Öxfjord, and Talvik.

The Murman Sea: Russian Lapland and the coast of Cisuralian Samoyede-land; the west coast of Novaya Zemlya according to Gobi; the west coast of Waygats, only a single specimen known.

The White Sea: rather common and abundant.

Baffin Bay: the west coast of Greenland at Fiskernæs, Sukkertoppen, Claushavn, and Godhavn.

Cladophora Diffusa (Roth) Harv.

Phyc. Brit. t. 130. Conferva diffusa Roth, Cat. Bot. 3, p. 207.

Descr. et Fig. Cladophora diffusa Harv. l. c.

Syn. Cladophora diffusa Kjellm. Spetsb. Thall. 2, p. 54; Algenv. Murm. Meer. p. 54.

Habitat. This species grows sublitoral in 2—5 fathoms water on gravelly and sandy bottom on exposed coasts. All the specimens from the Polar Sea that I have seen were sterile. They were collected during July, August, and September.

Geogr. Distrib. It is nowhere to be found in any considerable number within the Polar Sea. I have seen it more abundant on the coast of Finmarken than anywhere else, but even here it is rare. It is not known at all beyond that part of the Polar Sea which lies north of the Atlantic. Its northernmost point is Fairhaven on the northwest coast of Spitzbergen, Lat. N. 79° 49'.

Localities: The Norwegian Polar Sea: Finmarken local and scarce at Maasö.

The Greenland Sea: the north-west coast of Spitzbergen, rare.

The Murman Sea: the west coast of Novaya Zemlya, washed ashore at Matotshin Shar.

Cladophora Glaucescens (GRIFF.) HARV.

Phyc. Brit. t. 196. Conferva glaucescens GRIFF. in HARV. Man. p. 139.

Descr. Cladophora glaucescens HARV. Phyc. Brit. l. c.
Fig. » » KÜTZ Tab. Phyc. 4, t. 24.
Syn. Cladophora sericea Kleen, Nordl. Alg. p. 45.
Conferva glomerata β marina WG. Fl. Lapp. p. 513.

Remark on this species. The habit of the present species differs rather much at different stages of development and at different places of growth. However it is always to recognized and distinguished from cognate species by its considerable size, its comparatively short cells which are of about the same length in proportion to their thickness in different parts of the frond, its straight branches that spring out at acute angles and taper strongly towards the tip, and its long chains of sporangia (gametangia) produced of strongly metamorphosized branches.

Habitat. This species is litoral, growing partly in such places as are left exposed at low-tide, partly in rock-pools and other depressions in which there remains seawater even at that time. It is always attached at first to some object, usually small stones, and when living in localities exposed to a heavy surge it remains fastened during all its life. When on the contrary it occurs in quiet sheltered spots, especially in water-pools with gravelly bottom — where it is often one foot or more in length — it finally detaches itself from its substratum and afterwards forms masses of occasionally most considerable size which lie loose on the bottom or float on the surface. It grows both on exposed and sheltered coasts and on the north coast of Norway it is furnished with zoospores in July and August.

Geogr. Distrib. This species is known only from the north-west and north coast of Norway. Its most northern point is Gjesvær about Lat. N. 71°.

Localities: The Norwegian Polar Sea: Nordlanden according to numerous specimens in KLEEN'S herbarium; Finmarken, abundant at Gjesvær and Talvik.

Cladophora Gracilis (GRIFF.) HARV.

Phyc. Brit. t. 18. Conferva gracilis GRIFF. sec. HARV. l. c.

Descr. Cladophora gracilis FARL. New Engl. Alg. p. 55.
» » HARV. l. c.
Fig. » » KÜTZ Tab. Phyc. 4, t. 23.
» vadorum » » » 4, t. 20.
Syn. Cladophora gracilis KLEEN, Nordl. Alg. p. 45.

Remark on the determination of this species. I hold it to be beyond a doubt that the alga which I understand by the above name, is identical with that described and figured by the authors quoted under the names of *C. gracilis* and *C. vadorum*. Though approaching nearly to *C. glaucescens*, it differs from it by essential characteristics. Its relation to the plant commonly called *C. gracilis* by Scandinavian algologists and distributed by me under N:o 119 in Wittr. et Nordst. Alg. exsicc., is a question I cannot decide at present. That plant differs much in its habitat from the present species.

Habitat. On the polar coasts I have always found this alga in hollows filled with water on gravelly shores. It is attached at first and it then presents its typical aspect. When older, it forms loose intricate masses, exhibiting very little resemblance to a typical *C. gracilis*, as has been correctly pointed out by FARLOW l. c. In this condition it rather agrees with *C. crispata* as delineated by KÜTZING, Tab. Phyc. 4, pl. 40, fig. 1 A. KLEEN says that he found it attached to other algæ in the upper part of the sublitoral zone. This appears to be its mode of growth also on the coasts of England (Cp. HARV. Phyc. Brit. tab. 18). But on the coasts of France and New England its habitat is the same as at Finmarken, and this seems to be the usual case. KÜTZING says that it grows »in stagnis marinis et submarinis» (Cp. Spec. Alg. p. 403). The present species is gregarious. I have collected specimens with zoospores on the coast of Finmarken at the beginning of September.

Geogr. Distrib. It is known from the Norwegian Polar Sea and Baffin Bay. The most northerly point where it is certainly known to grow, is Gjesvær about Lat. N. 71°.

Localities: The Norwegian Polar Sea: Nordlanden rather common; Finmarken, local and not abundant at Gjesvær and Talvik.

Baffin Bay: Greenland according to specimens in the herbarium of the Copenhagen Museum.

Cladophora Crispata (ROTH) RABENH.

Fl. Eur. Alg. p. 333 et 336. Conferva crispata ROTH.

With regard to this species found in the White Sea by K. VON BAER, the reader is referred to GOBI, Algenfl. Weiss. Meer. p. 85.

Gen. **Rhizoclonium** KÜTZ.

Phyc. gener. p. 261.

Rhizoclonium Rigidum GOBI.

Algenfl. Weiss. Meer. p. 85.

Descr. Rhizoclonium rigidum GOBI, l. c. p. 86.

Exsicc. Conferva fracta f. longissima subsimplex ARESCH. Alg. Scand. exsicc. N:o 273.

Syn. Cladophora fracta KLEEN, Nordl. Alg. p. 45.

Habitat. This species grows in lagoons or depressions in the litoral zone, and is at first attached, afterwards floating on the surface as loose masses of indefinite shape. It is gregarious. I do not know anything with regard to its propagation.

Geogr. Distrib. It is known in the Polar Sea north of the Atlantic, occurring in large numbers luxuriantly developed even at Spitzbergen. Its northernmost point is Advent Bay on the west coast of Spitzbergen Lat. N. 78° 15'.

Localities: The Norwegian Polar Sea: Nordlanden common according to KLEEN; Finmarken, the south coast of Magerö abundant.

The Greenland Sea: in the lagoons at Advent Bay on the west coast of Spitzbergen, abundant.

The White Sea: Golaja-Koshka Isle.

Rhizoclonium Pachydermum KJELLM.

Algenv. Murm. Meer. p. 55.

f. *typica.*

Descr. et *Fig.* Rhizoclonium pachydermum KJELLM. l. c. et t. 1, fig. 26—28.

f. *tenuis nob.*

f. quam f. typica tenuior et ramosior; ramis cauloideis 30—40 μ. crassis, e cellulis membrana tenuiore contextis, rhizoideis crebris.

Remark on the forms of this species. In the herbarium of the Copenhagen Museum there is a considerable quantity of *Rhizoclinium* from Greenland coinciding so closely with the *Rh. packydermum* from Novaya Zemlya described by me, that it can hardly be regarded as specifically distinct from this. It is much branched, even more so than the typical *Rh. pachydermum,* with two sorts of branches, numerous rhizoid branches, usually formed of more than three cells, and pretty scarce cauloid branches that give rise to rhizoid branches and even to one or two cauloid branches of a higher order. The sympodial main axis is much thicker than the secondary axes and ends downwards in an obovate cell with discoid expanded extremity. The cells of the main axis have very thick, distinctly stratiform walls. It is chiefly distinguished from the typical form by its cauloid branches being more slender, usually only 30—40 μ. in diameter. Their cell-walls are somewhat thinner than in f. *typica,* although even in f. *tenuis,* especially in the lower portions of the cauloid branches, they are much thicker than in the majority of known *Rhizoclonia.*

Habitat. This species belongs to the litoral zone. At Novaya Zemlya I saw it forming a thin stratum on rocks at high-water mark. I have afterwards met with it on the north coast of Sibiria in lagoons with somewhat brackish water, forming mats lying loose on the bottom.

Only sterile specimens have been observed.

Geogr. Distrib. It seems to be widely distributed within the Arctic region, and is probably circumpolar. The most northern place where it is known to occur is Karmakul Bay on the west coast of Novaya Zemlya, about Lat. N. 72° 30'.

Localities: The Murman Sea: the west coast of Novaya Zemlya, scarce at Karmakul Bay.

The Siberian Sea: Pitlekay.

Baffin Bay: the west coast of Greenland, locality unrecorded.

Rhizoclonium Riparium (ROTH) HARV.

Phyc. Brit. t. 238. Conferva riparia ROTH, Cat. Bot. 3, p. 216.

Descr. Conferva implexa ARESCH. Phyc. Scand. p. 434.
Fig. Rhizoclonium riparium HARV. l. c.
Exsicc. Conferva implexa ARESCH. Alg. Scand. exsicc. N:o 136.
Syn. Conferva arenosa CROALL, Fl. Disc. p. 46 (?).
» obtusangula SOMMERF. Suppl. p. 195.
Rhizoclonium litoreum ZELLER, Zweite d. Polarf. p. 82 (?).
» » DICKIE, Alg. Cumberl. p. 239.
» » KLEEN, Nordl. Alg. p. 46.

Remark on this species. The plant here called *Rh. riparium* seems to agree most nearly with that which ARESCHOUG l. c. has described under the name of *Conferva implexa.* The boreal and arctic form is somewhat coarser and has somewhat thicker cell-walls than that distributed by ARESCHOUG in Alg. Scand. exsicc.; it is somewhat less curved and in general has fewer rhizoid branches than is shown by the quoted figure in Harv. Phyc. Brit. However these characteristics vary much even in the same specimens.

Habitat. It grows in the uppermost part of the litoral region, or even somewhat above its upper margin, forming strata over rocks above tide-mark, sometimes alone, sometimes together with several other algæ, usually *Rhodochorton Rothii.* It is gregarious and flourishes in sheltered as well as exposed localities. I have found specimens with zoospores on the coast of Finmarken during the latter part of August.

Geogr. Distrib. It is known from the Polar Sea north of the Atlantic, and reaches its maximum of frequency on the coast of Norway.

The northernmost point where it has been taken, is Fairhaven on the north-west coast of Spitzbergen Lat. N. 79° 49'.

Localities: The Norwegian Polar Sea: Nordlanden, at several places at Saltenfjord abundant according to SOMMERFELT; Tromsö amt, rather plentiful at Tromsö; Finmarken, local, but plentiful at Maasö, Gjesvær, and Öxfjord.

The Greenland Sea: the west and north coasts of Spitzbergen, local, but pretty abundant; the east coast of Greenland (?).

The Murman Sea: the west coast of Novaya Zemlya, local, but pretty abundant.

Baffin Bay: Cumberland Sound, rather common; the west coast of Greenland at Frideriksdal and Tessarmiut. Also at Jakobshavn, in case *Conferva arenosa* CROALL is the present species.

Gen. **Chætomorpha** KÜTZ.

Phyc. Germ. p. 203.

Chætomorpha Melagonium (WEB. et MOHR.) KÜTZ.

l. c. p. 204. Conferva melagonium WEB. et MOHR. Reise p. 194.

f. *typica.*

Descr. Conferva melagonium WEB. et MOHR. l. c.
Fig. Chætomorpha Picquotiana KÜTZ. Tab. Phyc. 3, t. 58.
Exsicc. » melagonium WITTR. et NORDST. Alg. exsicc. N:o 415.

f. *rupincola* ARESCH.

Conferva melagonium var. rupincola ARESCH. Alg. Scand. exsicc. N:o 275, a.
Descr. Conferva melagonium HARV. Phyc. Brit. t. 99, A.
Fig. Chætomorpha melagonium KÜTZ. Tab. Phyc. 3, t. 61.
Exsicc. Conferva melagonium var. rupincola ARESCH. l. c.
Syn. Chætomorpha melagonium DICKIE, Alg. Cumberl. p. 239.
» » GOBI, Algenfl. Weiss. Meer. p. 87.
» » KJELLM. Algenv. Murm. Meer. p. 56.
» » KLEEN, Nordl. Alg. p. 45.
» » ZELLER, Zweite d. Polarf. p. 83.
Conferva linum POST. et RUPR. Ill. Alg. p. 11, Cfr. GOBI, l. c.
» melagonium J. G. AG. Progr. p. 2; Bidr. p. 11; Grönl. Alg. p. 110.
» » CROALL, Fl. Disc. p. 461.
» » DICKIE, Alg. Sutherl. 1, p. 143; 2, p. 192; Alg. Walker, p. 86.
» » EATON. Liste, p. 44.
» » KJELLM. Vinteralgv. p. 65; Spetsb. Thall. 2. p. 56; Kariska hafvets Algv. p. 29.
» » LINDBL. Bot. Not. p. 158.
» » LYNGB. Hydr. Dan. p. 148.
» » NYL. et SÆL. Herb. Fenn. p. 75.
» » WITTR. in Heugl. Reise, p. 284.

Habitat. The form *rupincola* is litoral, growing in rock-pools or on rocks covered with sand. The typical form, on the contrary, is always sublitoral, occurring in 2—15 fathoms water, chiefly on gravelly bottom. It grows sometimes on exposed coasts, sometimes in sheltered localities, attached to stones or (more seldom) to algæ. I have myself always found it scattered, but the large quantities of it brought home from Greenland seem to indicate that it occurs gregarious there, at least in certain places. On the north coast of Spitzbergen it develops during the whole winter. I have seen specimens with zoospores in July at Finmarken.

Geogr. Distrib. The present species is circumpolar and apparently attains its maximum of frequency in Baffin Bay. The most northerly locality where it has been found, is Discovery Bay in Smith Sound Lat. N. 81° 41'.

Localities: The Norwegian Polar Sea: Nordlanden (f. *typica* and f. *rupincola*) local and scarce; Finmarken at Maasö, f. *rupincola* rather common and abundant, f. *typica* local and scarce.

The Greenland Sea: (f. *typica*) commonly distributed, but not abundant on the coast of Spitzbergen; Sabine Island on the east coast of Greenland.

The Murman Sea: Russian Lapland and the coast of Cisuralian Samoyede-land; commonly spread, but scarce on the west coast of Novaya Zemlya and Waygats (f. *typica.*)

The Kara Sea: Uddebay, Cape Palander, and Actinia Bay (f. *typica*).

The American Arctic Sea: Port Kennedy, Assistance Bay, Beechy Island.

Baffin Bay: Cumberland Sound, common; Cape Boven; the west coast of Greenland, probably commonly distributed and abundant. I have seen f. *typica* from Tessarmiut, Friderikshaab, Neuherrnhut, and Godhavn. Besides, it is reported from Jacobshavn, Rittenbenk, Whale Sound, and Discovery Bay.

Chætomorpha Wormskioldii (Fl. Dan.) nob.

Conferva Wormskioldii Fl. Dan. t. 1547.

Descr. et *Fig.* Conferva Wormskioldii Lyngb. Hydr. Dan. p. 158; t. 55.

Habitat. It grows according to labels by J. Vahl »ad saxa et rupes in summo refluxu maris».

Geogr. Distrib. It is known only from Baffin Bay.

Locality: In the herbarium of the Copenhagen Museum there are found specimens taken at Godthaab. It is said by Fabricius to occur »sat vulgaris at litora maris Groenlandii» (Cp. Lyngb. l. c.).

Chætomorpha Linum (Roth) Kütz.

Phyc. Germ. p. 204. Conferva linum Roth, Cat. Bot. 3, p. 257.

Descr. Chætomorpha linum Kütz. Spec. Alg. p. 378.
Fig. » » Tab. Phyc. 3, t. 55.
Syn. Chætomorpha linum Gobi, Algenfl. Weiss. Meer. p. 87.

Habitat. Nothing is known in this respect.

Locality: The White Sea: at Solowetzki Isles according to Gobi l. c.

Chætomorpha Tortuosa (Dillw.) Kleen.

Nordl. Alg. p. 45. Conferva tortuosa Dillw. Brit. Conf. Syn. p. 46.

Descr. et *Fig.* Conferva tortuosa Aresch. Phyc. Scand. p. 433 et t. 3. f. G.
Exsicc. » » Aresch. Alg. Scand. exsicc. N:o 29.
Syn. Chætomorpha tortuosa Dickie, Alg. Cumberl. p. 240? Nonne Rhizoclonii spec.
» » Kleen, Nordl. Alg. p. 45; fide spec.
Conferva » Sommerf. Suppl. p. 196 (?) Nonne Rhizoclonii spec.

Habitat. I have found this species only in lagoons on sheltered coasts, forming dense masses of considerable size that lie loose on the bottom or float on the surface of the water. On this account it seems to me improbable that Sommerfelt's *Conferva tortuosa*, which is said to grow »supra arenam in rimis rupium», should be the present species. Just as *Chætomorpha tortuosa* Dickie (l. c.), it may possibly be some species of *Rhizoclinium*. The specimens of *Chætomorpha tortuosa* in Kleen's herbarium, according as I understand this species, have grown in the same kind of localities as those in which I have found the plant. Only sterile specimens are known from the Polar Sea.

Geogr. Distrib. It is found in the Norwegian Polar Sea and in Baffin Bay. The northernmost point where it is certainly known to occur, is Talvik in Altenfjord, about Lat. N. 70°.

Localities: The Norwegian Polar Sea: Nordlanden at Lofoten; Finmarken, local, but plentiful at Talvik.

Baffin Bay: The west coast of Greenland at Julianeshaab and Godhavn; Kikerton Islands (?)

Chætomorpha Septemtrionalis FOSLIE.

Arct. Havalg. p. 10.

Descr. et *Fig.* Chætomorpha septemtrionalis FOSLIE l. c. et t. 2, fig. 13.
Exsicc. » » FOSLIE in Wittr. et Nordst. Alg. exsicc. N:o 416.

Habitat. It occurs according to FOSLIE in the sublitoral region on dead bottom in 8—12 fathoms water.

Locality: The Norwegian Polar Sea: Finmarken at Gjesvær.

Gen. **Ulothrix** KÜTZ.

Alg. Dec. N:o 144; sec. Spec. Alg. p. 345.

Ulothrix (?) Sphacelariæ (FOSLIE) nob.

Chætomorpha Sphacelariæ FOSLIE, Arct. Havalg. p. 11.

Habitat. It grows epiphytic on *Sphacelaria arctica.*

Locality: The Norwegian Polar Sea: Finmarken at Honingsvaag.

Ulothrix Submarina KÜTZ.

Spec. Alg. p. 349.
Syn. Ulothrix submarina GOBI, Algenfl. Weiss. Meer. p. 88.

Locality: The White Sea: at Golaja-Koshka (Cp. GOBI l. c).

Ulothrix Discifera KJELLM.

Spetsb. Thall. 2, p. 52.
Descr. et *Fig.* Ulothrix discifera KJELLM. l. c. et t. 5, fig. 10—14.

Habitat. Litoral, attached to rocks, in company with *Enteromorpha compressa, Calothrix scopulorum,* and *Chætomorpha maritima,* hitherto found in small numbers and sterile.

Geogr. Distrib. Known only from the eastern part of the Greenland Sea. Its northernmost point is Fairhaven on the north-west coast of Spitzbergen, Lat. N. 79° 49'.

Localities: The Greenland Sea: Fairhaven and Duympoint on the coast of Spitzbergen, local, very scarce.

Gen. **Urospora** Aresch.

Obs. Phyc. 1, p. 15.

Urospora Penicilliformis (Roth) Aresch.

Obs. Phyc. 2, p. 4. Conferva penicilliformis Roth, Cat. Bot. 3, p. 271.

Descr. Urospora penicilliformis Aresch. l. c.

Fig. Lyngbya Carmichaelii Harv. Phyc. Brit. t. 186, A.

» Cutleriae » » » t. 336.

» speciosa » » » t. 186, B.

Conferva Youngeana » » » t. 328.

Exsicc. Conferva speciosa Aresch. Alg. Scand. exsicc. N:o 132, 185.

» hormoides » » » » » 133, 186.

Hormiscia flacca » » » » » 342.

Urospora mirabilis » » » » » 340.

» penicilliformis Wittr. et Nordst. Alg. exsicc. N:o 417, 418.

Syn. Conferva hormoides J. G. Ag. Spetsb. Alg. Progr. p. 2; Bidr. p. 11.

» » Lindbl. Bot. Not. p. 158.

» » Wittr. in Heugl. Reise, p. 284.

» Youngeana Croall, Fl. Disc. p. 461.

Lyngbya Carmichaelii Croall, l. c. p. 461.

» flacca » » » 462.

» speciosa » » » »

Urospora mirabilis Kleen, Nordl. Alg. p. 46.

» penicilliformis Kjellm. Spetsb. Thall. 2, p. 55; Algenv. Murm. Meer. p. 56.

Habitat. This alga is litoral, usually growing on stones and rocks above tide-mark, sometimes in rock-pools within or somewhat above the litoral zone. It grows gregarious in considerable masses both on exposed and sheltered coasts, but exposed localities are preferred by it. I have found specimens with zoospores at Spitzbergen in the middle of Jyly, at Novaya Zemlya at the end of Jyly, on the north coast of Norway in August.

Geogr. Distrib. It is probably circumpolar. However it has not been found as yet in the Siberian and American Arctic Seas. Its maximum of frequency is on the north coast of Norway. The most northern place where it has been taken is Treurenberg Bay on the north coast of Spitzbergen, Lat. N. 79° 56'.

Localities: The Norwegian Polar Sea: Nordlanden, common; Finmarken, local, but abundant at Maasö.

The Greenland Sea: the north-east and south-east coasts of Spitzbergen, local and scarce.

The Kara Sea: Kjellman's Iles, scarce.

Baffin Bay: the coast of Greenland at Frideriksdal, Nanortalik, Sermilik, Julianehaab, Godthaab. Besides, it has been found floating in the sea at several places off the coast. (Cp. Croall l. c. and Dickie, Alg. Sutherl. 1, p. 143).

(?) Gen. **Bulbocoleon** PRINGSH.
Morph. Meeresalg. p. 8.

Bulbocoleon Piliferum PRINGSH.
l. c.
Descr. et *Fig.* Bulbocoleon piliferum l. c. et t. 1.
Syn. Bulbocoleon piliferum CIENK. Bericht. p. 24.

Habitat. It lives endophytic in other algæ. I have found it in *Dumontia filiformis* and *Myrionema strangulans.* I have seen only sterile individuals.

Geogr. Distrib. Found within the Polar Sea at Nordlanden on the coast of Norway and in the White Sea. I cannot state any exact localities.

Fam. DERBESIACEÆ THUR.
in LE JOL. Liste Alg. Cherb. p. 14.

Gen. **Derbesia** SOLIER.
Ann. d. Sc. p. 158.

Derbesia Marina (LYNGB.) SOLIER.
l. c. Vaucheria marina LYNGB. Hydr. Dan. p. 79.
Descr. et *Fig.* Derbesia marina SOLIER, l. c. et t. 9. fig. 1—17.

Habitat. It grows in the Polar Sea in the deeper parts of the sublitoral zone, in 10—20 fathoms, attached to *Lithothamnion soriferum* and corals. I have found it only on an exposed coast, in small number. It bore almost ripe zoosporangia at the end of August.

Geogr. Distrib. Known only from the Norwegian Polar Sea.

Locality: Finmarken at Gjesvær.

Fam. BRYOPSIDEÆ THUR.
In LE JOL. Liste Alg. Cherb. p. 14.

Gen. **Bryopsis** LAMOUR.
Ess. p. 281.

Bryopsis Plumosa (HUDS.) AG.
Spec. Alg. 1, p. 448. Ulva plumosa HUDS. Fl. Angl. p. 571.
Descr. et *Fig.* Bryopsis plumosa HARV. Phyc. Brit. t. 3.
Exsicc. » » ARESCH. Alg. Scand. exsicc. N:o 422.

Syn. Bryopsis plumosa ASHM. Alg. Hayes, p. 96.
» » CROALL. Fl. Disc. p. 460.
» » KLEEN, Nordl. Alg. p. 40.

Habitat. On the north coast of Norway it grew in rock-pools in the litoral zone, scattered, attached to stones or *Lithothamnion polymorphum.* The Norwegian specimens collected in July and August are sterile and very feebly developed.

Geogr. Distrib. Known from the Norwegian Polar Sea and Baffin Bay. In the latter region it goes far northwards, being reported by ASHMEAD from Smith Sound between Lat. N. 78° and 82°.

Localities: The Norwegian Polar Sea: Nordlanden, rather rare at Fleinvær and Röst.

Baffin Bay: The west coast of Greenland, exact locality unrecorded; Smith Sound.

Fam. CHARACIACEÆ (NÄG.) WITTR.

Gotl. och Öl. Alg. p. 32; NÆG. Gatt. einz. Alg. p. 64, excl. gen.

Gen. **Characium** AL. BRAUN.

in KÜTZ. Spec. Alg. p. 208.

Characium Marinum nob.

Descr. et *Fig.* Characium spec. KJELLM. Spetsb. Thall. 2, p. 57 et t. 4, fig. 10.

Habitat. Of this alga, for which I think fit now to propose a special name, I have found only one individual. This was attached to *Pylaiella litoralis* and furnished with ripe zoospores, when taken in the month of October.

Locality: The Greenland Sea: Musselbay on the north coast of Spitzbergen.

Gen. **Codiolum** Al. BRAUN.

Alg. unic. p. 19.

Codiolum Longipes FOSLIE.

Arct. Havalg. p. 11.

Descr. et *Fig.* Codiolum longipes Foslie l. c. et t. 2, fig. 4.
Exsicc. » » » in WITTR. et NORDST. Alg. exsicc. N:o 458.

Addition to the description of the species. The stipe does not always pass into the club-head so without a limit, as appears in the figures of FOSLIE. Among the specimens kindly communicated to me by FOSLIE there are several that accord nearly with *C. gregarium* AL. BRAUN in regard to the shape of the club-head. I think I ought to point out this fact, because it shows that *C. longipes* is a species very slightly differentiated from *C. gregarium.*

Habitat. It has been found attached to an iron column exposed to the waves (investiens columnam ferream maris undis expositam: Foslie). I have not seen it with zoospores, but the specimens distributed by FOSLIE in the above-mentioned collection of exsiccatæ are near the stage when the zoospores are to be developed. These specimens having been collected in the beginning of September, the season for producing zoospores may be assumed to be the latter part of this month.

Locality: The Norwegian Polar Sea: Finmarken at Gjesvær.

Codiolum Pusillum (LYNGB.) KJELLM.

in FOSLIE, Arct. Havalg. p. 12. Vaucheria pusilla LYNGB. Hydr. Dan. p. 791.
Descr. Codiolum pusillum FOSLIE l. c.
Fig. » » » » t. 2, fig. 1.
Vaucheria pusilla LYNGB. l. c. t. 22, fig. B.
Exsicc. Codiolum pusillum FOSLIE in WITTR. et NORDST. Alg. exsicc. N:o 457.

Remark on this species. The Norwegian specimens differ on some points from those which are preserved in the herbarium of the Copenhagen Museum under the name of *Vaucheria pusilla* by the determination of LYNGBYE. In the former the club-head is longer in proportion to the shaft than in the latter. For although these often have the club-head longer than the shaft, it is never 1 ½—3 times longer, as is stated by FOSLIE to be the case in Norwegian specimens; it is also often about as long as the shaft or even shorter. The present species is recognized from the other species of the genus by its considerable length, by the shape of the club-head, and by the slight thickness of the club-head in proportion to its length.

Habitat. This alga clothes with a thin cover rocks exposed to the surge in the lower part of the litoral zone. Like other species of this genus it grows gregarious in large masses. I know only sterile specimens collected at the end of August.

Locality: The Norwegian Polar Sea: Finmarken at Gjesvær.

Codiolum Nordenskiöldianum KJELLM.

Spetsb. Thall. 2, p. 56.
Descr. et *Fig.* Codiolum Nordenskiöldianum KJELLM. l. c. et t. 5. fig. 1—9.
Exsicc. » » » in ARESCH. Alg. Scand. exsicc. N:o 425 et WITTR. et NORDST. Alg. exsicc. N:o 51.

Remark on the species. When I described the present alga from specimens collected at Spitzbergen, only one species of this genus was known with certainty, viz. *C. gregarium* described by the founder of the genus, AL. BRAUN. The species from Spitzbergen was so essentially different from this, that it could not but be regarded as an independent species. However, another species of the same generic type was described and delineated long ago, alhough this circumstance had escaped both AL. BRAUN and myself. After the publication of my treatise on the algæ of Spitzbergen I began to suspect that LYNGBYE's *Vaucheria pusilla* was in fact a *Codiolum*, and this suspicion was confirmed when some years ago I had the opportunity of examining LYNGBYE's collections of algæ preserved in the Copenhagen Museum. This *Codiolum* agrees more closely with *C. Nordenskiöldianum* than with *C. gregarium*. FOSLIE afterwards found the same plant at Finmarken. He detected there yet another *Codiolum*, which has been described by him under the name of *C. longipes* and is intermediate between *C. gregarium* and *C. Nordenskiöldianum*. It resembles *C. gregarium* in size and with regard to the proportion in length between the club-head and the stipe, *C. Nordenskiöldianum* with regard to the shape of the club-head. After I had described *C. Nordenskiöldianum*, I have also myself found a *Codiolum* on the coast of Finmarken, which I have distributed under this name in the above-quoted collection of exsiccatæ. It does not however completely coincide with the Spitzbergen form, being larger than this and not always exhibiting the same proportion as this between the length of the club-head and the shaft. Probably it is to be considered an intermediate form between *C. Nordenskiöldianum* from Spitzbergen and *C. longipes* from the north coast of Norway. A fourth form of the genus has been lately observed on the north-east coast of North America. According to what FOSLIE has communicated to me by letter, it may be identified with *C. longipes*, although it differs somewhat from this. It appears to me to agree more nearly than *C. longipes* with *C. gregarium*. It results from these facts that the genus *Codiolum* has shown itself of late to possess in the northern seas a considerable number of forms that are only slightly differentiated and should possibly be justly regarded as forms of one and the same species. It is evidently a genus in course of developing species. Nevertheless the species distinguished ought to be kept up, until more forms shall have been discovered at other places. This will no doubt happen now, since more attention has been directed to these small and easily overlooked algæ. The last finds prove that they are no rarities and have an extensive range on the coasts of the old as well as the new world. In the present work I understand *C. Nordenskiöldianum* as including both the form from Spitzbergen and that from Finmarken. Its length does not in general exceed 600 μ. The thickness of the club-head usually amounts to 25—50 μ., sometimes 70 μ. in specimens from Finmarken. The head of the club is in most cases longer than the shaft, sometimes as long as or somewhat shorter than the shaft, and elongated-obovate in shape.

Habitat. It forms a thin cover on stones within the litoral zone on exposed coasts, together with *Urospora penicilliformis*. It is gregarious, occurring in considerable masses. I have found specimens with zoospores in July on the north coast of Spitzbergen, towards the end of August on the coast of Finmarken.

Geogr. Distrib. Known from the Norwegian Polar Sea and the eastern Greenland Sea. Its northernmost point is Duympoint on the north coast of Spitzbergen Lat. N. 79° 30'.

Localities: The Norwegian Polar Sea: Finmarken at Maasö, local, but abundant.

The Greenland Sea: the north coast of Spitzbergen, Duympoint, local, not abundant.[1])

[1]) It has been stated by mistake, that the *C. Nordenskiöldianum* distributed under N:o 425 in Alg. Scand. exsicc. has been collected at *Insulæ Spetsbergenses*. This is not true. It has been taken on the north coast of Norway at Maasö in Finmarken, at the same place and time as the specimens distributed in Wittr. et Nordst. Alg. exsicc. N:o 51.

Gen. **Chlorochytrium** Cohn.

Biol. Pflanz, 1,2. p. 102.

Chlorochytrium Inclusum nob.

Chl. in statu vegetativo sphæricum vel subsphæricum in planta gestatrice omnino inclusum, evolutione zoosporarum instante paullulo prolongatum, depresso-conicum, ampullæforme, ovoideum vel ellipsoideum, demum vertice apiculato telam corticalem plantæ gestatricis penetrante nudum, ostiolo formato zoosporas emittens. Tab. 31, fig. 8—17.

Description. This alga lives endophytic in *Sarcophyllis arctica*, being placed in most cases near the surface of the nurse-plant, sometimes in the middle of it (fig. 8). In its vegetative state it is completely enclosed in the nurse-plant, being covered at least by its cortical layer, but sometimes surrounded by its middle layer which is formed of branching cell-filaments (fig. 9). It then has a spherical or almost spherical shape, 80—100 μ. in diameter. Its colour is yellowish green. The thin cell-wall is of equal thickness. The chromatophore is thin, spread along the whole of the wall. At that period when the zoospores are to be produced, the cell is elongated in the direction of the nearest surface of the nurse-plant, and becomes ovoid, ellipsoidical, short cone-shaped, or bottle-shaped (fig. 12—15). The membrane grows thicker especially at the side facing outwards, and there is formed here a short cone-shaped outgrowth of cellulosa, which contributes probably to the piercing of the cortical layer. The alga assumes a more intense yellowish-green colour, the coloured plasma increases in mass and is finally divided into a large number of densely packed zoospores. As included in the nurse-plant, the alga, both in its vegetative and in its fructiferous stage, possesses a greater or less number of different bulgings, evidently caused by the surrounding tissue impeding its equal growth (fig. 9, 11, 16). The zoospores issue through an opening formed by the dissolution of the cell-wall at the top of the cell beyond the nurse-plant (fig. 16). As to the structure, germination and further development of the zoospores I know nothing, because I have only had the opportunity of examining dried specimens. Those individuals which lie in the central part of the nurse-plant are usually developed to a far greater size than the others. Their longest diameter can attain even 275 μ. Their membrane is much thickened equally. This may possibly be a state of rest which is interrupted when at the dissolution of the nurse-plant those individuals are liberated. I have hesitatingly referred the present alga to the genus *Chlorochytrium*, with the other species of which it has much in common. The question as to what genus it rightly belongs to, can only be determined with certainty, when the history of its development shall be known.

Habitat. All the specimens of *Sarcophyllis edulis* that I have examined, at whatever degree of longitude and latitude and whatever season of the year they have been taken, have been found to contain a greater or less number of this endophyte. I have found it most plentiful and most strongly developed in specimens of *Sarcophyllis arctica* collected in the winter months, especially in December. It was then so numerous as to be counted in hundreds in a couple of square millimeters of the surface of the nurse-plant. At this season it is richly provided with zoospores. However I think I have found specimens with zoospores also in summer.

The present alga has the same geographical distribution as *Sarcophyllis arctica*. In all the localities where I have met with the last-mentioned species, I have also found *Chl. inclusum*.

Fam. PALMELLACEÆ (Näg.) Wittr.

Pl. Scand. p. 19; Näg. Gatt. einz. Alg. p. 61; lim. mut.

Gen. **Chlorangium** Cienk.

Bericht. p. 23.

Chlorangium Marinum Cienk.

l. c. et t. 1, fig. 7—9.

This species is reported from the White Sea by Cienkowsky (Cp. l. c.).

Ser. **NOSTOCHINEÆ** (Ag.) Näg.

sub. nom. Nostocaceæ Algensyst. p. 132; Ag. Syst. Alg. p. XV; lim. mut.

Fam. RIVULARIACEÆ Harv.

in Engl. Fl. 5, p. 262.

Syn. Calotricheæ Thur. Nostoch. p. 10.

Gen. **Rivularia** (Roth) Thur.

Nostoch. p. 5; Roth, Cat. Bot. 3, p. 332; char. emend.

Rivularia Hemisphærica (L.) Aresch.

Alg. Scand. exsicc. Ser. 1, N:o 47. Tremella hemisphærica L. Spec. Pl. 2, p. 1158; sec. Aresch. Phyc. Scand. p. 437.

Descr. et *Fig.* Rivularia atra Harv. Phyc. Brit. t. 239.

Exsicc. » hemisphærica Aresch. l. c.

Syn. Rivularia hemisphærica Foslie, Arct. Havalg. p. 9.
Linkia atra Sommerf. Suppl. p. 201.

Habitat. It grows in the upper part of the litoral zone on sheltered coasts, gregarious, but in small numbers. I have taken specimens with hormogonia in course of development at Finmarken in the beginning of September.

Geogr. Distrib. Known from the north coast of Norway and the west coast of Greenland. The most northern point where it has been found is Talvik in Finmarken, about Lat. N. 70°.

Localities: The Norwegian Polar Sea: Nordlanden according to specimens in KLEEN'S herbarium; Finmarken at Talvik, local, but rather abundant.

Baffin Bay: The west coast of Greenland at Tessarmiut, Ameralik, and Pikitsok, according to specimens in the herbarium of the Copenhagen Museum.

Rivularia Microscopica DICKIE.

Alg. Sutherl. 2, p. 193.

Descr. Rivularia microscopica DICKIE, l. c.

Habitat. Found growing on *Enteromorpha compressa*.

Locality: The American Arctic Sea: Assistance Bay, »and other localities», according to DICKIE, l. c.

Gen. **Calothrix** (AG.) THUR.

Nostoch. p. 5; AG. Syst. Alg. p. XXIV; char. mut.

Calothrix Harveyi nob.

Descr. et *Fig.* Calothrix fasciculata HARV. Phyc. Brit. t. 58, A.

Remark on this species. I share the opinion of ARESCHOUG, pronounced in Phyc. Scand. (p. 439), that AGARDH'S *C. scopulorum* and *C. fasciculata* are forms (or developments) of the same species, *C. scopulorum* WEB. et MOHR. *C. fasciculata* HARV. is no doubt specifically distinct from this. I have found the last-named alga at Finmarken, and I propose for it the name of *C. Harveyi*. HARVEY'S description of it is satisfactory. Only it should be observed that *C. Harveyi* is sometimes almost simple, sometimes richly branched, just as there is to be found of *C. scopulorum* WEB. et MOHR one form that is luxuriant and branching and another that is less luxuriant and only little branching.

Habitat. This alga forms dense mats several feet in extent on mouldering woodwork at high-water mark. In the Polar Sea it has been found only in sheltered localities. When I collected it in the beginning of September, the hormogonia were in development.

Locality: The Norwegian Polar Sea: Finmarken at Talvik, local, but abundant.

Calothrix Scopulorum (WEB. et MOHR) AG.

Syst. Alg. p. 70. Conferva scopulorum WEB. et MOHR, Reise, p. 195.

Descr. Calothrix scopulorum ARESCH. Phyc. Scand. p. 438.

» » » Alg. Scand. exsicc. N:o 139.

Syn. Schizosiphon scopulorum KJELLM. Spctsb. Thall. 2, p. 58. Algenv. Murm. Meer. p. 57.

Habitat. The present species in the Polar Sea forms, in conjunction with several other algæ, *Rhodochorton Rothii*, *Chætophora maritima*, *Urospora penicilliformis*, species of *Codiolum* a. o., a thin stratum over stones within the upper part of the litoral zone. It is scarce here, and nowhere occurs in great masses. It has been found both on exposed and sheltered coasts. I have collected specimens with hormogonia in development on the coast of Spitzbergen in July.

Geogr. Distrib. Known from the Polar Sea north and north-east of the Atlantic. Its maximum of frequency is on the coast of Norway. Its most northern point is Fairhaven on the north-west coast of Spitzbergen, Lat. N. 79° 49'.

Localities: The Norwegian Polar Sea: Finmarken at Maasö and Giesvær, rather common and abundant.

The Greenland Sea: rather common, but in small numbers on the west coast of Spitzbergen.

The Murman Sea: local and scarce on the west coast of Novaya Zemlya and Waygats.

Calothrix Confervicola AG.

Syst. Alg. p. 70.

Descr. et *Fig.* Calothrix confervicola BORN. et THUR. Not. algol. 1, p. 8 et t. 3..
Syn. Calothrix confervicola GOBI, Algenfl. Weiss. Meer. p. 88.

Habitat. Found attached to *Phlœospora tortilis.*

Locality: The White Sea: Solowetzki Isles (Cfr. GOBI, l. c.)

Fam. OSCILLARIACEÆ (AG.) NORDST.

Pl. Scand. p. 53. Oscillatorinæ AG. Syst. Alg. p. XXIV; lim. mut.

Gen. **Lyngbya** (AG.) THUR.

Nostoch. p. 4; AG. Syst. Alg. p. XXV; char. emend.

Lyngbya Semiplena (HARV.) nob.

Descr. et *Fig.* Calothrix semiplena HARV. Phyc. Brit. t. 309. Cfr. Symploca Harveyi THUR. Nostoch. p. 8.

Habitat. Found growing on wood-work within the litoral zone in a sheltered locality, in small numbers.

Locality: The Norwegian Polar Sea: Finmarken at Talvik, local, scarce.

Gen. **Oscillaria** (BOSC.) THUR.

Nostoch. p. 4 et 7; Bosc. sec. THUR. l. c.

Oscillaria Subsalsa AG.

Oscillatoria subsalsa AG. Syst. Alg. p. 66.

Descr. Oscillaria subsalsa KÜTZ. Spec. Alg. p. 246.
Fig. » » » Tab. Phyc. 1, t. 42, fig. 5.

Habitat. According to a label, it grows »in fossis submarinis».

Locality: Baffin Bay: Tessarmiut on the west coast of Greenland according to specimens in the herbarium of the Copenhagen Museum.

Gen. **Spirulina** TURP.

Sec. THUR. Nostoch. p. 7.

Spirulina Tenuissima KÜTZ.

Phyc. gener. p. 183.

Descr. Spirulina tenuissima FARL. New. Engl. Alg. p. 31.
Fig. » » HARV. Phyc. Brit. t. 105, C.

Habitat. The specimens that I have had the opportunity to examine, have grown in company with the preceding alga in cavities with brackish water among mouldering algæ.

Locality: Baffin Bay: Tessarmiut on the west coast of Greenland, according to specimens taken by Wormskiold.

Uncertain Species.

Halosaccion dumontioides DICKIE, Alg. Cumberl. p. 239.
Ceramium tenuissimum J. G. AG. Spetsb. Alg. Bidr. p. 11.
Ectocarpus Durkeei (?) DICKIE. Alg. Sutherl. I, p. 142.
Cladophora Inglefieldii » » » 1, p. 143.
Conferva ærea » » » 2, p. 192.
Gleocapsa spec. CIENK. Bericht. p. 22,

List of Literature.

AGARDH, C. A. Dispositio Algarum Sueciæ. Lundæ 1810—12. — (AG. Disp. Alg.)

» Synopsis Algarum Scandinaviæ. Lundæ 1817. — (AG. Syn. Alg.)

» Systema Algarum. Lundæ 1824. — (AG. Syst. Alg.)

» Species Algarum. Gryphiswaldiæ 1821—1828. — (AG. Spec. Alg.)

AGARDH, J. G. Novitiæ Floræ Sueciæ ex Algarum familia. Lundæ 1836. — (J. G. AG. Nov.)

» In historiam Algarum symbolæ. — Linnæa von Schlechtendal. Vol. 15. Halle 1841. — (J. G. AG. Symb.)

» Algæ maris Mediterranei et Adriatici. Parisiis 1842. — (J. G. AG. Alg. Med.)

» In systemata Algarum hodierna adversaria. Lundæ 1844. — (J. G. AG. Advers.)

» Nya alger från Mexico. Öfversigt af Kongl. Vetenskaps-Akademiens Förhandlingar 1847. Stockholm 1848. — (J. G. AG. Alg. Liebm.)

» Species, genera et ordines Algarum. Lundæ 1848—1863. — (J. G. AG. Spec. Alg.)

» Om Spetsbergens Alger. Akademiskt program. Lund 1862. — (J. G. AG. Spetsb. Alg. Progr.)

» De Laminarieis symbolas offert. — Lunds Universitets Årsskrift. Tome 4. Lund 1867. — (J. G. AG. Lam.)

» Bidrag till kännedomen af Spetsbergens Alger. — Kongl. Svenska Vetenskaps-Akademiens Handlingar. Band 7, N:o 8. Stockholm 1868. — (J. G. AG. Spetsb. Alg. Bidr.)

» Bidrag till kännedomen af Spetsbergens Alger. Tillägg till föregående afhandling. — Anf. st. — (J. G. AG. Spetsb. Alg. Till.)

» Algæ Spetsbergenses. — (J. G. AG. Enum.)

» Alger insamlade på Grönland 1870 af dr Sv. Berggren och P. Öberg, bestämda af prof. J. G. AGARDH. — Redogörelse för en expedition till Grönland år 1870 af A. E. Nordenskiöld in Öfversigt af Kongl. Vetenskaps-Akademiens Förhandlingar 1870. Stockholm 1871. — (J. G. AG. Grönl. Alg.)

» Bidrag till kännedomen af Grönlands Laminarieer och Fucaceer. — Kongl. Svenska Vetenskaps-Akademiens Handlingar. Band 10. N:o 8. Stockholm 1872. — (J. G. AG. Grönl. Lam. och Fuc.)

» Epicrisis systematis Floridearum. Lipsiæ 1876. — (J. G. AG. Epicr.)

» Till Algernes Systematik. Nya bidrag. — Lunds Universitets Årsskrift. Tome 17. Lund 1880—1881. — (J. G. AG. Alg. Syst. 2.)

AHLNER, K. Bidrag till kännedomen om de Svenska formerna af Algslägtet Enteromorpha. Akademisk afhandling. Stockholm 1877. — (Ahln. Enterom.)

ARESCHOUGH, J. E. Algæ Scandinavicæ exsiccatæ. Fasc. 1—3. Gotoburgi 1840—1841. (ARESCH. Alg. Scand. exsicc. ser. 1.)

» Algarum (Phycearum) minus rite cognitarum pugillus secundus. — Linnæa von Schlechtendal. Tome 17. Halle 1843. — (ARESCH. Pugill.)

ARESCHOUG, J. E. Phycearum, quæ in maribus Scandinaviæ crescunt, enumeratio. — Nova Acta regiæ Societatis scientiarum Upsaliensis. Vol. 13 & 14. Upsaliæ 1847, 1850. — (ARESCH. Phyc. Scand.)

» Algæ Scandinavicæ exsiccatæ. Ser. nova. Fasc. 1—9. Upsaliæ 1861—1879. — (ARESCH. Alg. Scand. exsicc.)

» Observationes Phycologicæ 1—3. — Acta regiæ Societatis scientiarum Upsaliensis. Ser. 3, Vol. 6, 9. 10. Upsaliæ 1866, 1874, 1875. — (ARESCH. Obs. Phyc.)

» Slägtena Fucus (L.) DECAISNE et THURET och Pycnophycus KÜTZ, jemte tillhörande arter. — Botaniska notiser 1868. — (ARESCH. Fuc. et Pycnoph.)

» Om de skandinaviska Algformer, som äro närmast beslägtade med Dictyosiphon foeniculaceus eller kunna med denna lättast förblandas. — Botaniska Notiser 1873. — (ARESCH. Bot. Not. 1873.)

» De Algis nonnullis maris Baltici et Bahusiensis. — Botaniska Notiser 1876. — (ARESCH. Bot. Not. 1876)

ASHMEAD, S. Algæ. — Enumeration of Arctic Plants collected by dr J. J. Hayes in his exploration of Smith's Sound between parallels 78th and 82d during the months of July, August and beginning of September 1861. — Proceedings of the Academy of Natural Sciences of Philadelphia 1863. Philadelphia 1864. — (ASHM. Alg. Hayes.)

BERGGREN, S. Musci et Hepaticæ Spetsbergenses. — Kongl, Svenska Vetenskaps-Akademiens Handlingar. Band 13, N:o 7. Stockholm 1875. (BERGGREN, Musci Spetsb.)

BERTHOLD, G. Die Bangiaceen des Golfes von Neapel und der angrenzenden Meeres-Abschnitte. — Fauna und Flora des Golfes von Neapel und der angrenzenden Meeres-Abschnitte herausgeben von der zoologischen Station zu Neapel. Leipzig 1882. — (BERTH. Bangiaceen.)

BORNET, ED. et THURET, G. Notes Algologiques. Fasc. 1, 2. Paris 1876, 1880. — (BORN. et THUR. Not. algol.)

» » Études Phycologiques. Paris 1878. — (BORN. et THUR. Etud. Phycol.)

BORY DE SAINT VINCENT. Dictionnaire classique d'Histoire naturelle. Paris 1822—1831. — (BORY, Dict. Class.)

BRAUN, AL. Algarum unicellularium genera nova et minus cognita. Lipsiæ 1855. — (AL. BRAUN. Alg. unic.)

BROWN, R. Catalogue of Plants found in Spitsbergen. — An Account of the arctic Regions with a History and description of the Nothern Whale-Fishery by W. Scoresby. Appendix 5. Vol. 1. Edinburgh 1820. — (R. BR. in Scoresby, Account.)

BÖRGEN, C. Aräometer-Beobachtungen. — Die Zweite Deutsche Nordpolarfahrt in den Jahren 1869 und 1870 unter Führung des Kapitain Karl Koldeway. Herausgegeben von dem Verein für die Deutsche Nordpolarfahrt in Bremen. Band 2. Leipzig 1874. — (BÖRGEN Zweite d. Polarf.)

CHAUVIN. J. et ROBERGE, M. Algues de la Normandie. Caen 1826—1831. — (CHAUV. Alg. Norm.)

CIENKOWSKY, L. Bericht über die Excursion zum Weissen Meer 1880. Russian text. — (CIENK. Bericht.)

Contributions to our knowledge of the Meteorology of the arctic regions. Published by the Authority of the meteorological council. Part II. London 1880. — (Contrib. Arct. Meteor.)

COOLEY, W. D. Physical Geography. London 1876. — (Cool. Phyc. Geogr.)

COHN, F. Beiträge zur Biologie der Pflanzen. — Band. 1, Del. 2. Breslau 1872. — (COHN, Biol. Pflanz.)

CROALL, A. Marine Algæ. — The Florula Discoana. Contributions to the Phyto-Geography of Greenland within the parallels of 68° and 70° North Latitud by R. BROWN. — Transactions of the Botanical Society. Vol. 9. Edinburgh 1868. — (CROALL, Fl. Disc.)

CROUAN P. L. et H. M. Algues marines du Finistère classées. Brest 1852. — (CROUAN, Alg. Finist.)

» » Florule du Finistère. Paris 1867. — (CROUAN, Flor.)

» » Observations sur le genre Peyssonelia. — Annales des Sciences naturelles. Ser. 3, Tome 2. Paris 1844. — (CROUAN. Ann. d. Sc.)

DECAISNE, J. Essai sur une classification des Algues et des Polypiers calcifères de Lamouroux. — Annales des Sciences naturelles. Ser. 2. Tome 17. Paris 1842. — (DCSNE. Class.)

» Plantes de l'Arabie heureuse receuillies par P. E. BOTTA. Paris 1841. — (DCSNE Pl. Arab.)

» et THURET, G. Recherches sur les anthéridies et les spores de quelques Fucus. — Annales des Sciences naturelles. Ser. 3, Tome 3. Paris 1845. — (DCSNE et THUR. Rech. Fuc.)

DE CANDOLLE, A. P. et DE LAMARK, J. B. Flore Française. Ed. 3. Tome 2. Paris 1815. — (DC. Fl. Fr.)

DE LA PYLAIE, M. Quelques observations sur les productions de l'île de Terre Neuve et sur quelques Algues de la côte de France, appartenant au genre Laminaire. — Annales des Sciences Naturelles Ser. 1, Tome 4. Paris 1824. — (DE LA PYL. Prod. Terre neuve.)

» Flore de l'île de Terre neuve et des îles S:t Pierre et Niclon. Paris 1829. (DE LA PYL. Fl. Terre neuve.)

DICKIE, G. Notes on the Algæ. — Journal of a Voyage in Baffin's Bay and Barrow Straits in the Years 1850—1851, by P. C. Sutherland. Vol. 2. London 1852 — (DICKIE, Alg. Sutherl. 1.)

» Notes on Flowering plants and Algæ collected during the Voyage of the »Isabel». — A Summer Search for Sir John Franklin with a Peep into the Polar Basin by Inglefield. London 1853. — (DICKIE, Alg. Sutherl. 2.)

» Algæ. — An Account of the Plants collected by dr Walker in Greenland and arctic America during the Expedition of Sir Francis M'Clintock in the Yacht »Fox». — Journal of the Proceedings of the Linnean Society. Botany. Vol. 5. London 1861. — (DICKIE, Alg. Walker.)

» Notes on a Collection of Algæ procured in Cumberland Sound by Mr James Taylor and Remarks on arctic species in general. — The Journal of the Linnean Society. Botany. Vol. 9. London 1867. — (DICKIE, Alg. Cumberl.)

» On the Algæ found during the arctic Expedition (under the command of Nares). From The Journal of the Linnean Society. Botany. Vol. 17. — (DICKIE, Alg. Nares.)

DILLWYN, L. W. British Confervæ. London 1809. —

DUNÉR, N. och NORDENSKIÖLD, A. E. Anteckningar till Spetsbergens Geografi. — Kongl. Svenska Vetenskaps-Akademiens Handlingar. Band 6, N:o 5. Stockholm 1865. — (DUNÉR, NORDENSKIÖLD, Spetsb. Geogr.)

EATON, A. E. A List of Plants collected in Spitzbergen in the Summer of 1873 with their localities. — The Journal of Botany british and foreign. New Series. Vol. 5. London 1876. — (Eaton, List.)

ENGLER, A. Versuch einer Entwicklungsgeschichte der Pflanzenwelt insbesondere der Florengebiete seit der Tertiärperiode. Theil 1. Leipzig 1879. — ENGLER, Pflanzenw.)

English Botany. London 1790—1814; Supplement. London 1831. (Engl. Bot.; Engl. Bot. Suppl.)

FARLOW, W. G. Marine Algæ of New-England and adjacent Coast. Washington 1881. — (FARL. New Engl. Alg.)

» ANDERSSON, C. L. and EATON, D. C. Algæ exsiccatæ Americæ borealis. Fasc. 3. Bostoniæ 1878. — (Farl. Eat. et Ands. Alg. Amer.)

Flora Danica. Tome 1—16. Hafniæ 1766.. — 1877.

FOSLIE, M. Om nogle nye arctiske havalger. — From Christiania Videnskabsselskabs Forhandlinger 1881. N:o 14. — (FOSLIE, Arct. Havalg.)

» Bidrag til kundskab om de til gruppen Digitatæ hörende Laminarier. — From Christiania Videnskabsselskabs Forhandlinger 1883. N:o 2. — (FOSLIE, Digitata-Lam.)

FRIES, E. Systema orbis vegetabilis. Lundæ 1825. — (Fries Syst. Veg.)

» Corpus Florarum provincialium Sueciæ. Flora Scanica. Upsaliæ 1835. — (FR. Fl. Scan.)

» Summa vegetabilium Scandinaviæ. Upsaliæ 1845. — (FR. Sum. Veg.)

FRIES, TH. M. Lichenes Spitsbergenses. — Kongl. Svenska Vetenskaps-Akademiens Handlingar. Band 7, N:o 2. Stockholm 1867. — (TH. FRIES, Lich. Spetsb.)

GMELIN, S. G. Historia Fucorum. Petropoli 1768. — (GMEL. Hist. Fuc.)

GOBI, CHR. Die Brauntange des Finnischen Meerbusens. — Mémoires de l'Académie impériale des Sciences de St.-Pétersbourg. Ser. 7. Tome 21, n:o 9. St.-Pétersbourg 1874. — (Gobi, Brauntange.)

» Die Algenflora des Weissen Meeres und der demselben zunächstliegenden Theile des nördlichen Eismeeres — Mémoires de l'Académie impériale des Sciences de St.-Pétersbourg, Ser 7. Tome 26, N:o 1. — (GOBI, Algenfl. Weiss. Meer.)

GOODENOUGH, S. and WOODWARD, T. J. Observations on the British Fuci with particular Descriptions of each Species. — Transactions of the Linnean Society. Vol. 3. London 1797. — (GOOD. et WOOD. Linn. Trans.)

GRAY, J. E. A natural arrangement of Brittish Plants. London 1821. — (GRAY Brit. Pl.)

GREVILLE, R. K. Descriptiones novarum specierum ex algarom ordine. — Nova acta Physico-medica Academiæ Cesareæ Leopoldino-Carolinæ naturæ curiosorum. Tome 14. Pars posterior. Bonnæ 1829. — (GREV. Act. Leop,)

» Algæ Britannicæ. Edinburgh 1830. — (GREV. Alg. Brit.)

GUNNERUS, J. E. Flora norvegica. Nidrosiæ 1766—1772. — (GUNN. Fl. Norv.)

» Om nogle norske Planter. — Det Kongelige Norske Videnskabers Selskabs skrifter. Del. 4. Kjöbenhavn 1768. — (GUNN. Act. Nidros.)

HARVEY, W. H. Algological illustrations N:o 1. — The Journal of Botany by W. J. Hooker. Vol. 1. London 1834. — HARV. in Hook. Journ. Bot.)

» A Manual of the British Algæ. Ed. 1. London 1841; Ed. 2. London 1849. — (HARV. Man.; HARV. Man. Ed. 2.)

» Algæ. — Flora of Western Eskimaux-Land. The Botany of the Voyage of H. M. S. Herald by B. Seeman. London 1852—1857. — (HARV. Fl. West-Esk.).

» Nereis Boreali-Americana. 1—2. — Smithsonian Contributions to knowledge. Vol. 3, 5. Washington 1852, 1853. — (HARV. Ner. Am.)

» Charactes of New Algæ, chiefly from Japan and adjacent Regions, collected by Charles Wright in the North Pacific Exploring Expedition under Captain John Rodgers. — From Proceedings of the American Academy Vol. 4. Oktober 1859. — (HARV. New Alg.)

» Notice of a Collection of Algæ made on North-West Coast of North America, chiefly at Vancouver's Island, by David Lyall. — Journal of the Proceedings of the Linnean Society. Botany. Vol. 6. London 1862. — (HARV. Alg. Vanc.)

» Phycologia Britannica or a History of British Sea-Weeds. New Edition. London 1871. — (HARV. Phyc. Brit.)

HAUCK, T. Die Meeresalgen. — Dr L. Rabenhorst's Kryptogamen-Flora von Deutschland, Oesterreich und der Schweiz. Band 2. Leipzig 1883. — (HAUCK, Meeresalg.)

HAUGHTON, S. Geological account of the arctic archipelago, drawn up principally from the specimens collected by Captain F. L. M'Clintock from 1849 to 1859. — The Voyage of the »Fox« in the Arctic Seas by Captain M'Clintock. Appendix 4. London 1860 — (HAUGHTON, Fox-Exped.)

HOHENACKER, R. F. Algæ marinæ siccatæ. Esslingen, Kirchheim 1852—1862. — HOHENACK. Alg. Mar.)

HILDEBRANDSSON, H. H. Observations météorologiques faites par l'Expédition de la Vega du Cap Nord a Yokohama par le détroit de Behring. — Vega-Expeditionens Vetenskapliga Iakttagelser bearbetade af deltagare i resan och andra forskare. Band. 1. Stockholm 1882. — (HILDEBRANDSSON, Obs. Météor.)

HOOKER, W. J. The British Flora. Vol. 2. London 1853. (Hook. Brit. Fl.)

HUDSON, G. Flora Anglica. Ed. 2. Londini 1778. (Huds. Fl, Angl.)

JENSEN, J. A. D. Astronomiske Observationer foretagne i en Del af Holstenborgs og Egedesmindes Distrikter og Undersögelser over Vandets Saltholdighed. — — Meddelelser om Grönland udgivne af Commissionen for Ledelsen af de geologiske og geographiske Undersögelser i Grönland. Hefte 2. Kjöbenhavn 1881. — (JENSEN, Grönl. Meddel.)

JOHNSTON, G. A. History of British Sponges and Lithophytes. Edinburgh 1842. — (JOHNSTON, Brit. Spong. Lith.)

KIEPERT, H. Uebersichts-Karte der Nordpolar-Länder. Neue berichtige Ausgabe. Berlin 1874.

KJELLMAN, F. R. Bidrag till kännedomen om Skandinaviens Ectocarpeer och Tilopterider. Stockholm 1872. — (KJELLM. Skand. Ect. och Tilopt.)

» Förberedande anmärkningar om algvegetationen i Mosselbay enligt iakttagelser under vinterdraggningar, anstälda af Svenska polarexpeditionen 1872—1873. — Öfversigt af Kongl. Vetenskaps-Akademiens Förhandlingar 1875, N:o 5. Stockholm. (KJELLM. Vinteralgv.)

» Om Spetsbergens marina klorofyllförande Thallophyter 1, 2. Bihang till Kongl. Svenska Vetenskaps-Akademiens Handlingar, Band 3, N:o 7, Band 4, N:o 6. Stockholm 1875, 1877. (KJELLM. Spetsb. Thall.)

» Ueber die Algenvegetation des Murmanschen Meeres an der Westküste von Nowaja Semlja und Wajgatsch. — Nova acta regiæ Societatis scientiarum Upsaliensis. Ser. 3. — Upsaliæ 1877. — (KJELLM. Algenv. Murm. Meer.)

» Redogörelse för Prövens färd från Dicksons hamn till Norge samt för Kariska hafvets växt- och djurverld. — Redogörelse för en Expedition till mynningen af Jenissej och Sibirien år 1875 af A. E. Nordenskiöld. — Bihang till Kongl. Svenska Vetenskaps-Akademiens Handlingar. Band 4, N:o 1, Stockholm 1877. — (KJELLMAN, Pröven.)

» Bidrag till kännedomen af Kariska hafvets Algvegetation. — Öfversigt af Kongl. Vetenskaps-Akademiens Förhandlingar 1877, N:o 2. Stockholm. — (KJELLM. Kariska hafvetsAlgv.)

» Bidrag till kännedomen om Islands hafsalgflora. — Botanisk tidskrift. Ser. 3. Band 3. Köpenhamn 1879. (KJELLM. Isl. Alg.)

» Rhodospermeæ et Fucoideæ. — Enumerantur Plantæ Scandinaviæ. 4. Lund 1880. — (KJELLM. Pl. Scand.)

KLEEN, E. Om Nordlandens högre Hafsalger. — Öfversigt af Kongl. Vetenskaps-Akademiens Förhandlingar 1874, N:o 9. Stockholm. — (KLEEN, Nordl. Alg.)

KOLDEWEY, K. Meerestemperaturen und Strömungen; Ebb- und Flutbeobachtungen. — Die Zweite Deutsche Polarfahrt etc. Cp. Börgen.) — (KOLDEWEY, Zweite deutsche Polarf.)

KORNERUP, A. Notices générales sur la nature du Grönland, — Meddelelser om Grönland etc. (Cp. JENSEN. Hefte 3. Kjöbenhavn 1880 — KORNERUP, Grönl. Meddel. I.)

KÜTZING, F. T. Ueber Ceramium AG. LINNÆA VON SCHLEETENDAL. Vol. 5. Halle 1841. — (KÜTZ. Linnæa).

» Phycologia generalis. Leipzig 1843. — (KÜTZ. Phyc. gener.)

» Tabulæ Phycologicæ 1—19. Nordhausen 1845—1869. — (KÜTZ. Tab. Phyc.)

» Diagnosen und Bemerkungen zu neuen oder kritischen Algen. — Botanische Zeitung 1847. (KÜTZ. Bot. Zeit. 1847.)

» Species Algarum. Lipsiæ 1849. — (KÜTZ. Spec. Alg.)

LAGERSTEDT, N. G. W. Om algslägtet Prasiola. Försök till en Monographi. Upsala 1869. — (LAGERST. Monogr.)

LAMARK. Histoire naturelle des animaux sans vertèbres. Tome 2. Paris 1816. — (LAM. Hist. Anim.)

LAMOUROUX, J. V. F. Essai sur les genres de la famille des Thalassiophytes non articulées. — Annales du Muséum d'Histoire naturelle. Tome 20. Paris 1813. — (LAMOUR. Ess.)

» Histoire des Polypiers coralligènes flexibles, vulgairement nommés Zoophytes. Caen 1816. (LAMOUR. Hist. Polyp.)

LEPECHIN, J. Quattuor Fucorum species descriptæ. — Novi Commentarii Academiæ Scientiarum imperialis Petropolitanæ. Tome 19. Petropoli 1775. — (LEPECH. Comment. Petrop.)

LE JOLIS, A. Examen des espèces confondues sous le nom de Laminaria digitata auct., suivi de quelques observations sur le genre Laminaria. — Nova acta Academiæ Cesareæ Leopoldino-Carolinæ naturæ curiosorum Vol. 26. Pars posterior. Vratislaviæ et Bonnæ 1856. — LE JOL. Exam.)

» Liste des Algues marines de Cherbourg. Cherbourg 1863. — (LE JOL. Liste Alg. Cherb.)

LIGHTFOOT, J. Flora Scotica. London 1777. — (LIGHTF. Fl. Scot.)

LINDBLOM, A. E. Förteckning öfver de på Spetsbergen och Beeren Eiland anmärkta växter. — Botaniska Notiser 1840. — (LINDBL. Bot. Not.)

LINK, H. F. Epistola de Algis aquaticis in genera disponendis. — C. G. Nees ab Esenbeck. Horæ phyceæ berolinensis. Bonnæ 1820. — (LINK, Epist.)

LINNÉ, C. VON. Species Plantarum. Ed. 1, 2. Holmiæ 1753, 1763. — (L. Spec. Pl.)

» Systema naturæ. Ed. 10. Holmiæ 1758 (L. Syst. Nat.); Ed. 12. Holmiæ 1763. — (L. Syst. Nat. Ed. 12.)

» Fauna Svecica. Ed. 2. Stockholmiæ 1761. — (L. Fauna Svec.)

» Mantissa Plantarum. Holmiæ 1767. — (L. Mant.)

LYNGBYE, H. G. Tentamen Hydrophytologiæ Danicæ. Hafniæ 1819. — (LYNGB. Hydr. Dan.)

MAGNUS, P. Die Botanischen Ergebnisse der Nordseefahrt von 21. juli bis 9. september 1872. — Jahresbericht der Kommission zur Untersuchung der deutschen Meere in Kiel, II. Berlin 1874, — (MAGNUS, Nordseef.)

MARKHAM, C. R. On the Threshold of the Unknown Region. Ed. 2. London 1873. — (MARKHAM, Threshold.)

MARTIN, A. R. Meteorologiska observationer gjorde på en resa till Spitsbärgen. — Kongl. Svenska Vetenskaps-Akademiens Handlingar 1758. — (MARTIN, Met. Observ.)

MARTENS, F. Journal d'un Voyage au Spitsbergen et au Groenlandt. — Recueil de Voyages au Nord. Tome 2. Amsterdam 1715. — (MARTENS, Voyage Spitsb.)

MOHN, H. VON. Die Temperatur Verhältnisse im Meere zwischen Norwegen, Schottland, Island und Spitsbergen. — Petermann's Mittheilungen über wichtige neue Erforschungen auf dem Gesammtgebiete der Geographie. Band. 22. Gotha 1876. — (MOHN, Temp. Verhältn.)

MONTAGNE, J. F. C. Sixième centurie de plantes cellulaires nouvelles, tant indigènes qu'exotiques. — Annales de Sciences naturelles. Ser. 3. Tom. 11. Paris 1849. — (MONT. Ann. d. Sc. 9.)

» Sylloge generum specierumque cryptogamarum quas in variis operibus descriptas iconibusque illustratas nunc ad diagnosim reductas multasque novas interjectas ordine systematico disposuit. Parisiis 1856. — (MONT. Syll.)

NARDO. De novo genere Algarum cui nomen est Hildbrandtia protypus. — Isis von Oken. Leipzig 1843. — (NARDO, Isis.)

NORDENSKIÖLD, A. E. Redogörelse för den Svenska Polarexpeditionen år 1872—1873, — Bihang till Kongl. Svenska Vetenskaps-Akademiens Handlingar, Band 2, N:o 18. Stockholm 1875. — (NORDENSKJÖLD, Spetsb.-Exp.)

» Karta öfver Prövens färd till Jenisej och åter 1875.

» Redogörelse för en expedition till mynningen af Jenissej och Sibirien år 1875. — Bihang till Kongl. Svenska Vetenskaps-Akademiens Handlingar. Band 4, N:o 1. Stockholm 1877. — (NORDENSKIÖLD, Pröven.)

» Rapporter skrifna under loppet af Vegas expedition till D:r Oscar Dickson. — Vega-expeditionens Vetenskapliga Iakttagelser etc. Band. 1. — (NORDENSKIÖLD, Vega-exp.)

» Om möjligheten att idka sjöfart i det Sibiriska Ishafvet. — l. c. — (NORDENSKIÖLD, Vega-exp.)

NORDSTEDT, O. Nostocaceæ. — Enumerantur Plantæ Scandinaviæ 4 Lund 1880. — (NORDST. Pl. Scand.)

NYLANDER, W. och SÆLAN, TH. Herbarium Musei Fennici. Helsingfors 1859. — NYL. et SÆL. Herb. Fenn.)

NÄGELI, C. Die neuern Algensysteme und Versuch zur Begründung eines eigenen Systems der Algen und Florideen. Neuenburg 1847. — (NÄG. Algensyst.)

» Gattungen einzelliger Algen. Zürich 1849. — (NÄG. Gatt. einz. Alg.)

NÄGELI, C. Morphologie und Systematik der Ceramiaceen. Sitzungsberichte königl. bayer. Akademie der Wissenschaften zu München 1861. Band 2. München 1861. — (NÄG. Ceram.)

PALLAS, P. S. Reise durch verschiedene Provinzen des Russischen Reichs. 3. S:t Petersburg 1776. — (PALL. Sib. Reise.)

PARRY, W. E. Zweite Reise zur Entdeckung einer nordwestlichen Durchfahrt aus dem atlantischen in das stillen Meer. Hamburg 1822. — (PARRY, Zweite Reise.)

PHILIPPI. Beweis dass die Nulliporen Pflanzen sind. — Archiv für Naturgeschichte von A. F. A. Wiegman. Jahrg. 3. Band 1. Berlin 1837. — (PHIL. Wiegm. Arch.)

POSTELS, A. et RUPRECHT, F. Illustrationes Algarum Oceani Pacifici imprimis septemtrionalis. Petropoli 1840. — POST. et RUPR. Ill. Alg.)

PRINGSHEIM, N. Beiträge zur Morphologie der Meeres-Algen. From Abhandlungen der königl. Akademie der Wissenschaften zu Berlin 1861. Berlin 1862, — (PRINGSH. Morph. Meeresalg.)

RABENHORST, L. Flora Europæa Algarum aquæ dulcis et submarinæ. Sectio 3. Lipsiæ 1868. — (RABENH. Fl. Eur. Alg.)

» Die Algen Sachsens respective Mittel-Europas. — (RABENH. Alg. Eur.).

RINK. H. Grönland geographisk og statistisk beskrevet. Kjöbenhavn 1857. — (RINK, Grönland.)

ROSANOFF, S. Recherches anatomiques sur les Mélobésiées. — Mémoires de la société impériale des Sciences naturelles de Cherbourg. Tome 12. Cherbourg 1866. — (ROSAN. Melob.)

ROTH, A. W. Tentamen Floræ germanicæ. Lipsiæ 1788—1800. — (ROTH, Fl, Germ.)

» Catalecta Botanica 1—3. Lipsiæ 1797—1806. — (ROTH, Cat. Bot.)

RUPRECHT, F. Tange des Ochotischen Meeres. — Reise in den äussersten Norden und Osten Sibiriens von A. Th. v. Middendorff. Band 1, Theil 2. S:t Petersburg 1848. — (RUPR. Alg. Och.)

SCHÜBELER. Algæ (of Novaya-Zemlya.) — Reisen nach dem Nordpolarmeer in den Jahren 1870 und 1871 von M. Th. von Heuglin. Theil 3. Braunschweig 1874. — (SCHÜBELER in Heugl. Reise.)

SCHRENK, A. G. Reise durch die Tuudren der Samojeden. Theil 2. Dorpat 1854. — (SCHRENK, Ural Reise.)

SCORESBY, W. An Account etc. (Cp. BROWN.)

SOLIER, A. J. J. Mémoire sur deux Algues zoosporées devant former un genre distinct, le genre Derbesia. — Extrait des Annales des Sciences naturelles. Tome 7. 1847. — (SOLIER, Ann. d. Sc.)

SOLMS-LAUBACH, GRAF ZU. Die Corallinenalgen des Golfes von Neapel und der angrenzenden Meeres-Abschnitte. — Fauna und Flora des Golfes von Neapel etc. (Cp. BERTHOLD — Leipzig 1881.)

SOMMERFELT, CHR. Supplementum Floræ lapponicæ quam edidit D:r G. Wahlenberg. Christianiæ 1826. — (SOMMERF. Suppl.)

» Bidrag till Spitsbergens og Beeren-Eilands Flora efter Herbarier, medbragte af M. Keilhau. — Magazin for Naturvidenskaberne. 2 Rækkes 1 Bind. Christiania 1832. — (SOMMERF. Spitsb. Fl.)

SPÖRER J. Novaja Semlä in geographischer, naturhistorischer und Volkwirthschaftlicher Beziehung, Ergänzungsheft N:o 21 zu Petermann's Geographische Mittheilungen. Gotha 1867. — (SPÖRER, Nov. Semlä.)

STACKHOUSE, J. Description of Ulva punctata. — Transactions of the Linnean Society. Vol. 3. London 1797. — (STACKH. Linn. Trans.)

» Nereis Britannica. Ed. 2. Oxonii 1816. — (STACKH. Ner. Brit.)

STUXBERG, A. Evertebratfaunan i Sibiriens Ishaf. Förelöpande Meddelande. — Vega-expeditionens vetenskapliga iakttagelser etc. Cp. HILDEBRANDSSON. — (STUXBERG, Vega-exp.)

SUHR, J. N. VON. Beiträge zur Algenkunde. — Flora. Jahrg. 23. Band 1, N:o 19. — (SUHR, Flora 1840.)

THURET, G. Essai de Classification des Nostochinées. — Extrait des Annales des Sciences naturelles. Ser. 6, Tome 1. — (THUR. Nostoch.)

TOURNEFORT, J. P. Institutiones rei Herbariæ. Tome 3. Paris 1719. — TOURN. Inst. Herb.)

TURNER, D. Description of four new Species of Fucus. — Transactions of the Linnean Society. Vol. 6. London 1802. — (TURN. Linn. Trans.)

TURNER, D. Fuci, sive plantarum Fucorum generis a botanicis ascriptarum icones, descriptiones et historia. Vol. 1—4. Londini 1808—1819. — (TURN. Hist. Fuc.)

UNGER, F. Beiträge zur näheren Kenntniss des Leithakalkes namentlich der vegetabilischen Einschlüsse und der Bildungsgeschichte desselben. — Denkschriften der kaiserlichen Akademie der Wissenschaften. Band. 14. Wien 1858. — (UNGER, Leitha-Kalk.)

WAHLENBERG, G. Flora lapponica. Berolini 1812. — (WG. Fl. Lapp.)

WEBER F. und MOHR, D. M. H. Naturhistorische Reise durch einen Theil Schwedens. Göttingen 1804. — (WEB. et MOHR, Reise.)

WIJKANDER, A. Observations météorologiques de l'Expédition arctique Suédoise 1872—1873. — Kongl. Svenska Vetenskaps-Akademiens Handlingar Band 12, N:o 7. Stockholm 1875. — (WIJKANDER. Obs. Météor.)

WITTROCK, V. B. Försök till en Monographi öfver Algslägtet Monostroma. Stockholm 1866. — (WITTR. Monostr.)

» Om Gotlands och Ölands sötvattensalger. — Bihang till Kongl. Svenska Vetenskaps-Akademiens Handlingar. Band 1, N:o 1. Stockholm 1872. — (WITTR. Gotl. och Öl. Alg.)

» Algæ (of Spitzbergen.) — Reisen etc. von Heuglin. (Cp. SCHÜBELER.) — WITTR. in Heugl. Reise.)

» On the development and systematic arrangement of the Pithophoraceæ, a new order of Algæ. — Nova acta regiæ Societatis scientarium Upsaliensis. Ser. 3. Upsaliæ 1877. — (WITTR. Pithoph.)

» Chlorophyllophyceæ. — Enumerantur Plantæ Scandinaviæ 4. Lund 1880. — (WITTR. Pl. Scand.)

» et NORDSTEDT, O. Algæ aquæ dulcis exsiccatæ præcipue Scandinavicæ, quas adjectis algis marinis chlorophyllaceis et phycochromaceis distribuerunt. 1—10. Upsaliæ et Holmiæ 1877—1880.

ZANARDINI, G. Saggio di classificazione naturale della Ficee, aggiunti nuovi studii sopra l'Androsace degli antichi con tavola miniata ed enumerazione di tutte le specie scoperte e raccolte dall' autore in Dalmazia. Venezia 1843. — (ZANARD. Sagg.)

ZELLER, G. Algen. — Die Zweite Deutsche Polarfahrt etc. (Cp. BÖRGEN.) — (ZELLER, Zweite d. Polarf.)

Explanation of the figures.

PLATE 1.

Lithothamnion soriferum.

Fig. 1. Young specimen. $^1/_1$.
» *2.* Somewhat older specimen, seen from above. $^1/_1$.
» *3.* The same specimen, seen from below. $^1/_1$.
» *4.* Full-grown specimen. $^1/_1$.
» *5.* Half of a younger specimen, seen from the cleaving-surface. $^1/_1$.
» *6—10.* Branch-systems of different development. $^1/_1$.
» *11.* Branch with conceptacles of sporangia. $^2/_1$.
» *12.* Part of transversal thin section[1]) of a branch. $^{40}/_1$.
» *13.* Part of median thin section of a branch. $^{40}/_1$.
» *14.* Part of one of the outer concentric layers of a transversal thin section. $^{400}/_1$.
» *15.* Part of the central layer of a transversal thin section. $^{400}/_1$.
» *16.* Part of a superficial tangential thin section. $^{400}/_1$.
» *18.* Part of the roof of a conceptacle of sporangia, delivered from lime. $^{400}/_1$.
» *19.* Sporangium. $^{300}/_1$.

PLATE 2.

Lithothamnion glaciale.

Fig. 1. Older, full-grown specimen, seen from above. $^1/_1$.
» *2.* The same, seen from below. $^1/_1$.

PLATE 3.

Lithothamnion glaciale.

Fig. 1. Young specimen, surrounding part of a stone. $^1/_1$.
» *2.* Somewhat older specimen, completely inclosing a stone. $^1/_1$.

[1]) Thin section = slice made thin by grinding.

Fig. 3. Part of an older specimen in transverse section. $^{1}/_{1}$.
» *4.* Branch with conceptacles of sporangia, seen from above. $^{1}/_{1}$.
» *5.* Part of thin transverse section of a branch. $^{40}/_{1}$.
» *6.* Part of median thin section of a branch. $^{40}/_{1}$.
» *7.* Part of one of the outer concentric layers of a thin transverse section. $^{400}/_{1}$.
» *8.* Part of the central layer in a thin transverse section. $^{400}/_{1}$.
» *9.* Part of layer of tissue in a thin median section. $^{400}/_{1}$.
» *10.* Part of superficial, tangential thin section. $^{400}/_{1}$.
» *11.* Part of the roof a conceptacle of sporangia, delivered from lime. $^{400}/_{1}$.
» *12—14.* Sporangia. $^{400}/_{1}$.

PLATE 4.

Lithothamnion intermedium.

Fig. 1. Full-grown specimen. $^{1}/_{1}$.
» *2.* Half of an older specimen, seen from the cleaving-surface. $^{1}/_{1}$.
» *3.* Branch with conceptacles of sporangia. $^{2}/_{1}$.
» *4.* Part of thin transverse section of a branch. $^{40}/_{1}$.
» *5.* Part of thin median section of a branch. $^{40}/_{1}$.
» *6.* Part of one of the outer concentric layers of a thin transverse section. $^{400}/_{1}$.
» *7.* Part of the central layer of a thin transverse section. $^{400}/_{1}$.
» *8.* Part of layer of tissue in a thin median section. $^{400}/_{1}$.
» *9.* Part of superficial, tangential thin section. $^{400}/_{1}$.
» *10.* Sporangium. $^{400}/_{1}$.

PLATE 5.

Lithothamnion alcicorne. *Fig. 1—8.*

Fig. 1. Older, fully developed specimen. $^{1}/_{1}$.
» *2.* Branch with conceptacles of sporocarps and of sporangia. $^{2}/_{1}$.
» *3.* Part of thin median section of a branch. $^{40}/_{1}$.
» *4.* Part of thin transverse section of a branch. $^{40}/_{1}$.
» *5.* Part of one of the outer concentric layers of a thin transverse section. $^{400}/_{1}$.
» *6.* Part of central layer of thin transverse section. $^{400}/_{1}$.
» *7.* Part of layer of tissue in a thin median section. $^{400}/_{1}$.
» *8.* Sporangium. $^{400}/_{1}$.

Lithothamnion norvegicum. *Fig. 9—10.*

Fig. 9. Specimen from the north-west coast of Norway. $^{1}/_{1}$.
» *10.* Specimen from the south-west coast of Norway. $^{1}/_{1}$.

Lithothamnion foecundum. *Fig. 11—19.*

Fig. 11. Part of specimen incrusting a stone. $^{1}/_{1}$.
» *12.* Part of the crustaceous frond bearing conceptacles of sporangia, seen from above. $^{2}/_{1}$.

Fig. 13. Vertical, tangential thin section. $^{40}/_{1}$.
» *14.* Part of the basal system of the frond in thin radial section. $^{400}/_{1}$.
» *15.* Part of the thickening system of the frond. $^{400}/_{1}$.
» *16.* Part of horizontal, tangential, superficial thin section. $^{400}/_{1}$.
» *17.* Section of a part of the frond provided with a conceptacle of sporangia. $^{100}/_{1}$.
» *18.* Part of the roof of a conceptacle of sporangia, delivered from lime. $^{400}/_{1}$.
» *19.* Sporangium. $^{400}/_{1}$.

PLATE 6.

Lithothamnion flavescens. *Fig. 1—7.*

Fig. 1. Part of frond covering Lithothamion compactum. $^{1}/_{1}$.
» *2.* Conceptacle of sporangia, seen from above. $^{40}/_{1}$.
» *3.* Thin radial section with a grown-in conceptacle of sporangia. Of the sporangia only the gelinated tips remain. $^{40}/_{1}$.
» *4.* Thin radical section of the lower part of the frond. $^{400}/_{1}$.
» *5.* Part of horizontal, tangentical, superficial thin section. $^{400}/_{1}$.
» *6.* Part of the roof of a conceptacle of sporangia, delivered from lime. $^{400}/_{1}$.

Lithothamnion compactum. *Fig. 8—12.*

Fig. 8. Part of a crust-complex with a young crust on its surface. $^{1}/_{1}$.
» *9.* Thin vertical section of an older crust-complex. $^{40}/_{1}$.
» *10.* Part of the basal system of the frond in thin radial section. $^{400}/_{1}$.
» *11.* Part of the thickening system of the frond. $^{400}/_{1}$.
» *12.* Part of superficial, horizontal, tangential thin section. $^{400}/_{1}$.

PLATE 7.

Rhodomela virgata.

Fig. 1. First year's plant in summer habit. $^{1}/_{1}$.
» *2.* First year's plant in autumnal habit. $^{1}/_{1}$.
» *3.* Part of a branch with stands of sporocarps developed during the winter. $^{1}/_{1}$.
» *4.* Stand of sporocarps. $^{40}/_{1}$.
» *5.* Stand of antheridia. $^{40}/_{1}$.
» *6.* Stand of tetrasporangia. $^{40}/^{1}$.
» *7.* Part of cross section of the main axis of an older plant near the base. $^{80}/_{1}$.
» *8.* Part of median longitudinal section of the main axis of an older plant near the base. $^{80}/_{1}$.

PLATE 8.

Rhodomela subfusca.

Fig. 1. First year's plant in summer- and autumn-habit. $^{1}/_{1}$.
» *2.* First year's plant in winter habit. $^{1}/_{1}$.

Fig. 3. Part of plant, bearing sporocarps. $^{1}/_{1}$.
» *4.* Part of cross section of the main axis of an older plant near the base. $^{80}/_{1}$.
» *5.* Part of median longitudinal section of the main axis of an oder plant near the base. $^{80}/_{1}$.

PLATE 9.

Fig. 1. Rhodomela lycopodioides f. typica. β. laxa. $^{1}/_{1}$.
» *2.* Rhodomela lycopodioides f. typica. γ. tenera. $^{1}/_{1}$.
» *3.* Rhodomela lycopodioides f. setacea. $^{1}/_{1}$.

PLATE 10.

Rhodomela lycopodioides f. flagellaris. *Fig. 1—2.*

Fig. 1. Habit-figure. $^{1}/_{1}$.
» *2.* Part of a transverse section of the main axis of the frond near the base. $^{80}/_{1}$.

Delesseria Corymbosa. *Fig. 3.*

Fig. 3. Habit-figure. $^{1}/_{1}$.

PLATE 11.

Hæmescharia polygyna.

Fig, 1. Vertical section of a vegetative part of the frond. $^{80}/_{1}$.
» *2.* Vertical section of a part of the frond, bearing cystidia. $^{80}/_{1}$.
» *3.* Part of the vegetative system. $^{400}/_{1}$.
» *4—6.* Parts of the frond, bearing cystidia. $^{400}/_{1}$.
» *7.* Branch-system, one axis of which ends in a trichogyn (?). $^{400}/_{1}$.
« *8.* Branches with numerous lateral trichogyns (?). $^{500}/_{1}$.
« *9.* Branch-system with undeveloped tetrasporangia (?). $^{400}/_{1}$.

PLATE 12.

Chantransia efflorescens f. tenuis. *Fig. 1—2.*

Fig. 1. Part of sporiferous specimen. $^{40}/_{1}$.
» *2.* Sporiferous branch. $^{400}/_{1}$.

Rhodophyllis dichotoma. *Fig. 3.*

Fig. 3. Form occurring in the interior of deep bays. $^{1}/_{1}$.

Halosaccion ramentaceum f. robusta. *Fig. 4.*

Fig. 4. Specimen bearing tetrasporangia. $^{1}/_{2}$.

PLATE 13.

Halosaccion ramentaceum f. robusta. *Fig. 1—2.*

Fig. 1. Specimen, the »ramenta» of which have been dissolved for the greatest part after the tetrasporangia have ripened. $^{1}/_{1}$.
» *2.* Part of specimen with new ramenta shooting forth from the remains of the formerly dissolved ones. $^{1}/_{1}$.

Halosaccion ramentaceum f. subsimplex. *Fig. 3.*

Fig. 3. Habit-figure. $^{1}/_{1}$.

Halosaccion ramentaceum f. ramosa. *Fig. 4.*

Fig. 4. Habit-figure. $^{1}/_{1}$.

PLATE 14.

Sarcophyllis arctica. *Fig. 1—3.*

Fig. 1. Older, full-grown specimen. $^{1}/_{1}$.
» *2.* Transverse section of part of the frond, bearing procarps. $^{200}/_{1}$.
» *3.* Transverse section of part of the frond, bearing sporocarps. $^{40}/_{1}$.

Kallymenia septemtrionalis. *Fig. 4—6.*

Fig. 4. Young specimen. $^{1}/_{1}$.
» *5.* Older, probably full-grown specimen. $^{1}/_{1}$.
» *6.* Transverse section of the frond in an older specimen. $^{200}/_{1}$.

PLATE 15.

Ptilota pectinata f. integerrima. *Fig. 1.*

Fig. 1. Part of older, full-grown specimen. $^{1}/_{1}$.

Ptilota pectinata f. litoralis. *Fig. 2—5.*

Fig. 2. Habit-figure. $^{1}/_{1}$.
» *3.* Part of branch-system of the last order. $^{20}/_{1}$.
» *4.* Transverse section of a full-grown branch of the last order but one. $^{285}/_{1}$.
» *5.* Transverse section of one of the oldest and most robust axes of the frond. $^{125}/_{1}$.

Ptilota pectinata f. typica. *Fig. 6.*

Fig. 6. Transverse section of a full-grown branch of the last order but one. $^{125}/_{1}$.

Ceramium rubrum f. **squarrosa**. *Fig. 7.*

Fig. 7. Part of specimen bearing tetrasporangia. $^{1}/_{1}$.

Rhodochorton intermedium. *Fig. 8.*

Fig. 8. Part of full-grown specimen. $^{1}/_{1}$.

Rhodochorton Rothii f. **globosa.** *Fig. 9—13.*

Fig. 9. Specimen, seen from the surface. $^{5}/_{1}$.
» *10.* Part of the basal layer of the frond. $^{300}/_{1}$.
» *11.* Branch-system issuing from the basal layer. $^{40}/_{1}$.
» *12.* Lower part of the main axis of such a branch-system. $^{300}/_{1}$.
» *13.* Upper part of a branch of the last order but one. $^{300}/_{1}$.

PLATE 16.

Antithamnion Pylaisæi f. **norvegica.** *Fig. 1.*

Fig. 1. Upper part of a full-grown specimen. $^{80}/_{1}$.

Antithamnion boreale f. **typica.** *Fig. 2—3.*

Fig. 2. Lower part of a branch-system of the first order. $^{80}/_{1}$.
» *3.* Top of another. $^{80}/_{1}$.

Antithamnion boreale f. **corallina.** *Fig. 4—5.*

Fig. 4. Lower part of a branch-system of the first order. $^{80}/_{1}$.
» *5.* Top of a feeble branch-system of the first order. $^{80}/_{1}$.

Rhodochorton mesocarpum f. **penicilliformis.** *Fig. 6—7*

Fig. 6. Part of specimen bearing tetrasporangia. $^{80}/_{1}$.
» *7.* Part of branch with tetrasporangia. $^{300}/_{1}$.

PLATE 17.

Diploderma amplissimum. *Fig. 1—3.*

Fig. 1—3. The plant at different ages. $^{1}/_{1}$.

Porphyra abyssicola. *Fig. 4.*

Fig. 4. Full-grown specimen. $^{1}/_{1}$.

PLATE 18.

Diploderma amplissimum. *Fig. 1—8.*

Fig. 1. Part of the base of the frond, seen from the surface. $^{300}/_{1}$.
» *2.* Transverse section of the same part. $^{300}/_{1}$.
» *3.* The sterile part of the frond, seen from the surface. $^{300}/_{1}$.
» *4.* Transverse section of specimen 90 cm. long, at the middle of the frond. $^{300}/_{1}$.
» *5.* Transverse section at the middle of the specimen delineated pl. 17 fig. 3.
» *6.* Part of the plant, bearing sporocorps, seen from the surface. $^{300}/_{1}$.
» *7.* Part of the frond bearing sporocarps and antheridia, seen from the surface. $^{300}/_{1}$.
» *8.* Transverse section of the same. $^{300}/_{1}$.

Diploderma miniatum. *Fig. 9.*

Fig. 9. Transverse section of the sterile part of the frond. $^{300}/_{1}$.

Porphyra abyssicola. *Fig. 10—11.*

Fig. 10. Part of the frond with sporocarps in development. $^{300}/_{1}$.
» *11.* Transverse section of the same. $^{300}/_{1}$.

PLATE 19.

Fucus miclonensis. *Fig. 1—2.*

Fig. 1. Specimen from Nordlanden. $^{1}/_{1}$.
» *2.* Specimen from Finmarken. $^{1}/_{1}$.

Fucus filiformis f. Gmelini. *Fig. 3.*

Fig. 3. Specimen bearing receptacles, from Finmarken. $^{1}/_{1}$.

PLATE 20.

Alaria dolichorhachis.

Young specimen of the plant in natural size.

PLATE 21.

Alaria dolichorhachis.

Fig. 1. Older specimen. $^{1}/_{1}$.
» *2.* Sporophyll. $^{1}/_{1}$.

PLATE 22.

Alaria oblonga.

Fig. 1. Very young specimen. $^1/_1$.
» *2.* Older specimen, not yet bearing sporophyll. $^1/_1$.
» *3.* Old specimen with sporophylls bearing zoosporangia. $^1/_2$.
» *4.* Sporophyll with sorus developed. $^1/_1$.

PLATE 23.

Alaria elliptica.

Fig. 1—2. Specimens at different ages.

PLATE 24.

Phyllaria lorea.

Fig. 1. Young specimen. $^1/_1$.
» *2.* Somewhat older specimen. $^1/_1$.
» *3.* Full-grown specimen with sorus in development. $^1/_3$.

PLATE 25.

Phyllaria dermatodea. *Fig. 1—4.*

Fig. 1. Transverse section of the stipe of a young specimen with a ring of thick-walled, tubular cells on the limit between the middle and intermediate layers. $^5/_1$.
» *2.* Part of the same transverse section: im the thick-walled, tubular cells. $^{200}/_1$.
» *3, 4.* Parts of two such cells. $^{50}/_1$.

Phyllaria lorea. *Fig. 5—6.*

Fig. 5. Inner part of a transverse section of the stipe of an older specimen, with long, tubular, but thin-walled cells (m). $^{200}/_1$.
» *6.* Transverse section of the lamina with a cryptostoma. $^{100}/_1$.

Laminaria saccharina f. grandifolia. *Fig. 7.*

Fig. 7. Part of a transverse section of the lamina with lacuna mucifera (l) $^{100}/_1$.

Laminaria nigripes. *Fig. 8—10.*

Fig. 8. Outer part of a transverse section through the stipe with lacunæ muciferæ. $^{50}/_1$.
The specimen is from Spitzbergen, determined as L. nigripes by J. G. AGARDH.

Fig. 9. Part of the same transverse section with three lacunæ muciferæ (l.). $^{100}/_1$.
» *10.* Part of a transverse section of the lamina with a lacuna mucifera. $^{100}/_1$.

Alaria dolichorhachis. *Fig. 11—18.*

Fig. 11. Part of a transverse section of the stipe. $^{35}/_1$.
» *12.* Transverse section of the stipe near the base in an older specimen. $^1/_1$.
» *13.* Transverse section of the older part of the rhachis, not bearing sporophylls, of the same specimen. $^1/_1$.
» *14.* Transverse section of the part of the rhachis bearing sporophylls, of the same specimen. $^1/_1$.
» *15.* Transverse section of the rhachis where it passes into the costa. $^1/_1$.
» *16.* Transverse section of the costa at the middle of the lamina. $^1/_1$.
» *17.* Zoosporangium. $^{400}/_1$.
» *18.* Paraphysis. $^{400}/_1$.

Alaria esculenta f. typica. *Fig. 19.*

Fig. 19. Transverse section of the costa at the middle of the lamina. Specimen from the west coast of Norway. $^2/_1$.

Alaria membranacea. *Fig. 20.*

Fig. 20. Transverse section of the costa at the middle of the lamina, in a specimen from the coast of Norwegian Finmarken. $^1/_1$.

Alaria oblonga. *Fig. 21—24.*

Fig. 21. Transverse section of the stipe and
» *22.* Transverse section of the part of the rhachis bearing sporophylls, of the same specimen. $^1/_1$.
» *23.* Transverse section of the costa of a young and
» *24.* of an older specimen. $^1/_1$.

Alaria elliptica. *Fig. 25—26.*

Fig. 25. Transverse section of the costa of a young and
» *26.* of an older specimen. $^1/_1$.

PLATE 26.

Scytosiphon attenuatus. *Fig. 1—5.*

Fig. 1. Several specimens issuing from the same fastening-surface. $^1/_1$.
» *2.* Detached specimen. $^1/_1$.
» *3.* Transverse section of the wall of the frond with gametangia (zoosporangia multilocularia). $^{400}/_1$.
» *4.* Longitudinal section of the wall of the frond with gametangia and so-called paraphyses. $^{400}/_1$.
» *5.* Lower part of a hair. $^{400}/_1$.

Lithoderma fatiscens. *Fig. 6—7.*

Fig. 6. Transverse section of a part of the frond bearing zoosporangia. $^{400}/_1$.
» *7.* Stand of gametangia. $^{400}/_1$.

Lithoderma lignicola. *Fig. 8—11.*

Fig. 8. Piece of wood covered with the plant. $^1/_1$.
» *9.* Part of the basal layer of the frond. $^{400}/_1$.
» *10.* Radical transverse section of the frond. $^{400}/_1$.
» *11.* Horizontal tangential section of the frond. $^{400}/_1$.

Dictyosiphon corymbosus f. abbreviata. *Fig. 12—15.*

Fig. 12. Older specimen bearing zoosporangia. $^1/_1$.
» *13.* Part of transverse section of the lower part of the main axis of the frond. $^{200}/_1$.
» *14.* Part of branch, bearing zoosporangia, seen superficially. $^{100}/_1$.
» *15.* Part of transverse section of the same. $^{200}/_1$.

Chorda filum. f. crassipes. *Fig. 16.*

Fig. 16. Full-grown specimen. $^1/_1$.

Phloeospora pumila. *Fig. 17.*

Fig. 17. Lower part of the frond with rhizines. $^{80}/_1$.

PLATE 27.

Pylaiella varia. *Fig. 1—12.*

Fig. 1. Part of sterile individual. $^{40}/_1$.
» *2.* Zoosporangia developed in different manners. $^{225}/_1$.

Pylaiella nana. *Fig. 13—15.*

Fig. 13. Vegetative part of the frond. $^{225}/_1$.
» *14.* Part of vertical cell-row bearing two gametangia (zoosporangia multilocularia) one, of which is void. $^{225}/_1$.
» *15.* Parts of vertical cell-rows bearing zoosporangia (?). $^{225}/_1$.

PLATE 28.

Monostroma saccodeum. *Fig 1—10.*

Fig. 1—2. Young individuals, still cystiform. $^1/_1$.
» *3—4.* Older, full-grown individuals of somewhat different habit. $^1/_1$.
» *5.* Lowest part of the frond, seen superficially. $^{400}/_1$.
» *6.* Transverse section of the same part. $^{400}/_1$.
» *7.* Part of the vegetative portion of the frond. $^{400}/_1$.
» *8.* Transverse section of the same. $^{400}/_1$.
» *9.* Part of the zoosporiferous portion of the frond. $^{400}/_1$.
» *10.* Transverse section of the same. $^{400}/_1$.

Monostroma crispatum. *Fig. 11—13.*

Fig. 11. Habit-figure. $^{1}/_{1}$.
» *12.* Transverse section of the uppermost part of the frond. $^{400}/_{1}$.
» *13.* Longitudinal section of the lower part of the frond. $^{400}/_{1}$.

PLATE 29.

Monostroma angicava.

Fig. 1. Two specimens issuing from the same fastening surface, one (downwards to the right) young, the other full-grown bearing zoospores. $^{1}/_{1}$.
» *2.* Lowest part of the frond, seen superficially. $^{400}/_{1}$.
» *3.* Transverse section of the same. $^{400}/_{1}$.
» *4.* Part of the vegetative portion of the frond, seen superficially. $^{400}/_{1}$.
» *5.* Transverse section of the same. $^{400}/_{1}$.
» *6.* Part of the zoosporiferous portion of the frond. $^{400}/_{1}$.
» *7.* Transverse section of the same. $^{400}/_{1}$.

PLATE 30.

Monostroma cylindraceum.

Fig. 1. Full-grown specimen bearing zoospores. $^{1}/_{1}$.
» *2.* Lowest part of the frond, seen superficially. $^{400}/_{1}$.
» *3.* Transverse section of the same. $^{400}/_{1}$.
» *4.* Part of the vegetative portion of the frond, seen superficially. $^{400}/_{1}$.
» *5.* Transverse section of the same. $^{400}/_{1}$.
» *6.* Part of the zoosporiferous portion of the frond. $^{400}/_{1}$.
» *7.* Transverse section of the same. $^{400}/_{1}$.

PLATE 31.

Enteromorpha micrococca f. subsalsa. *Fig. 1—3.*

Fig. 1. Part of a specimen. $^{40}/_{1}$.
» *2.* Part of the wall of the frond, seen superficially. $^{400}/_{1}$.
» *3.* Transverse section of the same. $^{400}/_{1}$.

Chætophora pellicula. *Fig. 4—7.*

Fig. 4. Prostrate branch-system with branches issuing on one side. $^{400}/_{1}$.
» *5.* Ascending branch-system. $^{400}/_{1}$.
» *6.* Tip of branch bearing hairs. $^{400}/_{1}$.
» *7.* Part of branch with zoosporangia. $^{400}/_{1}$.

Chlorochytrium inclusum. *Fig. 8—17.*

Fig. 8. Transverse section of Sarcophyllis arctica including Chlorochytrium inclusum. $^{50}/_{1}$.
» *9.* Specimen in vegetative state, enclosed in frond of S. arctica. $^{200}/_{1}$.
» *10.* Specimen in the same state taken out of the same plant. $^{200}/_{1}$.
» *11.* Specimens bearing zoospores (gamets?), enclosed in the frond of the same plant; they have almost penetrated the cortical layer, but not yet opened themselves. $^{200}/_{1}$.
» *12—15.* Specimens bearing zoospores (gamets), that have been taken out. $^{200}/_{1}$.
» *16.* Specimen enclosed in S. arctica, which has penetrated the cortical layer and opened itself in order to discharge the zoospores (gamets?). $^{200}/_{1}$.
» *17.* Specimen that has lain deeply enclosed in the frond of S. arctica and probably entered in a state of rest.

Index.

Contents.

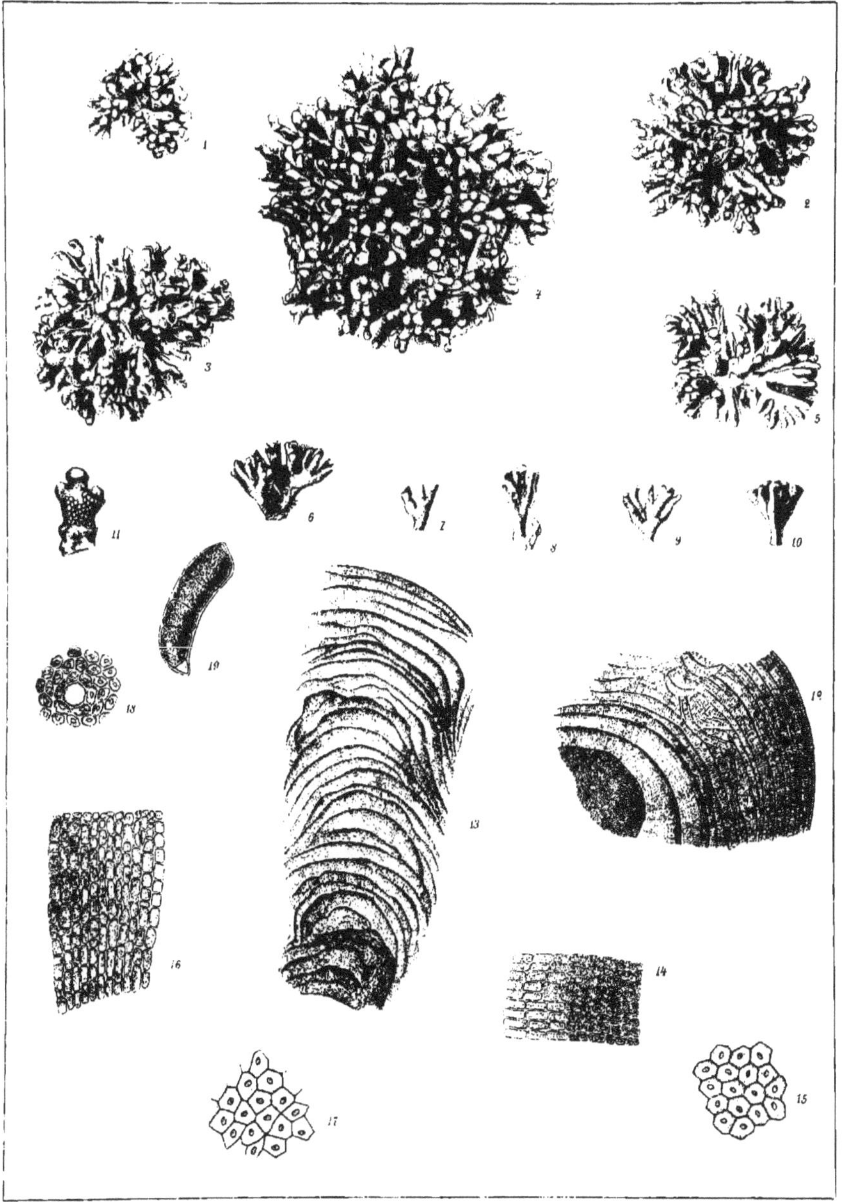
1
2
4
3
5
11
6
7
8
9
10
19
18
13
12
16
14
17
15

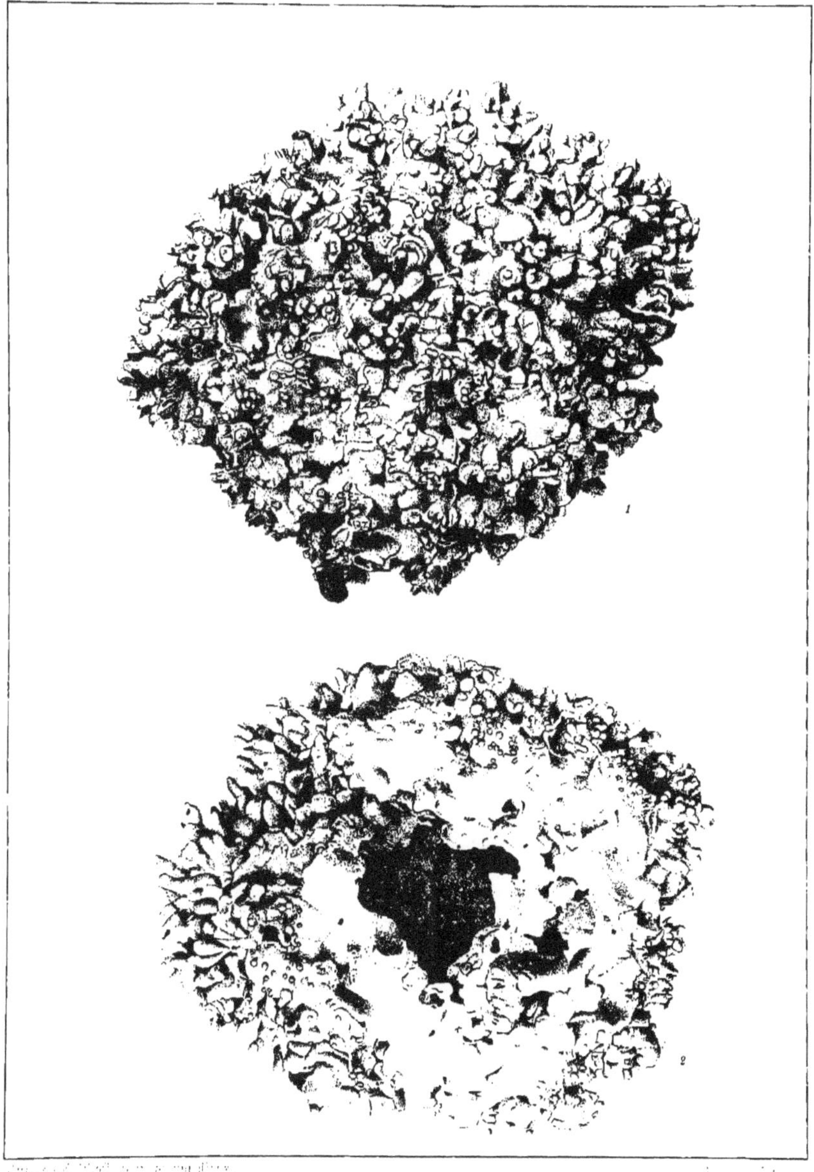
1
2

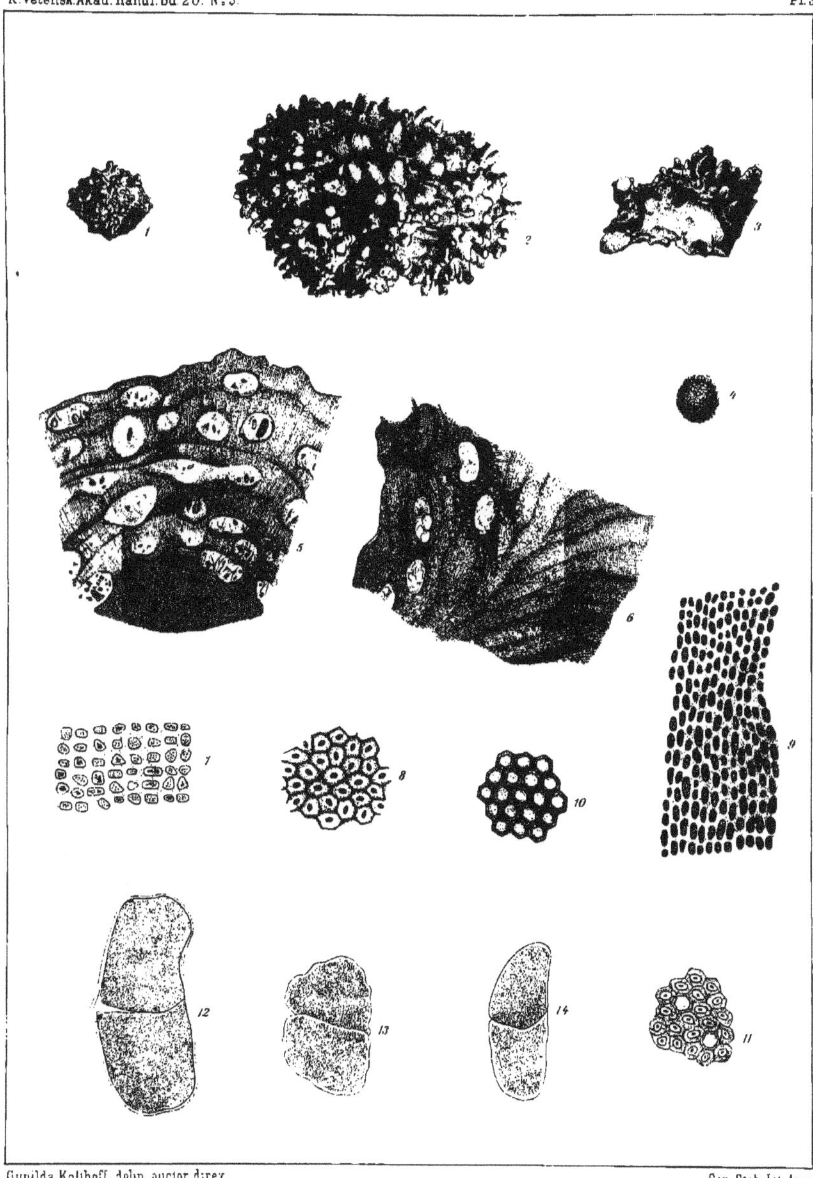

Gunilda Kolthoff delin, auctor direx. Gen. Stab. Lit. Anst.

Lithothamnion glaciale.

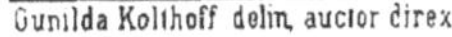

Gunilda Kolthoff delin. auctor direx

Gen. Stab. Lit. Anst.

Lithothamnion intermedium

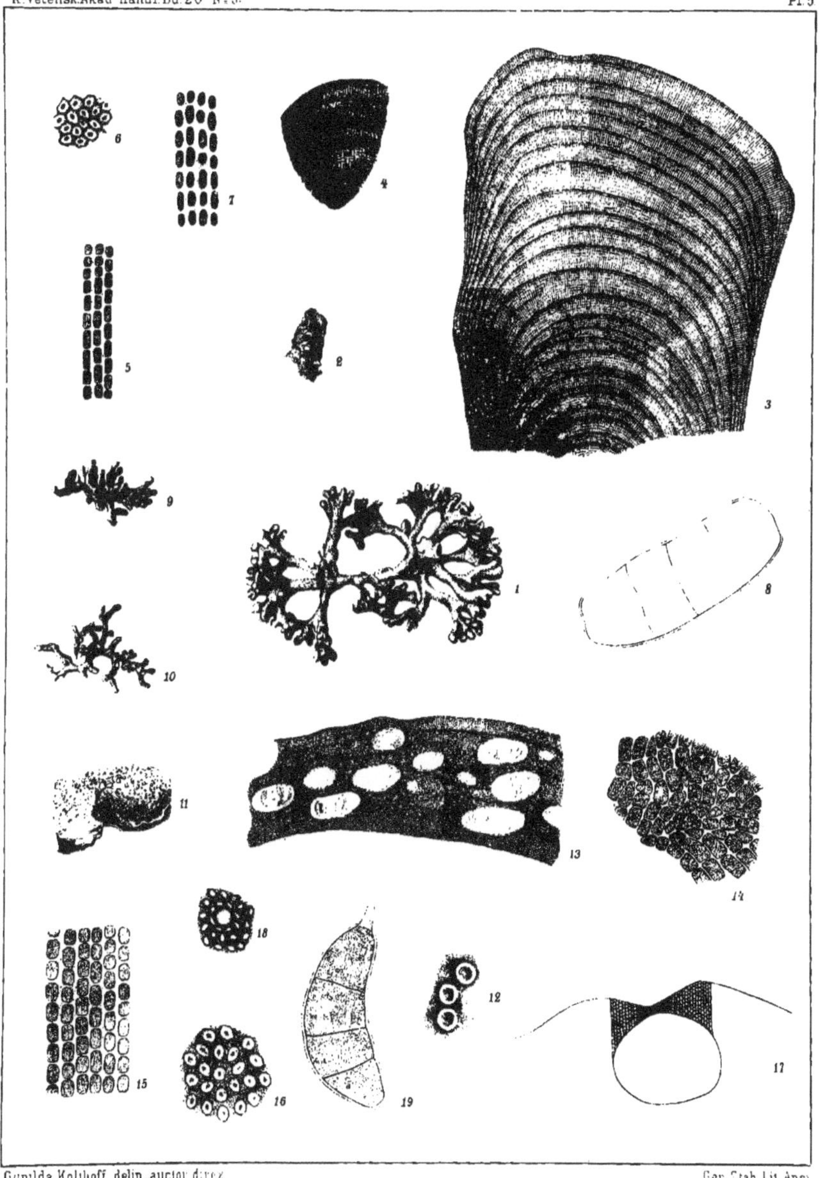

Gunilda Kolthoff delin. auctor direx. Gen. Stab. Lit. Anst.

1–8 Lithothamnion alcicorne. 9–10 Lithothamnion norvegicum

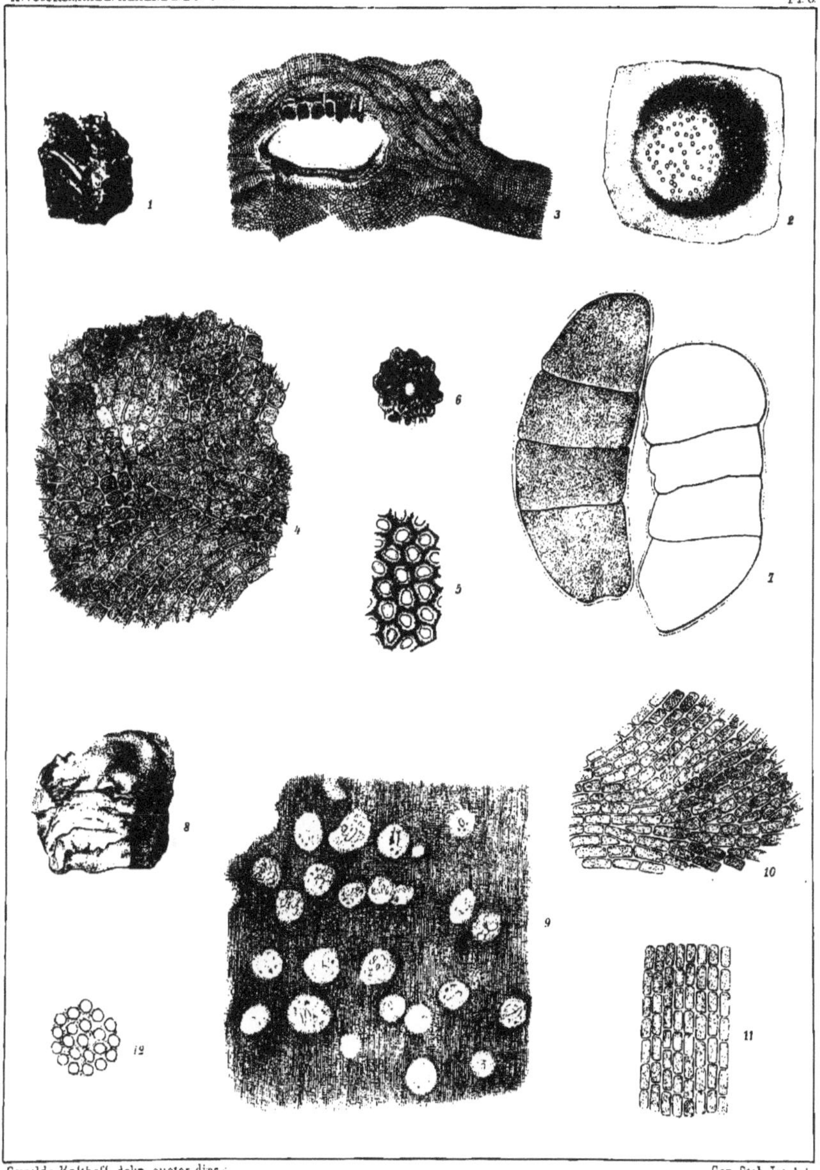

Gunilda Kolthoff delin. auctor direx. Gen. Stab. Lit. Anst.

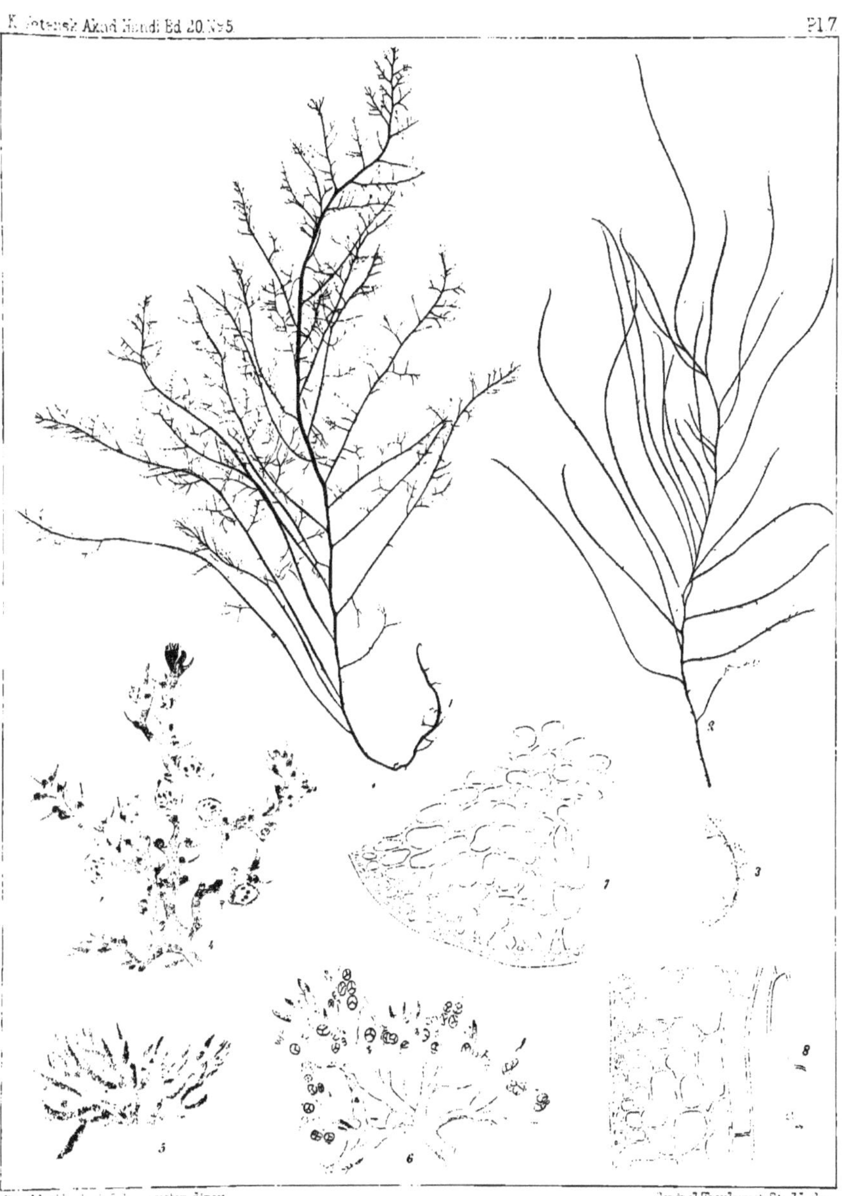

... auctor direx. Central-Tryckeriet, Stockholm.

Rhodomela virgata

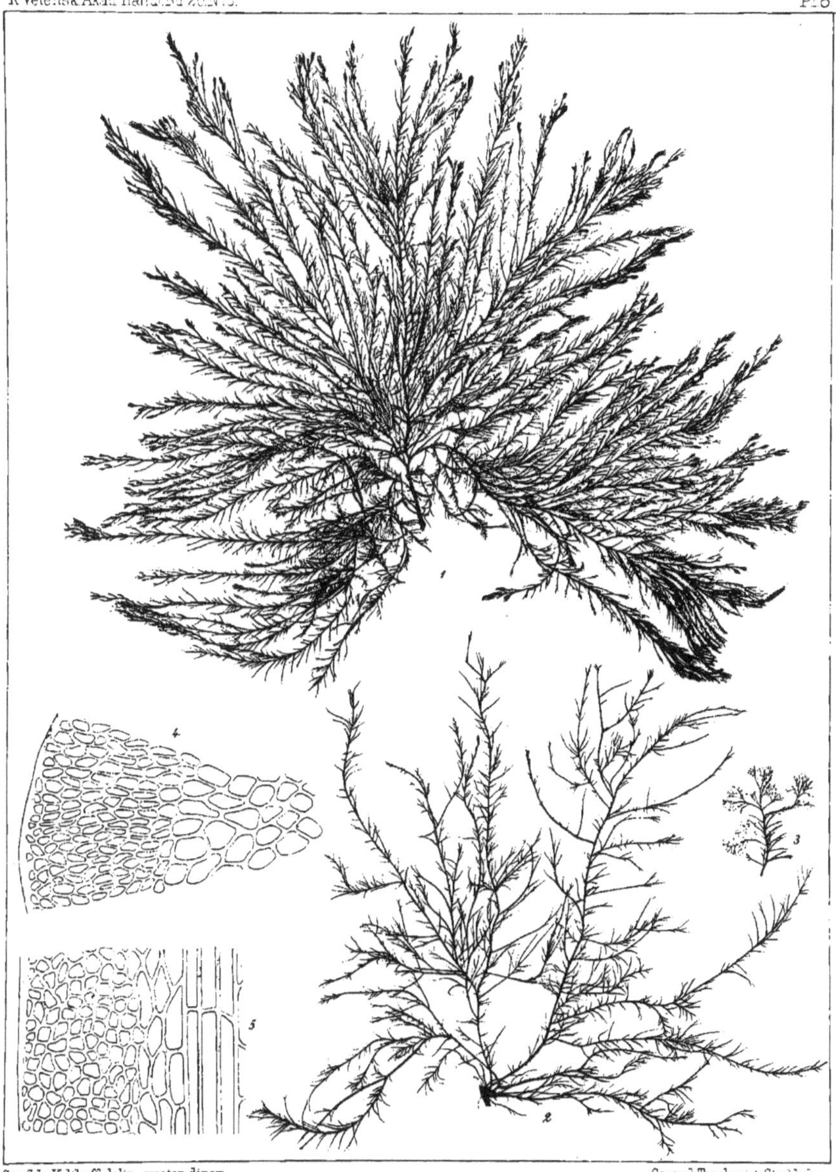

Gunilda Kolthoff delin, auctor direx. Central-Tryckeriet, Stockholm

Rhodomela subfusca.

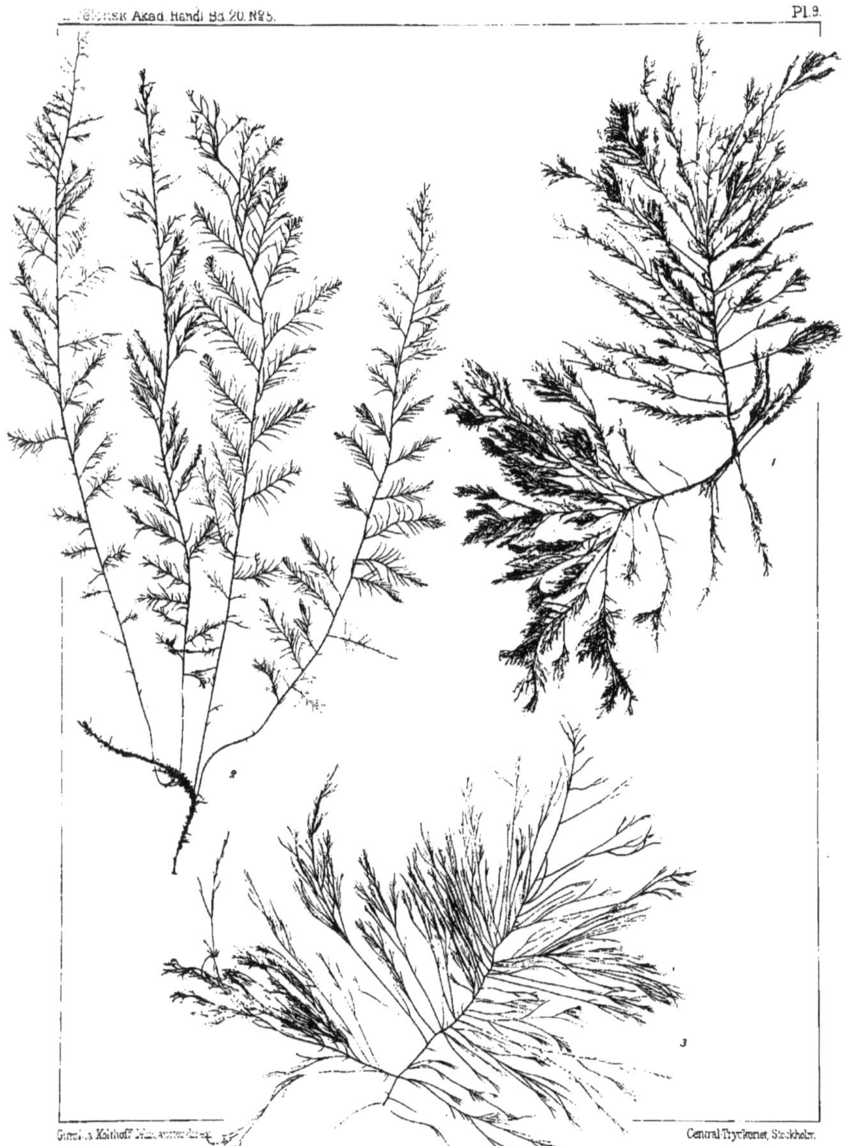

Gunnar & Kolthoff ... Central-Tryckeriet, Stockholm.

1 Rhodomela lycopodioides f. typica β laxa. 2 Rhodomela lycopodioides f. typica γ tenera.
3 Rhodomela lycopodioides f. setacea.

Gunilda Kolthoff delin., auctor direx. Central-Tryckeriet, Stockholm.

1-2 Rhodomela lycopodioides f flagellaris. 3 Delesseria corymbosa.

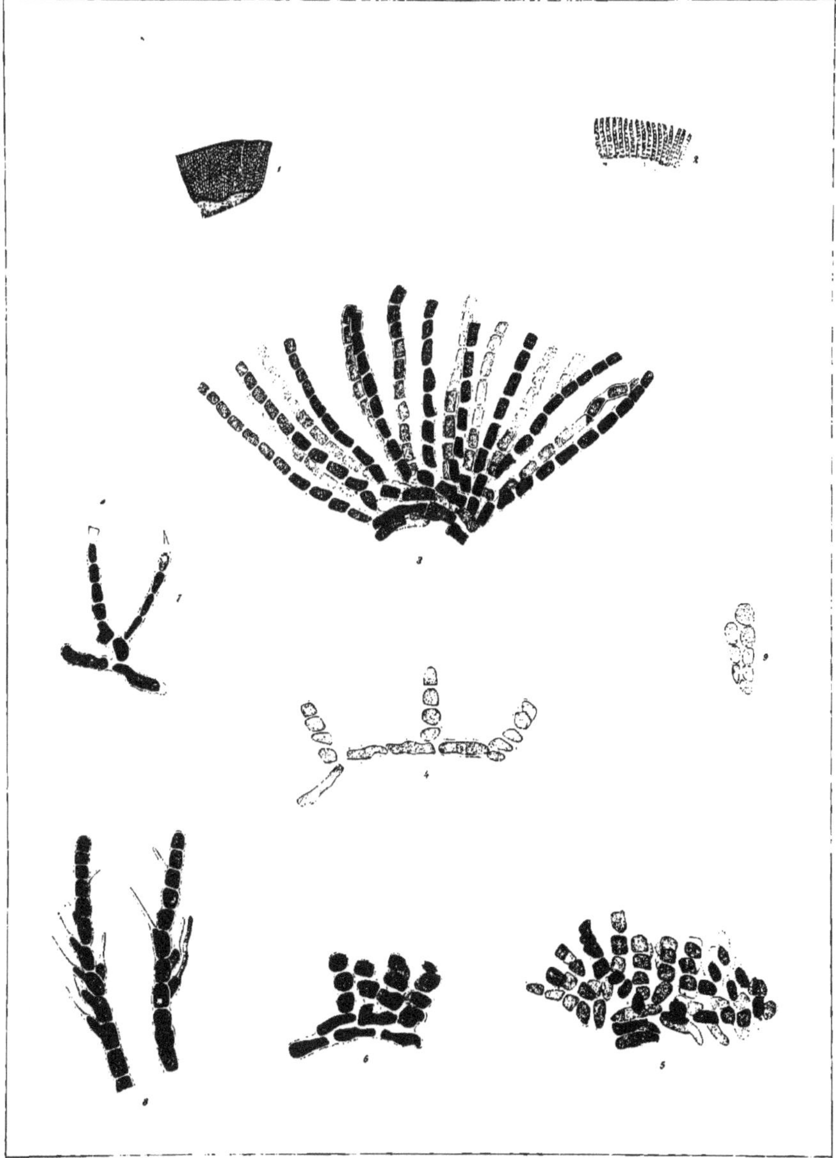

Gunilda Kolthoff delin., auctor direx

Central-Tryckeriet Stockh.

Gunilda Kolthoff delin. auctor direx. Central-Tryckeriet Stockholm

1-7 Chantransia efflorescens. 3. Rhodophyllis dichotoma
4 Halosaccion ramentaceum f. robusta

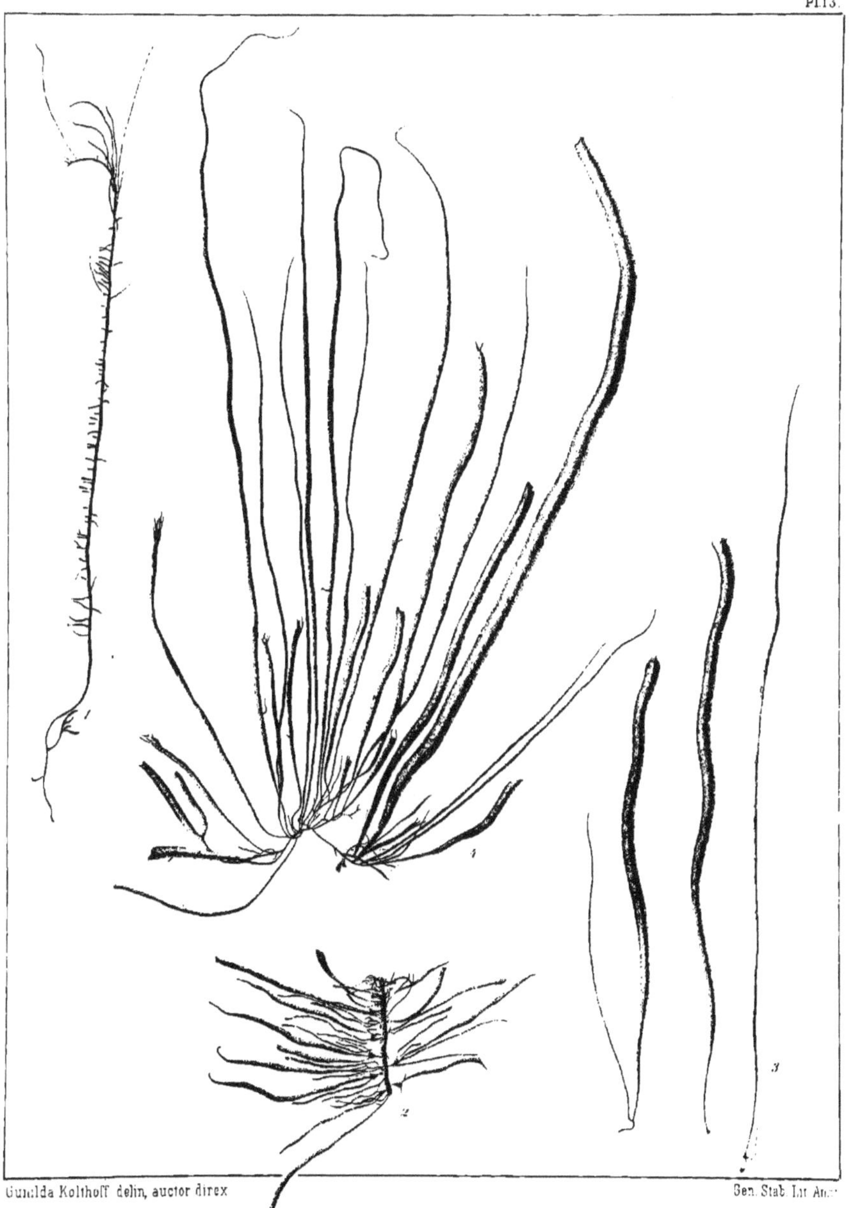

Gunilda Kolthoff delin, auctor direx

Gen. Stab. Lit. Anst.

1-2 Halosaccion ramentaceum f. robusta. 3 Halosaccion ramentaceum f. subsimplex.
4 Halosaccion ramentaceum f. ramosa.

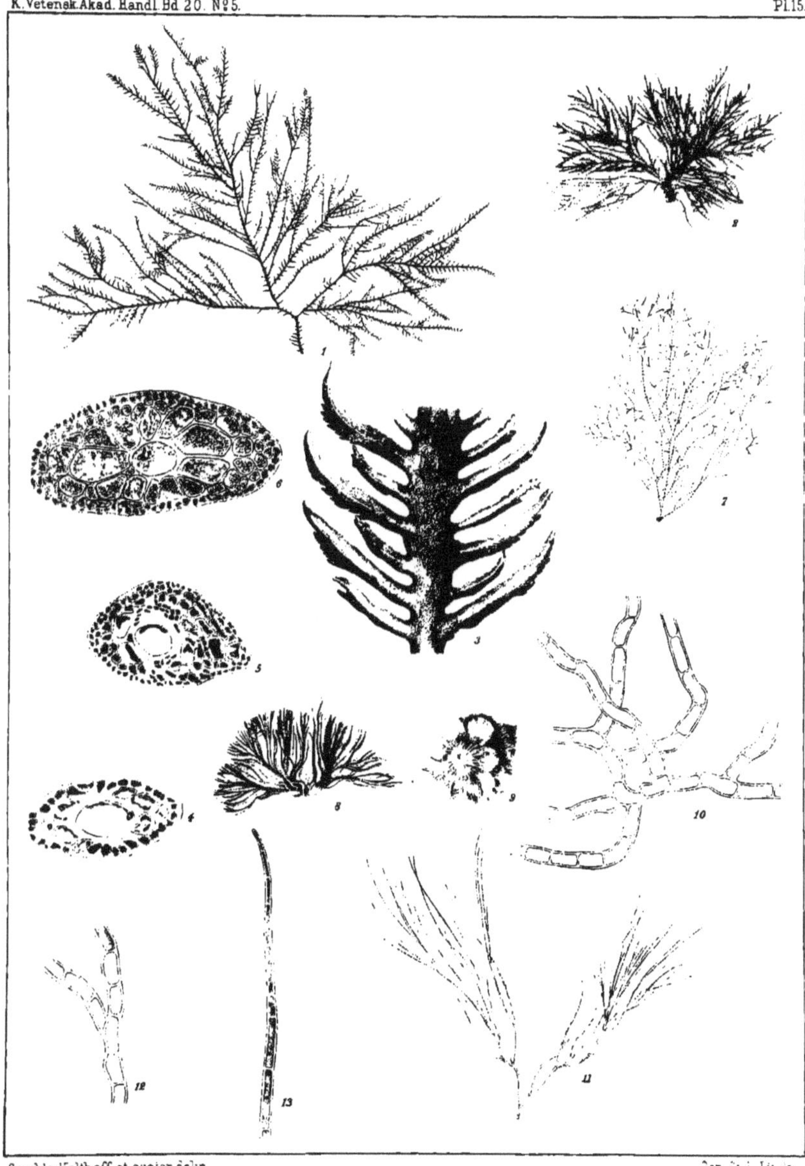

Gunilda Kolthoff et auctor delin. Gen. Stab. Lit. Anst.

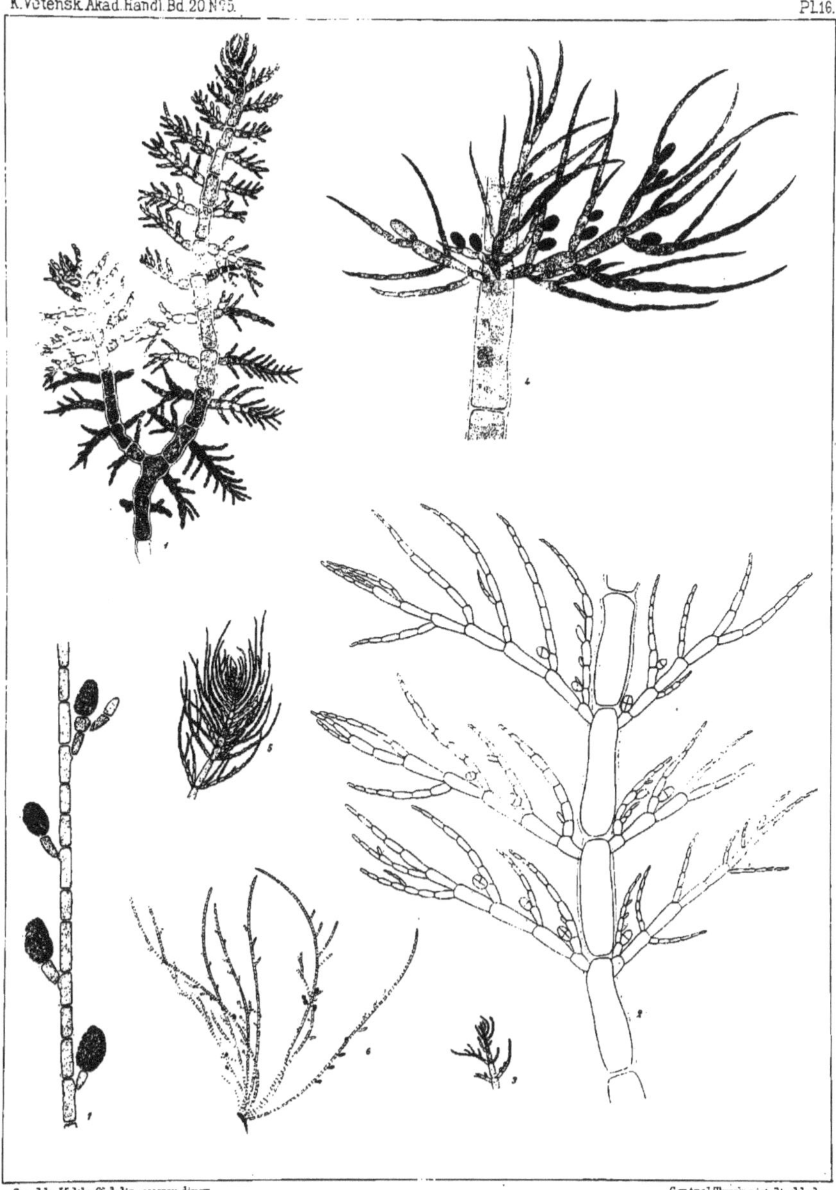

Gunilda Kolthoff delin, auctor direx Central-Tryckeriet Stockholm

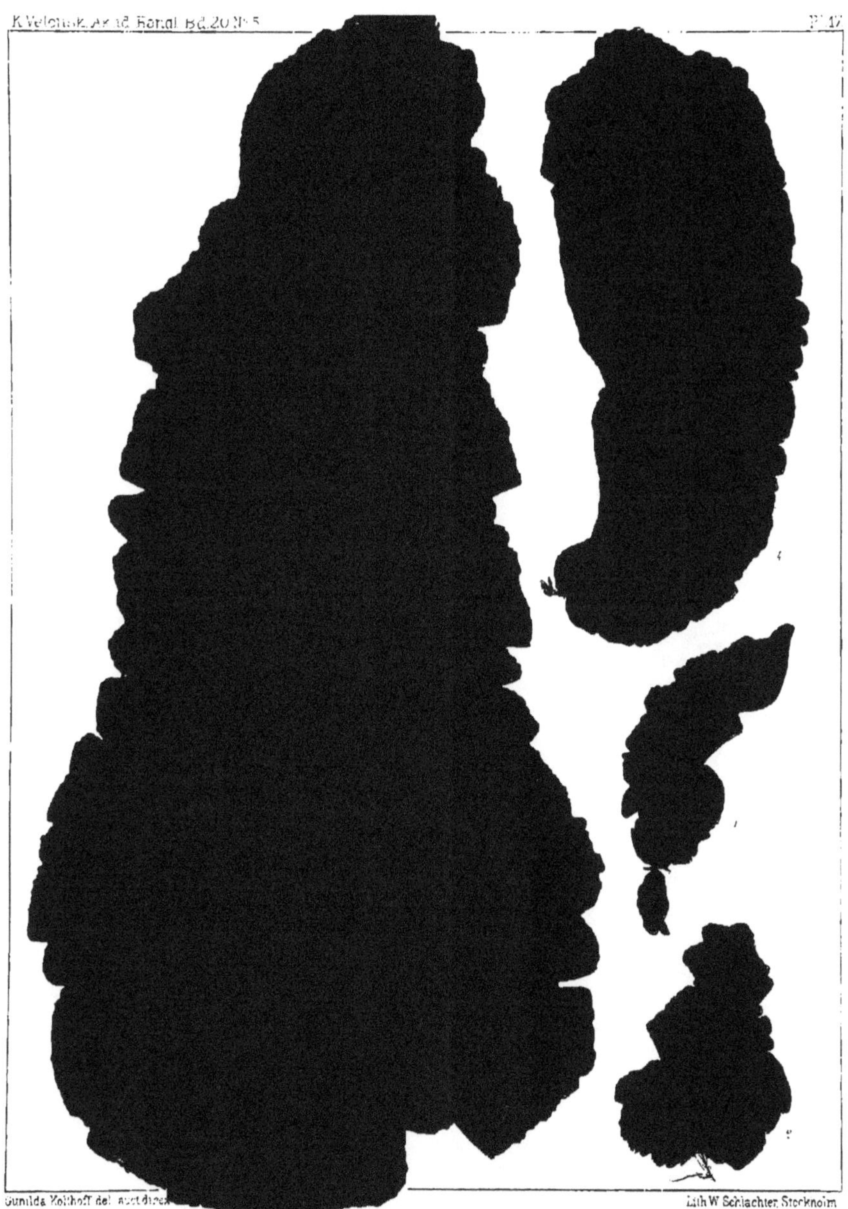

Gunilda Kolthoff del. auct. direx. Lith. W. Schlachter, Stockholm

1-3 Diploderma amplissimum 4 Porphyra abyssicola

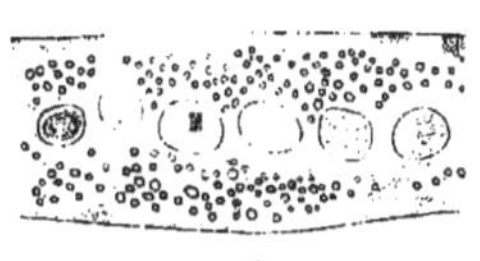

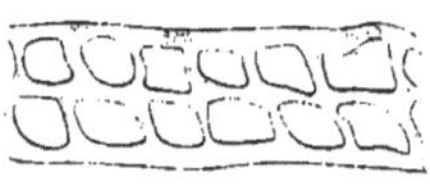

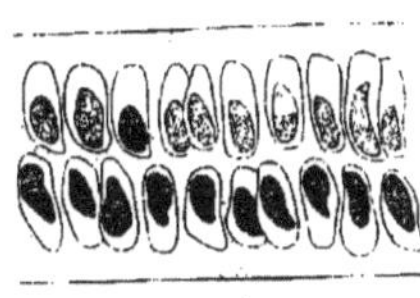

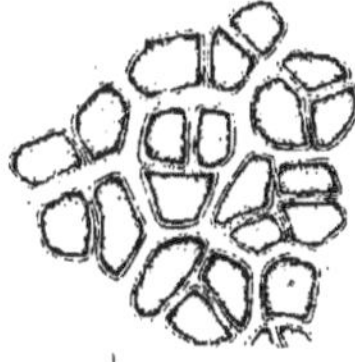

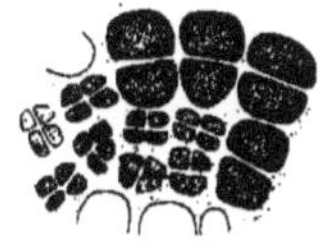

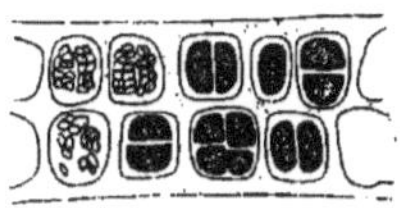

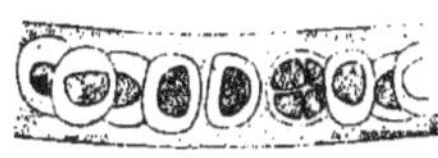

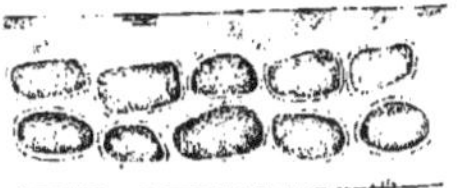

Gunilda Kolthoff delin, auctor direx. Centraltryckeriet, Stockholm.

1-8 Diploderma amplissimum. 9 Diploderma miniatum.
10-11 Porphyra abyssicola

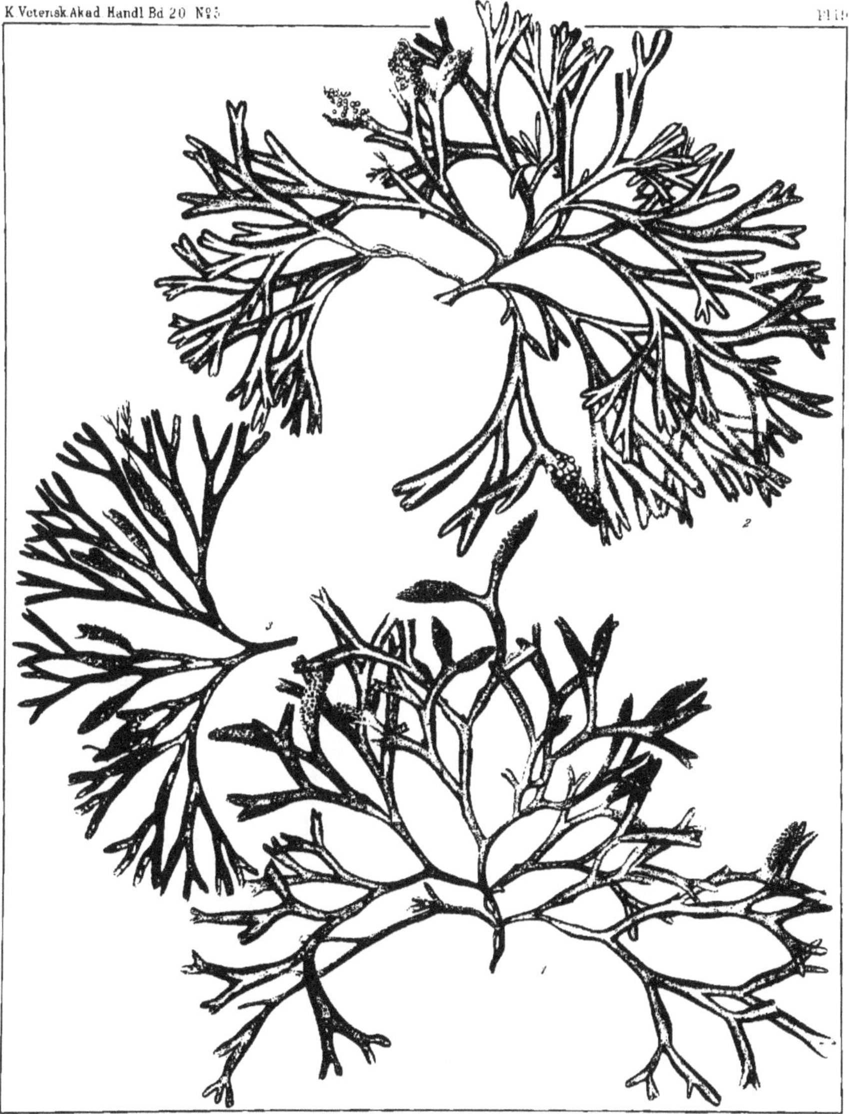

Gunilda Kolthoff delin, auctor direx.

Gen. Stab. lit. Anst.

Gunilda Kolthoff delin, auctor direx

Alaria dolichorhachis

Gunilda Kolthoff delin, auctor direx.

Centraltryckeriet Sthlm

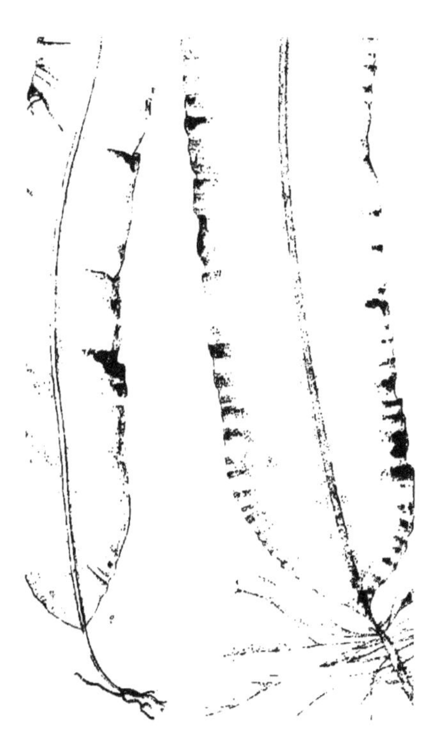

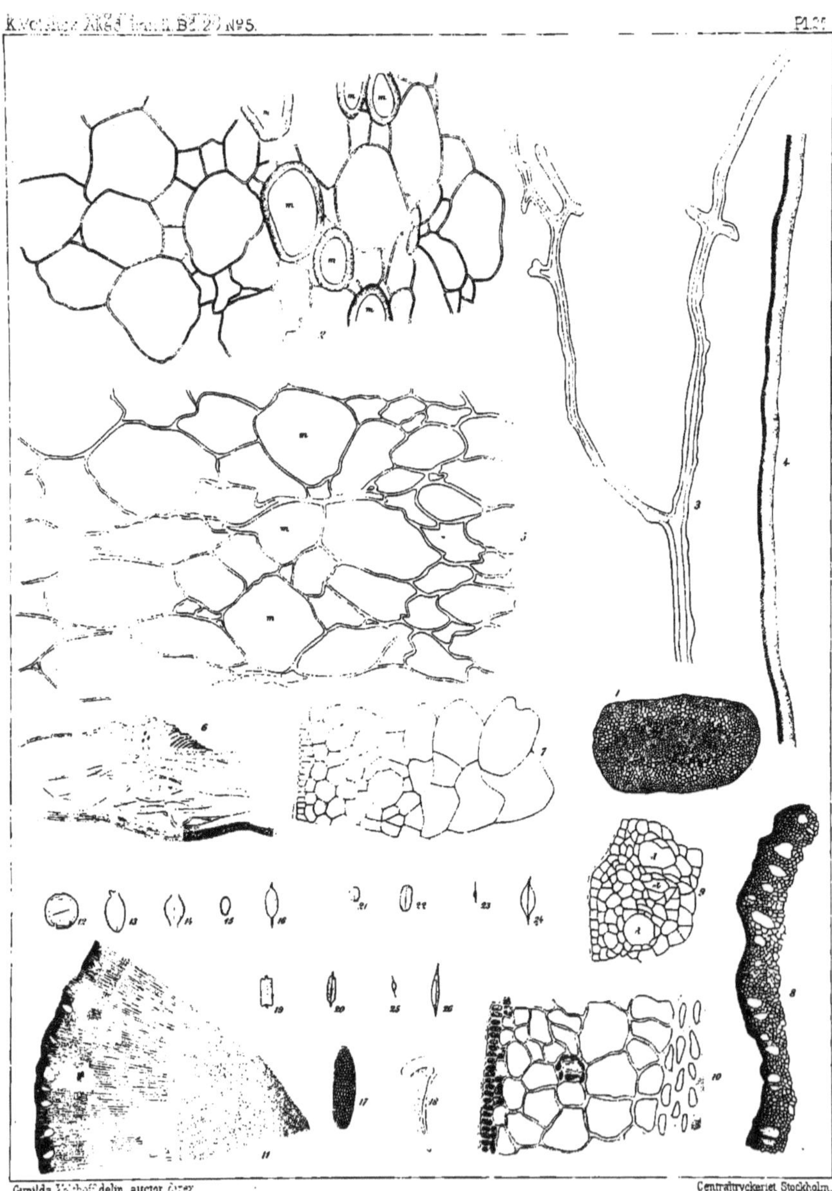

Gunilda Eichhoff delin, auctor direx Centraltryckeriet Stockholm.

1-4 Phyllaria dermatodea. 5-6 Phyllaria lorea. 7 Laminaria saccharina f. grandifolia. 8-10 Laminaria nigripes
11-18 Alaria dolichorhachis. 19 Alaria esculenta. 20 Alaria membranacea. 21-24 Alaria oblonga. 25-2? Alaria elliptica.

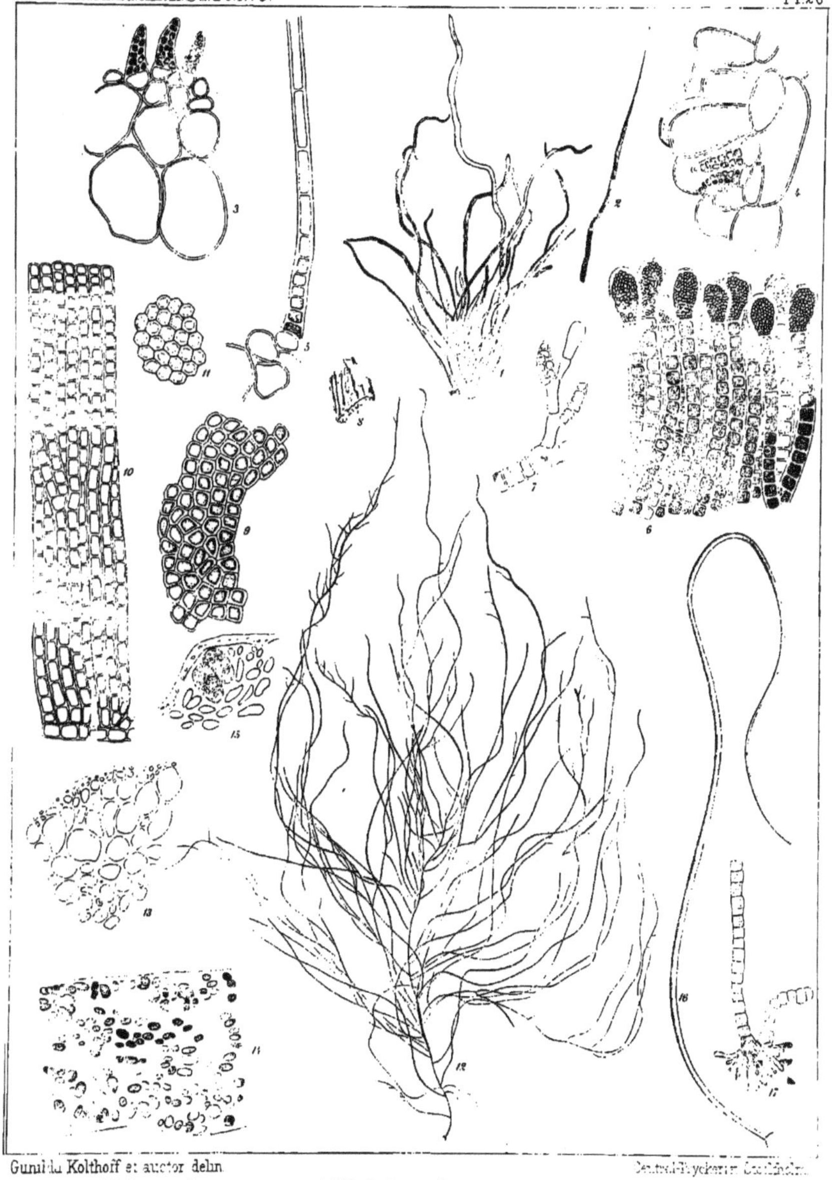

Gunilda Kolthoff et auctor delin. Central-Tryckeriet, Stockholm.

1-5 Scytosiphon attenuatus. 6-7 Lithoderma fatiscens. 8-11 Lithoderma lignicola.
12-15 Dictyosiphon corymbosus. 16 Chordaria filum f. crassipes. 17 Phloeospora pumila.

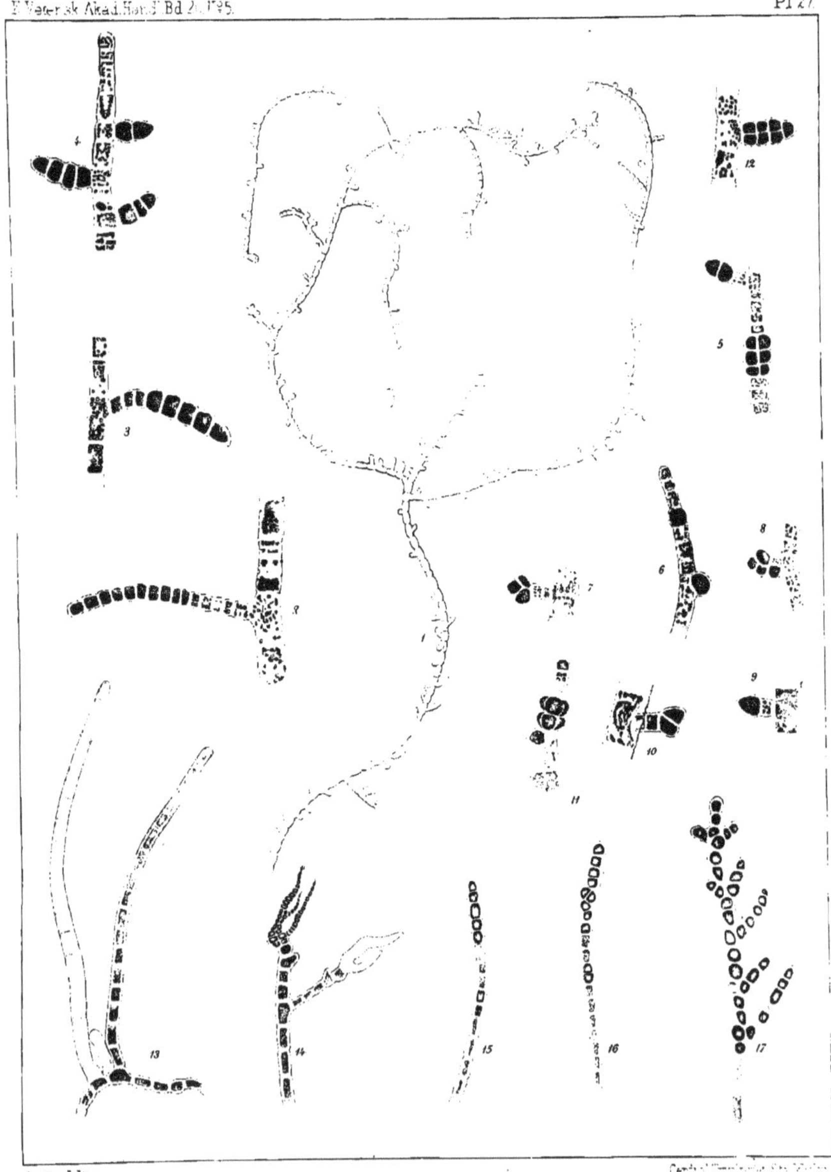

Auctor delin.

Cent[illegible]

1-12 Pylaiella varia 13-17 Pylaiella [illegible]

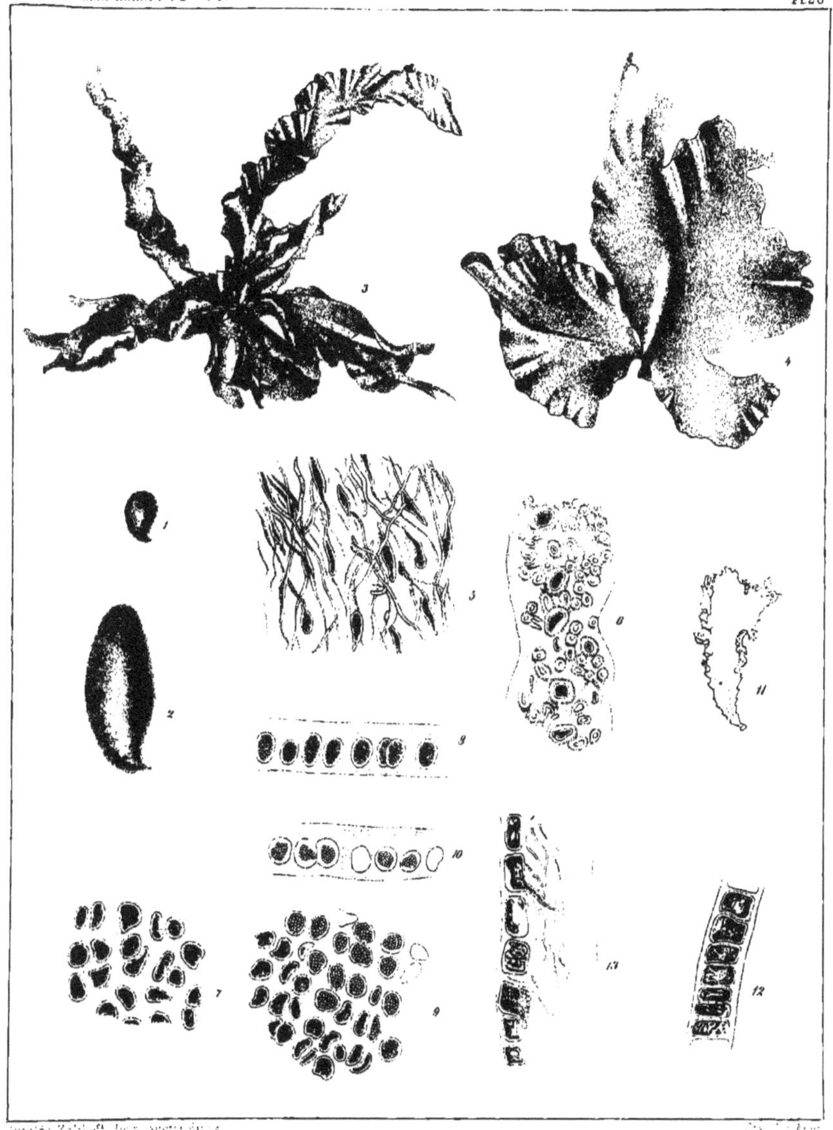
1
2
3
4
5
6
7
8
9
10
11
12
13

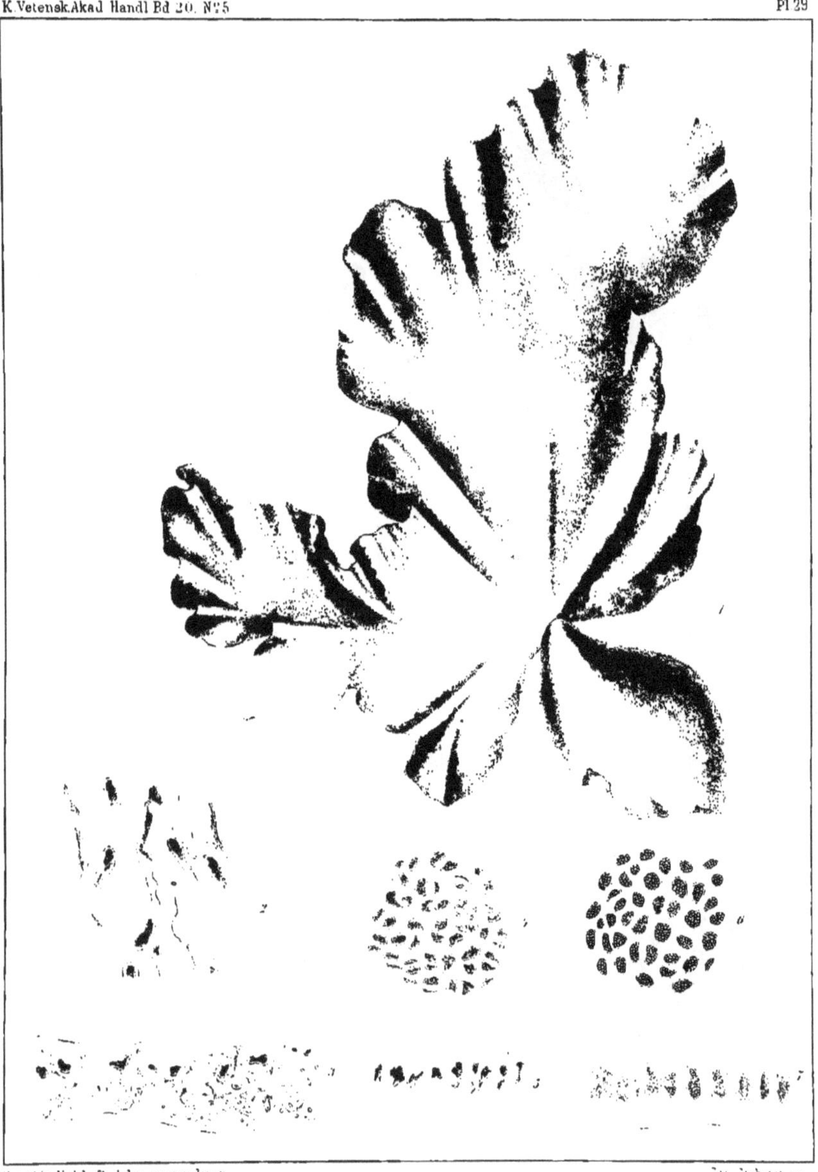

Gunilda Kolthoff delin. auctor direx. Gen. Stab. lit. an.

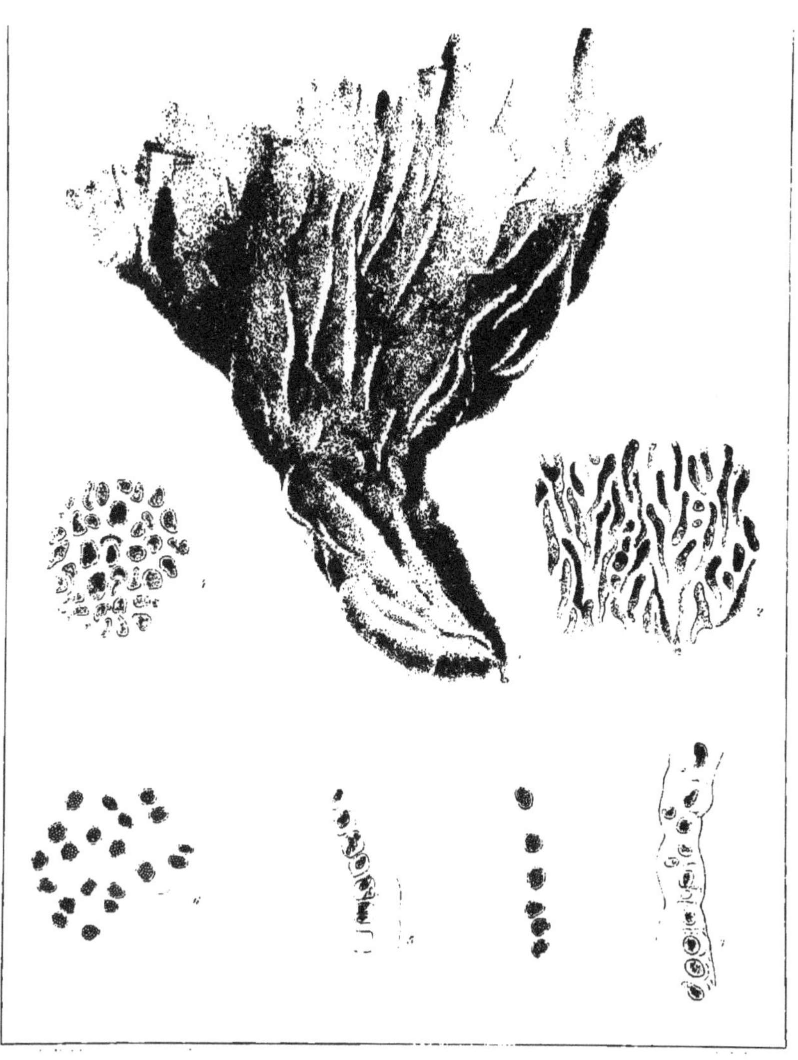

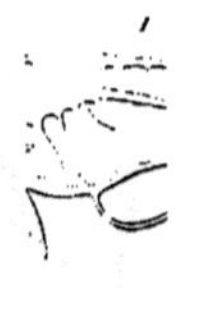